Agricultural Proteomics Volume 2

Ghasem Hosseini Salekdeh
Editor

Agricultural Proteomics Volume 2

Environmental Stresses

Springer

Editor
Ghasem Hosseini Salekdeh
Department of Systems Biology
Agricultural Biotechnology Research
Institute of Iran
Karaj
Iran

ISBN 978-3-319-43276-2 ISBN 978-3-319-43278-6 (eBook)
DOI 10.1007/978-3-319-43278-6

Library of Congress Control Number: 2016946323

Printed on acid-free paper

This Springer imprint is published by Springer Nature
The registered company is Springer International Publishing AG Switzerland

Preface

According to FAO's estimate, the number of people suffering from chronic hunger has increased to over a billion.

Because most of the extreme poor who suffer from hunger live in rural areas, the effort to enhance agricultural productivity will be a key element to reduce the numbers of the global population suffering hunger.

This goal will not be achieved unless we develop new genotypes of food crops and animals that will both improve production under suboptimal conditions. The discovery of genotypes with the capacity to cope with these problems suggests that increasing the support of breeding for fragile environments is a viable strategy for uplifting the rural poor. However, breeding for environmental stresses is a slow and inefficient process. Although several genotypes with good stress tolerance to environmental stresses have been identified or developed, it is difficult to transfer these traits to elite backgrounds because they are genetically very complex. One possibility currently being evaluated for enhancement of stress tolerance is to apply biomarkers in breeding programs to follow the inheritance of major genes that are difficult to phenotype, such as pyramids of disease resistance genes of similar effect. Proteomics is a powerful approach to identify proteins associated with stress tolerance. It offers an entry point for identifying possibly significant changes in protein levels against a background of unresponsive proteins.

The application of proteomics is usually initiated by detection of stress-responsive proteins through the comparison of proteomics data between stressed and control organisms. Identification of these expressional candidate proteins may then reveal that some of them have functions clearly consistent with the stress tolerance trait. Other relevant information including the expression pattern of mRNA and the metabolomics may help to further verify the correlation of these candidate proteins with desirable traits. The step forward from collecting proteomics data to functional prediction will pave the way for the sustainable agricultural production under unfavorable environmental conditions.

This book will cover several topics to elaborate how proteomics may contribute to our understanding of mechanisms involved in stress adaptation. The knowledge being accumulated through a wide range of proteomics technologies may eventually be utilized in breeding programs to enhance stress tolerance. This book presents a comprehensive review about the responses of crop and farm animals to environmental stresses. Challenges related to stress phenotyping and integration of proteomics and other omics data have also been addressed.

Karaj, Iran

Ghasem Hosseini Salekdeh

Contents

Contributors

Jose B. Abucay, Jr., Genetics and Biotechnology Division, Plant Molecular Biology Laboratory, International Rice Research Institute, Metro Manila, Philippines; College of Arts and Sciences, University of the Philippines Los Baños, Los Baños, Laguna, Philippines

Pooja R. Aggarwal National Institute of Plant Genome Research, Aruna Asaf Ali Marg, New Delhi, India

Owen K. Atkin ARC Centre of Excellence in Plant Energy Biology, Division of Plant Sciences, Research School of Biology, Australian National University, Canberra, ACT, Australia

Brian J. Atwell Department of Biological Sciences, Faculty of Science and Engineering, Macquarie University, Sydney, NSW, Australia

Elena Bencurova Department of Bioinformatics, Biocenter, University of Wuerzburg, Würzburg, Germany; Laboratory of Biomedical Microbiology and Immunology, Department of Microbiology and Immunology, University of Veterinary Medicine and Pharmacy in Kosice, Kosice, Slovakia

Niranjan Chakraborty National Institute of Plant Genome Research, Aruna Asaf Ali Marg, New Delhi, India

Subhra Chakraborty National Institute of Plant Genome Research, Aruna Asaf Ali Marg, New Delhi, India

Sixue Chen Department of Biology, Plant Molecular and Cellular Biology, Interdisciplinary Center for Biotechnology Research, Genetics Institute, University of Florida, Gainesville, FL, USA

Thomas Dandekar Department of Bioinformatics, Biocenter, University of Wuerzburg, Würzburg, Germany

Mehdi Dastranj Department of Systems Biology, Agricultural Biotechnology Research Institute of Iran, Agricultural Research, Education, and Extension

Organization, Karaj, Iran; Department of Plant Protection, College of Agriculture and Natural Resources, University of Tehran, Karaj, Iran

Peter David Eckersall Institute of Biodiversity, Animal Health and Comparative Medicine, College of Medical, Veterinary and Life Sciences, University of Glasgow, Glasgow, UK

Mohammad-Reza Ghaffari Systems Biology Department, Agricultural Biotechnology Research Institute of Iran, Agricultural Research, Education and Extension Organization (AREEO), Karaj, Iran

Javad Gharechahi Chemical Injuries Research Center, Baqiyatallah University of Medical Sciences, Tehran, Iran

Herra L. Grajo Genetics and Biotechnology Division, Plant Molecular Biology Laboratory, International Rice Research Institute, Metro Manila, Philippines; College of Arts and Sciences, University of the Philippines Los Baños, Los Baños, Laguna, Philippines

Mohammad-Reza Hajirezaei Department of Physiology and Cell Biology, Molecular Plant Nutrition, Institute of Plant Genetics and Crop Plant Research, Gatersleben, Germany

Yue Hao Institute of Apicultural Research, Chinese Academy of Agricultural Sciences, Beijing, China

Ante Jerkovic Department of Chemistry and Biomolecular Sciences, Macquarie University, Sydney, NSW, Australia

Mudassar Nawaz Khan Graduate School of Life and Environmental Sciences, University of Tsukuba, Tsukuba, Japan; National Institute of Crop Science, National Agriculture and Food Research Organization, Tsukuba, Japan

Ajay Kohli Genetics and Biotechnology Division, Plant Molecular Biology Laboratory, International Rice Research Institute, Metro Manila, Philippines

Setsuko Komatsu Graduate School of Life and Environmental Sciences, University of Tsukuba, Tsukuba, Japan; National Institute of Crop Science, National Agriculture and Food Research Organization, Tsukuba, Japan

Sheldon Lawrence II Department of Biology, Plant Molecular and Cellular Biology, University of Florida, Gainesville, FL, USA; Genetics Institute, University of Florida, Gainesville, FL, USA

Jianke Li Institute of Apicultural Research, Chinese Academy of Agricultural Sciences, Beijing, China

Mohsen Mardi Department of Systems Biology, Agricultural Biotechnology Research Institute of Iran, Agricultural Research, Education, and Extension Organization, Karaj, Iran

Mehdi Mirzaei Department of Chemistry and Biomolecular Sciences, Macquarie University, Sydney, NSW, Australia

Hans-Peter Mock Leibniz Institute of Plant Genetics and Crop Plant Research (IPK), Stadt Seeland, Gatersleben, Germany

Kanika Narula National Institute of Plant Genome Research, Aruna Asaf Ali Marg, New Delhi, India

Mohammad-Zaman Nouri Rice Research Institute of Iran, Mazandaran Branch, Agricultural Research, Education and Extension Organization (AREEO), Amol, Iran

Ehsan Oskoueian Agricultural Biotechnology Research Institute of Iran (ABRII), Mashhad, Iran

Jennifer Parker Department of Biology, Plant Molecular and Cellular Biology, Genetics Institute, University of Florida, Gainesville, FL, USA

Ghasem Hosseini Salekdeh Department of Systems Biology, Agricultural Biotechnology Research Institute of Iran, Agricultural Research, Education, and Extension Organization, Karaj, Iran

Andrew P. Scafaro ARC Centre of Excellence in Plant Energy Biology, Division of Plant Sciences, Research School of Biology, Australian National University, Canberra, ACT, Australia

Golandam Sharifi Department of Basic Sciences, Iranian Institute for Encyclopedia Research, Tehran, Iran

Hamid Sobhanian Department of Biology, Payame Noor Universtiy, Tehran, Iran

Jennylyn L. Trinidad Genetics and Biotechnology Division, Plant Molecular Biology Laboratory, International Rice Research Institute, Metro Manila, Philippines; College of Arts and Sciences, University of the Philippines Los Baños, Los Baños, Laguna, Philippines

Katja Witzel Leibniz Institute of Vegetable and Ornamental Crops (IGZ), Grossbeeren, Germany

Paul Worden Department of Chemistry and Biomolecular Sciences, Macquarie University, Sydney, NSW, Australia

Yunqi Wu Department of Chemistry and Biomolecular Sciences, Macquarie University, Sydney, NSW, Australia

Chapter 1
Well-Designed Experiments Make Proteomic Studies on Stressed Plants Meaningful

Brian J. Atwell

Abstract Analysis of the impact of abiotic stresses on plants is technically demanding. The cultivation of plants, application of treatments, choice of tissues and preparation of biological samples for proteomic analysis is as important as the subsequent identification of proteins. With appropriate precautions, proteomics will greatly improve our understanding of the mechanisms of abiotic stress tolerance. Hence, this chapter summarises some of the major design faults that can compromise the interpretation of 'stress experiments'. The examples of salt, drought, thermal stress and waterlogging are taken as representative of commonly encountered stresses, with recommendations for ways to avoid artefacts in design. The importance of interactions between these stresses is then discussed, pointing out the relevance of carefully constructed time courses and attendant physiological measurements to define the degree of stress. Tissue selection is also emphasised, recognising that stresses have differential impacts on different organs. Finally, the significance of choice of plant species is discussed, with recognition of the value of model species and the importance of expanding the range of taxa used if the full range of stress acclimation responses is to be identified through proteomics.

Keywords Experimental design · Abiotic stress

1.1 Introduction

Proteomic technologies have evolved rapidly in the past two decades, becoming an indispensable tool in the analysis of gene expression [1]. Because protein complements provide qualitatively different information from transcriptomes [2], proteomics will bring important new insights to plant phenomics under stress. However, the full extent of the disjunct between transcriptome and proteome is yet

B.J. Atwell (✉)
Department of Biological Sciences, Faculty of Science and Engineering, Macquarie University, Sydney, NSW 2109, Australia
e-mail: brian.atwell@mq.edu.au

G.H. Salekdeh (ed.), *Agricultural Proteomics Volume 2*,
DOI 10.1007/978-3-319-43278-6_1

to be revealed. Direct evidence for abiotic stresses modifying translation of mRNAs is scarce and deserves closer attention over a range of conditions. In hypoxic *Arabidopsis* plants, much of the mRNA population remains untranslated, leading to a proteome that is defined by the demands of the stressed cell [3].

In spite of great technical strides, the opportunities afforded by proteomics still have their limits, with detection of low-abundance proteins and post-translational modifications providing continuing challenges [2]. However, deep sequencing of DNA and extensive proteomic profiles are driving the concept of 'proteogenomics'—the marriage of proteomics with genomics to develop a deeper understanding of crop phenomics [4, 5]. Initial attempts will be based on the major, well documented crop species such as rice, from which they will extend to genetically complex species such as wheat and novel crop species.

This review does not set out to appraise these technologies but rather to analyse the methodology by which *biological samples are prepared* for subsequent proteomic analysis. Because 'agricultural proteomics' will make a major contribution to our understanding of the mechanisms of abiotic stress tolerance by quantifying gene expression levels under stress in high-performing hybrids [2], special care is required to avoid flawed experimental practices that could compromise interpretation of data and their application to breeding and targeted gene transfer. The sections that follow dissect the physiological, developmental and genetic factors that influence the results of gene expression analyses. They specifically address experimental design, particularly time courses of experiments and informed sampling of biological tissues from plants. Cautionary themes are presented under three headings (experimental design related to specific abiotic stresses, time frames and sampling). All three themes should be taken into consideration during the production of biological samples for proteomic experiments.

1.2 Designing Experiments to Mimic Abiotic Stress Observed in the Field

The environmental hazards that restrict agricultural productivity are either climatic (e.g. drought, salinisation, frost, light imbalance), chemical (e.g. inorganic nutrition, salt, herbicide residues) or biotic (invertebrate, fungal or bacterial attack). This section deals with the appropriate design of experiments required to mimic four of the major abiotic stresses on crops—salinity, drought, temperature and waterlogging. Through the precise application of these stresses in controlled conditions, we can gain confidence in proteomics as a tool to inform the genetic improvement of our major crop species. With sophisticated hardware (e.g. well-lit environmental cabinets) and software (e.g. ramping of conditions rather than simple day/night settings) now available, experimentalists can nuance the application of abiotic stress in a way not previously possible. Thereby, temperature, light, humidity and inorganic nutrition can all be very closely aligned with field observations.

Abiotic stresses do not act on plants independently—they interact, as we see at the tissue and cell level. This interaction is manifested in the phenotypic responses that we observe in experiments. For example, drought and salinity are mechanistically connected, with salt affecting land plants by perturbing cell water relations, as well as via the toxic effects of ions on cell metabolism. Therefore, 'osmotic drought' caused by salinity is likely to have gene expression responses in common with 'hydraulic drought', which is caused by soil water depletion, low air humidity and/or high wind speeds. However, longer term changes in the proteome will be specific to the toxic effects of sodium and chloride and will be manifested in expression of ion transporters that are required for compartmentation and efflux. Yet, surprisingly, many publications claim to reveal gene-level responses to salinity without designing experiments to discriminate between the dual effects of water relations and toxicity. If proteomics is to be effective, careful application of treatments (in this example, salt), time courses and environmental conditions must all be managed to lead us to the most likely explanation at the cell level for the responses seen in crop species.

A further dimension is the choice of species for gene expression studies: this is inseparable from the manner in which the stress is imposed, as seen in the specific examples referred to below. One must first look to the commonly used models such as *Arabidopsis*, *Chlamydomonas*, *Brachypodium*, *Nicotiana benthamiana* and the crop species *Oryza sativa* (monocotyledons) and *Medicago truncula*ta (legumes) because these species have contributed so much to our knowledge of gene-level responses to abiotic stress. However, generalising observations from these model genotypes to abiotic stress effects in all commercial crops is fraught because of the specific adaptations that might characterise particular species (Fig. 1.1). For example, the 'minimalist' deep tap-root of the dryland legume lupin contrasts with the expansive fibrous root system of wheat, in spite of both achieving efficient water use in identical dryland field conditions [6]; it is likely that each species employs some unique drought resistance strategies. Similar contrasts in root architecture can be seen for wheat and sugar beet in NMR images [7]. Dicotyledonous crop species are especially under-represented in studies aimed at identifying genes that respond to abiotic stress.

In summary, it behoves all those in the thrall of the technologies used to study gene expression to expand the range of taxa and improve the experimental designs that too often compromise abiotic stress studies. This will have the effect of creating ever larger and more reliable databases being applied to biological samples that genuinely mimic the physical constraints to yield in field crops. Relative to genomics, proteomics is a nascent science whose impact will be far deeper with rigorous application of the stresses applied (e.g. levels of stress, time courses, interaction effects). Naturally, experiments on biological extracts will always yield a proteomic profile—the challenge is to identify those protein changes that meaningfully reflect the system in which the plant normally grows. Modern proteomics based on well-executed experiments could obviate many of the criticisms that could be levelled at some earlier microarray studies.

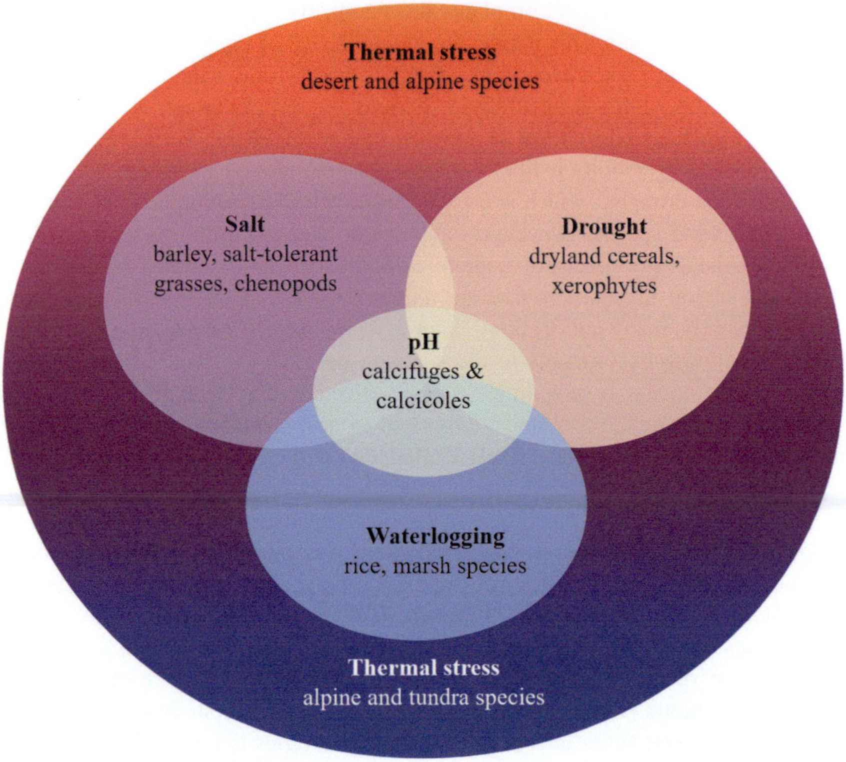

Fig. 1.1 Schematic to illustrate some ideal plant species or groups for proteomic studies on the range of abiotic stresses discussed in this chapter. The plant categories are not exclusive and the power of fully sequenced species (e.g. *Arabidopsis*, rice) as models for proteomic approaches is fully acknowledged. Calcifuges and calcicoles are plants naturally adapted to acid and alkaline soils respectively. Note that the effects of salt, drought and waterlogging interact and thus potentially produce unique proteomic responses. Extreme pH impinges especially on plants which are affected by other abiotic stresses. All these stresses are, in turn, subject to interaction with the experimental temperature regime

The following sections analyse the application of stresses individually and in combination, followed by the impact of temporal and spatial considerations in biological sampling.

1.2.1 Salt

Not uncommonly in the literature, sodium chloride is applied to plant roots in very high concentrations (100–200 mM), often in an instant, to mimic the effects of salinity on crop plants. However, only very rarely in nature is salinity visited on crops by a sudden rise from salt-free to inundation with highly saline solution and these are circumstances in which crops generally perish because of tidal surges or tsunamis. Salinity damage in crops is more likely to be subliminal and characterised by the gradual accumulation of salts in transpiring organs (mainly leaves), with associated osmotic and toxic impacts possible [8]. Furthermore, even though sodium and chloride are the dominant ionic species in sodic soils, the distortion of normal soil chemistry means that other ions can be present in concentrations far from their optimum [9]. Moreover, calcium also plays a key role in maintenance of membrane integrity and therefore it should ideally be present at millimolar concentrations in saline solutions around roots to prevent generalised membrane dysfunction and unrestrained passive uptake of sodium [8, 10]. Similarly, other macronutrients (e.g. N, P and K) must be sufficient to maintain an adequate steady-state supply to roots, and bathing solutions containing only sodium chloride must strictly be avoided as a mimic for salinity.

A further consideration must also be the inherent salt tolerance of the test species. It is unlikely that species poorly adapted to salinity will have a concentration of novel tolerance genes. The identification of high-affinity potassium transporters (HKTs) in wheat has led to a diverse family of genes from other species that confer salt tolerance [11], underlining the importance of experiments on a broad range of species.

Arabidopsis has relatively low physiological tolerance to salt and yet concentrations of sodium chloride up to almost half that of seawater are sometimes applied to this species in the laboratory to mimic saline conditions. Such experiments are more likely to educate us about the cell senescence and death than salt tolerance. The inclusion of halophytes and salt-tolerant grasses in proteomic experiments will go a long way to realising the full benefit of gene discovery technologies. Barley is clearly a case in point, where genomic and proteomic databases have the potential to reveal insights into mechanisms of salt tolerance. Furthermore, true halophytes such as the chenopods (family Amaranthaceae) and halophytic algae offer the opportunity to discover novel salt tolerance genes that have been lost in most land plants [12].

Recommendation: Apply relatively low sodium chloride concentrations in the presence of a full nutrient complement that includes calcium over relatively long periods (days, not hours) as a standard approach. The use of plant species that have at least moderate tolerance to salinity should also be encouraged but true halophytes are likely to be most informative.

1.2.2 Drought

Drought imposes itself on plants through a succession of processes that occur well before the common symptoms of wilting and death are observed: these events normally take place over the course of days or even weeks [13]. Impaired growth and diminished yield are the ultimate result of sustained drought but the physiological manifestations of withholding water (or dry atmospheres) are very complex. In those species which have been experimentally observed, acclimation to drought involves a suite of events in overlapping time frames—cell level changes associated with turgor loss are followed by accumulation of abscisic acid (ABA) and stomatal responses, and eventually, morphological adjustments such as thinner roots and altered root-to-shoot ratios [13]. With the benefit of many decades of information gathering on the expression of genes underlying these physiological observations, we now know that some drought responses are triggered directly by drought (e.g. ABA synthesis, biosynthesis of osmotic agents) while others are secondary or tertiary responses (e.g. accumulation of carbohydrates, senescence pathways, slower respiration).

Severe water deficits can be imposed effortlessly by removing a leaf from its parent plant, with wilting generally following quickly: naively, such an approach is sometimes thought to mimic drought. However, rapid dehydration involves little more than hydraulic shock and stomatal closure, with the more subtle adjustments to cell walls, hormone levels, tissue hydraulics and osmotic changes all masked. Thus, gradually withholding water is essential to elicit the full gamut of drought responses [13, 14] and therefore, to see changes to the proteome that represent all the acclimation processes in droughted plants. This is generally best achieved by using large soil volumes relative to plant size (see [7]), allowing soil water either to be depleted slowly [15, 16] or addition of very small volumes of water daily, enabling leaf hydration as plants acclimate to sub-optimal water supply [17, 18].

It is valuable in the analysis of abiotic stresses, including drought, to return plants to the non-stressed state by re-watering. Because re-watering immediately rehydrates plants, the pattern by which the proteomic profile returns to resemble that in continually watered plants can be re-assuring because the initial impacts of drought are likely to be reversed quickest (e.g. full cell hydration). For example, this might be seen in reduced levels of stress-inducible proteins [15]. Alternative approaches to manipulation of the proteome in response to drought ought to be employed where appropriate, including the classical split-root experiments [19]. This can effectively separate signals coming from a source (drying roots) from the hydraulic effects of drought in shoots. Another common technique used to impose drought is to add a non-permeating osmotic solute such as mannitol or polyethylene glycol to the root medium. While this achieves dehydration osmotically [20] it is unlikely to replicate the far more subtle acclimation responses of a true drought and can be hydrolysed and/or taken up by plant cells [21].

Recommendation: Drought is distinct from tissue dehydration and is most often imposed slowly in nature, leading to a wide range of acclimation responses. Therefore, in experiments drought should be mimicked by allowing plants to transpire water from large soil volumes. The effect of drought on gene expression can be further elucidated by re-watering to reverse the drought or splitting root systems into dry and wet compartments.

1.2.3 Thermal Stress

Many of the world's major crops grow and develop at temperatures outside the optimal diurnal range (say, 20–28 °C). While heat stress has frequently in the past been dismissed as little more than a subordinate of drought stress, it is a distinct phenomenon and in irrigated crops in the humid tropics, is likely to occur independently of drought. As with drought experiments, the artificial imposition of heat (and chilling) should be done using regimes that are guided by data from the field, such as those available from thermal loggers or meteorological observations. The imposition of drought and heat reported by Ashoub et al. [16] conforms to these general principles, with stress applied in graduated regimes. In that changes in the expression of stress-responsive genes are seen when temperate species are exposed to temperatures in the low thirties [22], extreme temperatures should only be imposed when justified by the habitat of the experimental species. Arguably, the most important metabolic changes occur within 5–10 °C of the optimal temperature range.

Similarly chilling must be imposed within physiological boundaries that are defined by field conditions, and at a rate that is plausible. Accordingly, chilling should be increased over timeframes of hours (simulating phenomena such as frost damage) or in some cases imposed over a period of days, as required for frost hardening in much colder environments [23]. Localised chilling of organs (e.g. roots) can be used to elicit release of mobile signals that trigger a change in the proteome of remote organs such as shoots [24]. Such an approach exploits proteomics to reveal the identity of either heat- and cold-inducible long-distance signals but has limited relevance to field plants outside those where rapid atmospheric heating accompanies evaporative cooling at the soil surface (e.g. irrigated rice in hot savanna).

Artificial growth conditions such as atmosphere-controlled glasshouses and growth cabinets have the capacity to heat and cool plants over a huge range in just minutes, further necessitating stepwise changes in temperature as a new steady-state is established. Ignoring the need for temperature ramping leads to experiments that measure how gene expression responds to thermal shock and provides no insights into acclimation to temperature shifts.

Recommendation: Impose heat stress by stepwise increases in temperature, generally during the daytime, and in accordance with the natural range of temperature stress that is likely to be experienced. Chilling should also be imposed gradually unless it is aimed at simulating sudden events such as frost in unhardened plants.

1.2.4 Waterlogging

As with drought, plants undergo a series of chemical, metabolic and structural changes during acclimation to flooding, with the primary impact being on roots, contrasting with impacts on shoots during drought and atmospheric fluctuations. The importance of care in the choice of tissues to be sampled for proteomics will be addressed in detail below (see '*The importance tissue sampling*').

Changing the oxygen supply to tissues abruptly is known to cause damage and even death of cells, especially in root apices, which are most metabolically active [25]. These authors showed that in the absence of internal ventilation in the form of aerenchyma, even flood-tolerant species such as rice are unable to withstand anoxia. In testing the effect of anoxia on plants, hypoxic pre-treatment is strongly recommended to alleviate damage from 'anoxic shock' (see [26]) as this qualitatively changes the tolerance of vulnerable tissues such as maize roots to anoxia [27]. The dissection of what constitutes shock versus steady-state stress is discussed in the final section.

Some experimenters advisedly test the recovery from low-oxygen stress by re-establishing aeration. However, just as the switch from normoxia to anoxia is very damaging, abrupt increases in oxygen supply to tissues are potentially deleterious, in this case because of the inadequacy of oxidative reactions to consume available oxygen, and subsequent release of deleterious reactive oxygen species [28]. Therefore, recovery treatments need to be applied with care, probably by hypoxic post-treatment.

Paradoxically, plant organs (e.g. roots, rhizomes) of highly flood-tolerant species largely owe their survival in low-oxygen environments to a system of aerenchyma which ventilate cells and re-supply surrounding medium with oxygen. This adaptation is highly developed in species such as rice and over-wintering wetland plants [29]. Furthermore, the rate at which oxygen diffuses out of roots varies with genotype [30]. Thus, while anoxia can be imposed on the root medium, the actual oxygen status of individual root zones from different genotypes might not be comparable at the time that they are sampled for proteomics because oxygen transport into these root systems varies with the proportion of aerenchyma and oxygen leakage rates [28]. This is particularly true for the stele of roots, which can be anoxic while the surrounding cortex is hypoxic [31]. Disparate anoxia tolerance in the dimorphic root systems of grasses [32] adds a further dimension that must be taken into account during sampling. Such subtleties require careful consideration and while in general, excision of organs should not be the first choice, there is a case where the confounding effects of long-distance transport of oxygen (or carbohydrates) make interpretation of data difficult in intact systems [33].

Choice of species is especially critical when probing the proteome of roots because some species are relatively tolerant to hypoxia/anoxia, while others are so intolerant that even hypoxia can kill them or at the least, inhibit all function [28].

This contrast is particularly pertinent when the pre-eminent plant model species (*Arabidopsis* vs. rice) represent extremes of tolerance to low oxygen, calling for low-oxygen treatments that recognise these tolerances. Broader taxonomic contrasts that include poplar and algae as well as *Arabidopsis* and rice, have been employed to identify common transcriptional responses [34] and similar metabolomic and expression profiles have also compared poplar with rice and *Arabidopsis* [35]. With gene expression having been studied in so few of the plant species which are adapted to marshes, wetlands and waterways, there is a powerful case for quantitative proteomics that encompasses more species and diverse oxygen treatments.

Anoxia severely impairs protein synthesis, even in rice seedlings [36] because most of the energy generated is used to synthesise new proteins [37]. It follows that tissues exposed to anoxia for short periods will reveal a proteomic profile dominated by proteins that were present prior to the low-oxygen treatment: this is obviously to be avoided. To discriminate the synthesis of novel proteins during the low-energy, low-oxygen period, quantitative proteins (e.g. enrichment of ^{15}N in proteins that were synthesised from labelled exogenous ammonium or amino acids) is a better approach [38].

Finally, the microbial populations that inhabit the rhizoplane of root systems that are not grown axenically are substantial; microbes have high protein concentration per unit biomass and rapid turnover rates [39]. These prokaryotic populations are clearly a confounding factor in proteome analysis and must be either eliminated or suppressed if the true root proteome is to be considered in gene expression studies. The advent of quantitative proteomics makes this even more pressing because the rates of incorporation of labelled precursor amino acids or ammonium into the microbial proteome will be so much faster that into the roots.

Recommendation: Lower (or raise) oxygen concentrations around root systems in one or more steps through the hypoxic range over at least 24-h periods in order to avoid tissue death and oxidative damage when anoxia (or normoxia) are reached. Roots should be sampled for proteomics with a clear knowledge of the actual oxygen status of the intact tissue, as well as its inherent tolerance to anoxia, developmental stage and the microbial populations that reside in the rhizosphere.

1.3 Managing Interactions Between Abiotic Stresses

Preceding sections describe how best to apply *individual* stresses to plants. However, appreciation of the more complex question of interacting abiotic events is also vitally important because the impact of one stress can exacerbate, or ameliorate, that of a second stress [40, 41]. Such interactions can be entirely abiotic, i.e. physical events external to the plant such as high temperature exacerbating oxygen deficiency. In reverse, low soil temperatures reduce root and microbial respiration and alleviate damage from waterlogging [42].

While the physiological manifestations of abiotic interactions might be obvious, there is far less certainty about the proteomic changes that are triggered as part of the biological response. Suzuki et al. [41] refer to signalling pathways that are common to particular stress combinations. As post-transcriptional modifications (e.g. RNA processing, protein phosphorylation) are revealed, greater complexity will necessarily be added to the gene expression patterns that are observed in response to interacting stresses [40].

A few common examples of stress interactions are listed below. This is not an exhaustive catalogue—see Suzuki et al. [41] for a more complete listing—but is an indication of some abiotic stresses that interact in a non-additive manner. While the impact of these interactions cannot be predicted at the gene or protein level, they should be foreseen using extensive knowledge of the whole-plant responses documented [43].

Drought and heat: Ambient temperature can exceed the actual leaf temperature by many degrees because of transpirational cooling [44, 45]. Thus, experimental protocols should take *actual* leaf temperature into account when assessing the impact of heat on leaves. The phenomenon of leaf 'self-cooling' adds complexity to the heat × drought interaction, with leaf temperatures rising close to the ambient atmospheric temperature as transpiration rates fall but the impacts of drought lessening as water losses are constrained by stomatal closure.

Drought and salinity: The introductory section raises a classical example of the complexity of salinity stress, where the dual impacts of hydraulics and toxicity can operate on separate time courses. To some degree, osmotic effects (leading to compromised hydraulics) and cell-level toxicity can be partly managed by sampling over rigorous time courses after stress application. For example, hydraulic effects become evident within minutes of adding salts to the root medium, with lower root water potential being transduced to the xylem, and subsequently the leaves [46]. Over a longer time course, salts can accumulate in the cell walls of leaves in non-halophytes, hastening the dehydration of mesophyll cells and initiating necrosis. Some of these salts are taken up by leaf cells, triggering biochemical and metabolic responses that are ultimately deleterious in the absence of compartmentation [47]. This chronological series of events is likely to elicit shifts in the proteome, with each tissue sampled minutes, hours, days and weeks after salinisation producing qualitatively distinct protein profiles. Well-designed experiments require time-course measurements of water and ionic status of tissues and aligning these data with the proteome at each time point. The proteomes of control plants should be reported alongside tissues of treated plants.

Temperature and low oxygen: Oxygen status is strongly dependent on temperature, with high temperature reducing soluble oxygen concentration and raising respiration rates, thus exacerbating the effects of inundation. However, this example amply reinforces the importance of time as an interacting factor with multiple stresses, with plants of the same *physiological* age not exhibiting a temperature × oxygen interaction while those of the same *chronological* age showed increased damage at high temperatures [48]. It is clearly a requirement that experiments on low oxygen responses in roots take careful account of temperature, developmental age and tissue type (see '*The importance tissue sampling*'). It is established above that temperature shifts produce major qualitative shifts in the proteome of rice leaves and cultured cells [49, 50] and low oxygen concentrations also cause a highly characteristic expression of anaerobically induced genes [51, 52]. However, the interaction of abiotic factors with oxygen supply must always be carefully considered if the full impact of stresses is to be revealed at the protein level. The best example of such an interaction comes from Waters et al. [53], who measured recovery of growth in wheat root apices as a way to assess the interaction of the various abiotic factors with oxygen deficits. Notably, root tip mortality rose dramatically as temperatures were increased from 15 to 25 °C, pH was lowered from 6 to 4 or carbohydrate supply was restricted, illustrating the importance of careful control of experimental conditions.

Low pH and various abiotic stresses: As shown above, oxygen deficits compromise the energy status of cells and in a low pH bathing medium, cell function is further impaired through cytoplasmic acidification [54]. Because regulation of proton transport, membrane potential and potassium retention have such profound implications for cell function [55], the protocols used when any abiotic stress is applied must take careful account of external pH. Moreover, as proteomics expands to tackle field-scale agricultural questions, the large range of pH observed in the natural environment must be considered, particularly for plants growing in the acid soils of many modern agricultural systems. External pH must be managed carefully in the laboratory, where acidification of the bathing medium around plant tissues is a risk if the volume of bathing solution is low and inadequately buffered.

The availability of metabolic energy lies at the core of the interaction between abiotic stresses and low external pH [55]. Specifically, metabolic energy is used to maintain membrane potential in living cells below −100 mV by extruding protons across the plasma membrane and tonoplast. Therefore, any abiotic factor that compromises ATP availability (e.g. anaerobiosis, thermal stress, phytotoxins) is likely to reduce cell membrane potentials and trigger the release of common stress sensors such as reactive oxygen species and Ca^{2+} [55]. These events are further amplified by acidification of the external medium because the free energy required for proton extrusion increases as the proton gradient becomes less favourable [56]. The expression of genes under these stress conditions is often coordinated by a series of transcription factors (e.g. AP2/ERF, B3, NAC, SBP and WRKY), many of which are common to multiple stresses such as cold, anoxia and dehydration (see [52, 57]). Transcription factors activate DNA-binding domains and trigger the transcription of a large array of proteins. Hence abiotic events, especially in combination with acidic conditions, will necessarily result in distinctive proteomes. One would expect that in acute stress, proteins typical of programmed cell death would be commonly observed [58]. It is therefore critical to control experimental conditions and the composition of bathing media very closely.

1.4 General Principles for the Design of 'Stress' Experiments

Two general principles should guide the design of experiments aimed at identifying the key processes in plant acclimation to abiotic stress—time and space. In short, one must first select a time course for the application of stress and recovery from it, compatible with the synthesis of proteins that are necessary for acclimation. Second, tissues which are sampled must be sufficient to provide a credible proteome but homogeneous enough to represent a tissue-specific response. This section is aimed at enunciating these general principles.

1.4.1 The Importance of Time

Decisions on time courses should be influenced by the intensity of stress and the rate of its imposition (Fig. 1.2). This should be guided as much as possible by whatever physiological literature is available for similar genotypes under the same stresses. For example, microarray data can be helpful in defining a physiologically meaningful time course for sampling tissues [59]. In this context, the general observation that protein turnover in plants has a half-time of 1–2 days [38] is germane; abiotic stresses applied for less than one day are unlikely to achieve a new steady state, with the proteome 'contaminated' with proteins that were present prior

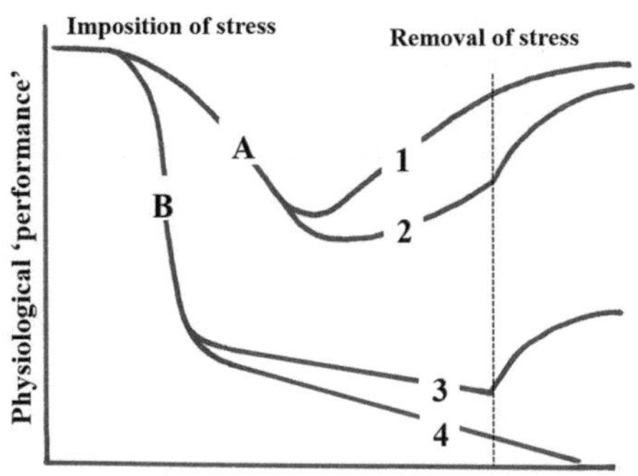

Fig. 1.2 Conceptual figure to show potential time courses of plant response to an arbitrary abiotic stress. Hence the y-axis is labelled *Physiological 'performance'* to indicate a process such as growth, development or, not of a metabolic function (e.g. photosynthesis). Scenarios *A* and *B* depict a mild and acute response to the stress; plants in Scenario *A* are likely to have been pre-conditioned to the stress (e.g. hypoxia prior to anoxia) while Scenario *B* is a shock treatment. After the physiological effects take hold, plants respond in at least four ways: (*1*) almost complete acclimation; (*2*) partial acclimation followed by rapid recovery after removal of stress; (*3*) severe decline under stress but not death—recovery on removal of stress; (*4*) irreversible damage and death—no recovery on removal of stress

to stress imposition. Therefore, sub-lethal levels of stress applied slowly enough to register a true acclimation is generally called for. There may also be cases for abrupt imposition of abiotic stress where the effect is not lethal and is believed to reflect a natural phenomenon such as flooding or frost. Furthermore, combining short and long-term stresses can help separate secondary (downstream) proteomic responses to stress from the primary effects, which might be the better targets for plant improvement.

Valuable information on appropriate sampling times under stress can often be found in transcriptomic and metabolomics studies, which can inform proteomics experiments. Alongside this, the power, efficiency and cost of the most contemporary proteomic techniques should enable far more intensive sampling and thus more detailed gene expression time courses. These promise to reveal important stages in the metabolic response to various abiotic stresses.

An excellent example of the importance of time courses is the distinct patterns of metabolite and expressed genes when oxygen was withheld from rice seedlings for up to 48 h [59]. After growing seedlings in anoxia, or aeration, some were switched to the opposite treatment and further tested up the 6 h later. This study clearly reveals the fact that gene products do not accumulate linearly over time, with a peak of 5000 transcripts being up- or down-regulated 3–12 h after imbibition but larger contrasts in transcript numbers between aerated and anoxic tissues appearing over the following 24 h. In another study on rice seedlings, Lasanthi-Kudahettige et al. [60] observed a similar disconnect between transcript levels for two isoforms of alcohol dehydrogenase, whereby one peaked at 3 h after anoxia and the other isoform after 7 h. This illustrates the distortion of gene expression data that can be caused by single, or too few, sampling times in non-steady state conditions after stress is imposed on plants [61]. Expression of genes that are induced by a variety of abiotic stresses are often analysed in detail over 24 h (e.g. [62]), revealing part of the acclimation response but almost certainly prior to the establishment of a new steady state. Moreover, changes in the proteome will generally become apparent in timeframes even slower than the transcriptional changes reported above. Processes such as carbohydrate accumulation, membrane properties and cell wall changes are typically observed over several days and ought to be more explicitly considered in experimental design.

1.4.2 The Importance of Tissue Sampling

Having designed a temporal regime for imposing abiotic stress that gives the best chance of identifying those proteins that are critical for acclimation and survival, it is then important to sample tissues judiciously in order to identify key proteins in subsequent proteomics analysis.

Higher plants differentiate into totally distinct tissue types: even apparently homogeneous tissues can have a high degree of heterogeneity (e.g. root apices, shoot apical meristems), while the functional specialisation in adjacent tissues (e.g. stele and cortex) is inevitably reflected in the genes expressed. One of the most convenient models for studying the spatial separation of function is in root apices, where adjacent zones of cell division, elongation/expansion and maturation have distinct functions and therefore proteomes (e.g. [63]). In preparing tissues for proteomics from these various root tissues, the proteome of the membrane fraction ought to be extracted alongside the soluble fraction because of the importance of transport in root function.

Tissue sampling is further complicated by the interaction between development and abiotic stress. One must question whether tissues at the same distance from common reference point (e.g. the apex of shoots or roots) in stressed and unstressed plants are necessarily at the same stage of development. In roots, for example, drought has been shown to qualitatively alter the dynamic of cell division and expansion [64], with the result that sampling the same length of tissues from contrasting drought regimes is almost certain to confound development with stress response.

Sampling is equally important in a number of other circumstances where stress is imposed. In the case of salt applied to roots, its accumulation in shoots is broadly proportional to the time for which leaves have been transpiring. This must be recognised during leaf sampling, where developmental age might be appropriate when a range of salt concentrations are to be compared. As in the previous example of roots in drought, the slowing of growth as a result of an abiotic stress complicates comparisons of tissue samples, which might alternatively be selected at a common chronological age or developmental stage.

In one of the earliest protein studies to be published, Sachs et al. [51] reported the major proteins that are synthesised when maize roots became anaerobic. This study has led over the years to a far more complete analysis of anaerobic gene expression, including in rice and *Arabidopsis*. Notably, a recent report on the relationship between the faster and slower growing regions of rice coleoptiles that were less than 20 mm long showed that fine-scale sampling within individual organs is rewarding and should be extended to the proteomic and metabolomics levels [52].

Plant survival during and after floods is a major agronomic question. For dryland species, little progress has been made and yet it has long been known that a major adaptation to inundation for many species, particularly monocotyledons, is the formation of aerenchyma—air channels that form in the root cortex through cell degradation. The cell-level events that lead to this phenomenon are critically important to breeding for greater flood tolerance in modern crops and therefore have captured the attention of researchers in recent years [65]. Because the proportion of root tissue that undergoes lysogeny is so small and close to the cell elongation zone, it is only now that proteomics has become a credible way to tackle the exact pattern of gene expression required to break down cortical cells in such an orderly fashion. This will require fine-scale tissue sampling which is guided by the anatomy of cortical cell breakdown and the molecular clues to when this degradation process is occurring [66] but promises great rewards if proteomics can lead us to targets for breeding programs.

Summary: The advent of mass spectrometry with higher sensitivity allows for physical samples of just a few tens of milligrams, enabling tissues with ever more highly defined physiological properties to be used in experiments. This is especially true where meristems are to be compared; arguably dividing cells have hitherto been ignored in proteomic studies and their response to abiotic stresses should be more deeply investigated as the opportunities for fine-scale sampling improve.

1.5 How Do Acclimation and Shock Differ?

A conceptual question in any discussion of experimental design is the line between stress (followed by acclimation) versus tissue shock, senescence and cell death (Fig. 1.2). This can never be satisfactorily resolved but the aim of the homily above is to design better experiments that inform us about acclimation and thereby, identify targets for genotypic improvement in subsequent breeding and biotechnology [4]. There is no single criterion for differentiating acclimation from damage due to shock. Markers for cell ageing or death might include caspases and other markers of programmed cell death, oxidative enzymes (polyphenol oxidases) and DNA repair enzymes. These molecular markers should be combined with physiological observations such as respiration rates, which should be sufficient to sustain cell function, and histochemical evidence (e.g. the use of vital stains—[53]). Recovery experiments are also vitally important because the failure of, not for a (healthy) steady state to be re-established indicates permanent tissue damage and is strong evidence that shock, senescence and cell death are taking precedence over acclimation. Comprehensive proteomic analyses promise to identify new markers for irreversible cell damage which might well become molecular signatures for over-zealous application of abiotic stress and a platform for design of meaningful experiments.

Acknowledgments I thank Aaron Phillips and Tom Roberts for careful editing of the manuscript.

References

1. Fritsche-Neto R, Borém A (2015) Phenomics: how next-generation phenotyping is revolutionizing plant breeding. Springer, Berlin
2. Vanderschuren H, Lentz E, Zainuddin I, Gruissem W (2013) Proteomics of model and crop plant species: status, current limitations and strategic advances for crop improvement. J Proteom 93:5–19
3. Sorenson R, Bailey-Serres J (2014) Selective mRNA translation tailors low oxygen energetics. In: Low-oxygen stress in plants. Springer, Berlin, pp 95–115
4. Guidetti-Gonzalez S, Labate MTV, De Santana Borges J, Budzinski IGF, Marques FG, Regiani T et al (2015) Proteomics and metabolomics as large-scale phenotyping tools. In: Phenomics. Springer, Berlin, pp 125–139

5. Low TY, Heck AJ (2016) Reconciling proteomics with next generation sequencing. Curr Opin Chem Biol 30:14–20
6. Hamblin A, Tennant D (1987) Root length density and water-uptake in cereals and grain legumes—how well are they correlated. Aust J Agric Res 38:513–527
7. Poorter H, Buhler J, Van Dusschoten D, Climent J, Postma JA (2012) Pot size matters: a meta-analysis of the effects of rooting volume on plant growth. Funct Plant Biol 39:839–850
8. Greenway H, Munns R (1980) Mechanisms of salt tolerance in nonhalophytes. Annu Rev Plant Physiol 31:149–190
9. Naidu R, Rengasamy P (1993) Ion interactions and constraints to plant nutrition in australian sodic soils. Aust J Soil Res 31:801–819
10. Cramer GR, Läuchli A, Polito VS (1985) Displacement of Ca^{2+} by Na^{+} from the plasmalemma of root cells a primary response to salt stress? Plant Physiol 79:207–211
11. Su Y, Luo W, Lin W, Ma L, Kabir MH (2015) Model of cation transportation mediated by high-affinity potassium transporters (HKTs) in higher plants. Biol Proc Online 17:1
12. Rozema J, Schat H (2013) Salt tolerance of halophytes, research questions reviewed in the perspective of saline agriculture. Environ Exp Bot 92:83–95
13. Hsiao TC (1973) Plant responses to water stress. Annu Rev Plant Physiol 24:519–570
14. Ludlow MM, Muchow RC (1990) A critical-evaluation of traits for improving crop yields in water-limited environments. Adv Agron 43:107–153
15. Mirzaei M, Pascovici D, Atwell BJ, Haynes PA (2012) Differential regulation of aquaporins, small GTPases and V-ATPases proteins in rice leaves subjected to drought stress and recovery. Proteomics 12:864–877
16. Ashoub A, Baeumlisberger M, Neupaertl M, Karas M, Bruggemann W (2015) Characterization of common and distinctive adjustments of wild barley leaf proteome under drought acclimation, heat stress and their combination. Plant Mol Biol 87:459–471
17. Atwell BJ, Henery ML, Rogers GS, Seneweera SP, Treadwell M, Conroy JP (2007) Canopy development and hydraulic function in *Eucalyptus tereticornis* grown in drought in CO_2-enriched atmospheres. Funct Plant Biol 34:1137–1149
18. Kelly JW, Duursma RA, Atwell BJ, Tissue DT, Medlyn BE (2015) Drought × CO_2 interactions in trees: a test of the low-intercellular CO_2 concentration (Ci) mechanism. New Phytol 209:1660–1612
19. Mirzaei M, Soltani N, Sarhadi E, Pascovici D, Keighley T, Salekdeh GH et al (2012) Shotgun proteomic analysis of long-distance drought signaling in rice roots. J Proteome Res 11: 348–358
20. Maksup S, Roytrakul S, Supaibulwatana K (2014) Physiological and comparative proteomic analyses of Thai jasmine rice and two check cultivars in response to drought stress. J Plant Interact 9:43–55
21. Dracup M, Gibbs J, Greenway H (1986) Melibiose, a suitable, non-permeating osmoticum for suspension-cultured tobacco cells. J Exp Bot 37:1079–1089
22. Hemantaranjan A, Patel PK, Singh R, Srivastava AK (2012) Heat stress responses of wheat and other plants. Adv Plant Physiol 13:279–313
23. Beck EH, Heim R, Hansen J (2004) Plant resistance to cold stress: mechanisms and environmental signals triggering frost hardening and dehardening. J Biosci 29:449–459
24. Neilson KA, Scafaro AP, Chick JM, George IS, Van Sluyter SC, Gygi SP et al (2013) The influence of signals from chilled roots on the proteome of shoot tissues in rice seedlings. Proteomics 13:1922–1933
25. Webb T, Armstrong W (1983) The effects of anoxia and carbohydrates on the growth and viability of rice, pea and pumpkin roots. J Exp Bot 34:579–603
26. Xia JH, Roberts J (1996) Regulation of H + extrusion and cytoplasmic ph in maize root tips acclimated to a low-oxygen environment. Plant Physiol 111:227–233
27. Bouny JM, Saglio PH (1996) Glycolytic flux and hexokinase activities in anoxic maize root tips acclimated by hypoxic pretreatment. Plant Physiol 111:187–194
28. Gibbs J, Greenway H (2003) Review: mechanisms of anoxia tolerance in plants. I. Growth, survival and anaerobic catabolism. Funct Plant Biol 30:353

29. Smith AM, Ap Rees T (1979) Pathways of carbohydrate fermentation in the roots of marsh plants. Planta 146:327–334
30. Colmer T (2002) Aerenchyma and an inducible barrier to radial oxygen loss facilitate root aeration in upland, paddy and deep-water rice (*Oryza sativa* L.). Ann Bot 91:301–309
31. Thomson CJ, Greenway H (1991) Metabolic evidence for stelar anoxia in maize roots exposed to low O_2 concentrations. Plant Physiol 96:1294–1301
32. Colmer TD, Greenway H (2011) Ion transport in seminal and adventitious roots of cereals during O_2 deficiency. J Exp Bot 62:39–57
33. Colmer TD, Huang S, Greenway H (2001) Evidence for down-regulation of ethanolic fermentation and K^+ effluxes in the coleoptile of rice seedlings during prolonged anoxia. J Exp Bot 52:1507–1517
34. Mustroph A, Lee SC, Oosumi T, Zanetti ME, Yang H, Ma K et al (2010) Cross-kingdom comparison of transcriptomic adjustments to low-oxygen stress highlights conserved and plant-specific responses. Plant Physiol 152:1484–1500
35. Narsai R, Rocha M, Geigenberger P, Whelan J, Van Dongen JT (2011) Comparative analysis between plant species of transcriptional and metabolic responses to hypoxia. New Phytol 190:472–487
36. Atwell BJ, Aprees T (1986) Distribution of protein synthesized by seedlings of *Oryza sativa* grown in anoxia. J Plant Physiol 123:401–408
37. Edwards JM, Roberts TH, Atwell BJ (2012) Quantifying ATP turnover in anoxic coleoptiles of rice (*Oryza sativa*) demonstrates preferential allocation of energy to protein synthesis. J Exp Bot 63:4389–4402
38. Nelson CJ, Li L, Millar AH (2014) Quantitative analysis of protein turnover in plants. Proteomics 14:579–592
39. Knief C, Delmotte N, Chaffron S, Stark M, Innerebner G, Wassmann R et al (2012) Metaproteogenomic analysis of microbial communities in the phyllosphere and rhizosphere of rice. ISME J 6:1378–1390
40. Atkinson NJ, Urwin PE (2012) The interaction of plant biotic and abiotic stresses: from genes to the field. J Exp Bot 63:3523–3543
41. Suzuki N, Rivero RM, Shulaev V, Blumwald E, Mittler R (2014) Abiotic and biotic stress combinations. New Phytol 203:32–43
42. Sojka R, Stolzy L, Kaufmann M (1975) Wheat growth related to rhizosphere temperature and oxygen levels. Agron J 67:591–596
43. Mittler R, Blumwald E (2010) Genetic engineering for modern agriculture: challenges and perspectives. Annu Rev Plant Biol 61:443–462
44. Matsui T, Kobayasi K, Yoshimoto M, Hasegawa T (2007) Stability of rice pollination in the field under hot and dry conditions in the Riverina region of New South Wales, Australia. Plant Prod Sci 10:57–63
45. Crawford AJ, Mclachlan DH, Hetherington AM, Franklin KA (2012) High temperature exposure increases plant cooling capacity. Curr Biol 22:R396–R397
46. Fricke W, Akhiyarova G, Wei W, Alexandersson E, Miller A, Kjellbom PO et al (2006) The short-term growth response to salt of the developing barley leaf. J Exp Bot 57:1079–1095
47. Munns R (2002) Comparative physiology of salt and water stress. Plant, Cell Environ 25:239–250
48. Trought M, Drew M (1982) Effects of waterlogging on young wheat plants (*Triticum aestivum* L.) and on soil solutes at different soil temperatures. Plant Soil 69:311–326
49. Gammulla CG, Pascovici D, Atwell BJ, Haynes PA (2010) Differential metabolic response of cultured rice (*Oryza sativa*) cells exposed to high- and low-temperature stress. Proteomics 10:3001–3019
50. Gammulla CG, Pascovici D, Atwell BJ, Haynes PA (2011) Differential proteomic response of rice (*Oryza sativa*) leaves exposed to high- and low-temperature stress. Proteomics 11:2839–2850
51. Sachs MM, Freeling M, Okimoto R (1980) The anaerobic proteins of maize. Cell 20:761–767

52. Narsai R, Edwards JM, Roberts TH, Whelan J, Joss GH, Atwell BJ (2015) Mechanisms of growth and patterns of gene expression in oxygen-deprived rice coleoptiles. Plant J 82:25–40
53. Waters I, Kuiper PJC, Watkin E, Greenway H (1991) Effects of anoxia on wheat seedlings. J Exp Bot 42:1427–1435
54. Roberts J, Callis J, Jardetzky O, Walbot V, Freeling M (1984) Cytoplasmic acidosis as a determinant of flooding intolerance in plants. Proc Natl Acad Sci USA 81:6029–6033
55. Shabala S, Bose J, Fuglsang AT, Pottosin I (2016) On a quest for stress tolerance genes: membrane transporters in sensing and adapting to hostile soils. J Exp Bot 67:1015–1031
56. Kulichikhin KY, Greenway H, Byrne L, Colmer TD (2009) Regulation of intracellular pH during anoxia in rice coleoptiles in acidic and near neutral conditions. J Exp Bot 60:2119–2128
57. Yamasaki K, Kigawa T, Seki M, Shinozaki K, Yokoyama S (2013) DNA-binding domains of plant-specific transcription factors: structure, function, and evolution. Trends Plant Sci 18:267–276
58. Cai YM, Yu J, Gallois P (2014) Endoplasmic reticulum stress-induced PCD and caspase-like activities involved. Front Plant Sci 5:41
59. Narsai R, Howell KA, Carroll A, Ivanova A, Millar AH, Whelan J (2009) Defining core metabolic and transcriptomic responses to oxygen availability in rice embryos and young seedlings. Plant Physiol 151:306–322
60. Lasanthi-Kudahettige R, Magneschi L, Loreti E, Gonzali S, Licausi F, Novi G et al (2007) Transcript profiling of the anoxic rice coleoptile. Plant Physiol 144:218–231
61. Kreps JA, Wu Y, Chang HS, Zhu T, Wang X, Harper JF (2002) Transcriptome changes for Arabidopsis in response to salt, osmotic, and cold stress. Plant Physiol 130:2129–2141
62. Seki M, Narusaka M, Ishida J, Nanjo T, Fujita M, Oono Y et al (2002) Monitoring the expression profiles of 7000 Arabidopsis genes under drought, cold and high-salinity stresses using a full-length cDNA microarray. Plant J 31:279–292
63. Mathesius U, Djordjevic MA, Oakes M, Goffard N, Haerizadeh F, Weiller GF et al (2011) Comparative proteomic profiles of the soybean (Glycine max) root apex and differentiated root zone. Proteomics 11:1707–1719
64. Sharp RE, Silk WK, Hsiao TC (1988) Growth of the maize primary root at low water potentials I. Spatial distribution of expansive growth. Plant Physiol 87:50–57
65. Rajhi I, Yamauchi T, Takahashi H, Nishiuchi S, Shiono K, Watanabe R et al (2011) Identification of genes expressed in maize root cortical cells during lysigenous aerenchyma formation using laser microdissection and microarray analyses. New Phytol 190:351–368
66. Takahashi H, Yamauchi T, Rajhi I, Nishizawa NK, Nakazono M (2015) Transcript profiles in cortical cells of maize primary root during ethylene-induced lysigenous aerenchyma formation under aerobic conditions. Ann Bot 115:879–894

Chapter 2
Cereal Root Proteomics for Complementing the Mechanistic Understanding of Plant Abiotic Stress Tolerance

Jennylyn L. Trinidad, Herra L. Grajo, Jose B. Abucay, Jr. and Ajay Kohli

Abstract Cereals are a staple food for four billion people globally with rice, wheat and maize making up 60 % of the energy intake by the world population. Climate change-mediated increase in the extent, frequency and unpredictability of the incidences of abiotic stresses frequently lead to decrease in the yield and grain quality of cereals. Additionally, demographic and socio-economic factors call for increase in the production of quality cereal grains. It is therefore crucial to generate stress tolerant cereal varieties and understand the underlying mechanisms so as to strategize the crop cultivation agro-physiology for long term benefits. Mechanistic understanding of plant responses to stress can best be elucidated through the omics tools and techniques and smart interpretation of their results. Proteomics forms an important aspect of the omics studies in relating the transcriptome to the metabolome. While most cereal proteomics studies dwell on the plants' overall tolerance strategies, proteomics studies either specifically on roots or comparing root responses to the aerial plant parts under stress have been somewhat limited. Root proteins are relatively difficult to extract and characterize, hence the lag in the identification of stress-specific proteins and transcription factors in the roots. However, with the advancements in protein identification and quantification, several important mechanisms have been determined to be at play during abiotic stresses. Root proteins with significant roles are mainly involved in ROS detoxification, energy metabolism, cell wall metabolism, and disease and defense responses. Plasma membrane proteins, regulators of signal transductions and ion channels also contribute to increased stress tolerance. This review brings together an understanding of stress response established by the proteomic studies on cereal

J.L. Trinidad · H.L. Grajo · J.B. Abucay Jr. · A. Kohli (✉)
Genetics and Biotechnology Division, Plant Molecular Biology Laboratory, International Rice Research Institute, DAPO 7777, Metro Manila, Philippines
e-mail: a.kohli@irri.org

J.L. Trinidad · H.L. Grajo · J.B. Abucay Jr.
College of Arts and Sciences, University of the Philippines Los Baños, College 4031, Los Baños, Laguna, Philippines

G.H. Salekdeh (ed.), *Agricultural Proteomics Volume 2*,
DOI 10.1007/978-3-319-43278-6_2

roots. Although intuitive to guess, the differences in proteomic responses of the roots can be very different from the aerial parts and in some cases diametrically opposite. As an integral component of the plants' aerial parts' response to stress, root response characteristics are important to be considered in the overall mechanistic understanding and engineering of plant response to stress, either through conventional breeding or modern biotechnological means.

Keywords Biotic · Abiotic · Stress · Root · Proteins · Drought · Salinity · Heat · Flooding · Nutrient deficiency · Heavy metal toxicity

2.1 Introduction

Cereals are the world's most important food source for humans. Their cultivation played a pivotal role in ushering human civilization. Billions of people around the world depend on rice, wheat, maize and to a minor extent, on barley, oat, sorghum and millet for their daily survival. With the increasing world population, the demand for cereals is expected to increase. The Food and Agriculture Organization [1] predicts that cereal crops must increase to about 3 billion tons to meet the future demand. In the developing countries, 60–80 % of the calories in the daily diet is derived from cereals [2]. It is also in these countries that the demand for cereals has exceeded the rate of production, hence, it is crucial to ensure growing production levels for a sustainable food supply. However, with the expected population to reach 9 billion by 2050 and the continuous loss of agricultural land, sustainable crop production must rely on increasing yield per unit area [3], and must be able to do so on less water, nutritional inputs, and pest and disease treatment chemicals [4].

Cereals can be severely affected by adverse environmental conditions. Being important food sources, it is imperative that they are able to withstand numerous biotic and abiotic stresses. Abiotic stresses such as drought, salinity, heat, cold and flooding cause molecular, biochemical, physiological and morphological changes in crop plants. These changes can severely affect cellular integrity and in turn cause major impairment in plant growth and reproduction, greatly reducing biomass and grain yield [5]. Plants, being sessile organisms, physically cannot escape harsh environmental conditions, hence must employ mechanisms to alleviate the effects of stress. Abiotic stresses in plants usually result in largely similar physiological effects. Plants respond to stress through altered calcium and reactive oxygen species (ROS) movement and metabolism and hormonal changes, generally leading to reduced photosynthesis and growth. If the plants are close to the reproductive phase during stress, it is generally fast tracked to produce propagules. In addition to the major environmental abiotic stresses, edaphic stresses such as heavy metal toxicity and nutrient deficiency can also severely affect cereal crops. Unsurprisingly, these two stresses are intertwined in their manifestation and effects. Heavy metal toxicity causes damaging effects and disturbances in the plants' metabolic functions such as transpiration and photosynthesis inhibition, disturbance of carbohydrate metabolism, nutritional stress and oxidative stress, ultimately affecting the plant's growth and development. Essential macro- and micronutrients play specific roles in different metabolic processes; hence deficiencies greatly affect crop growth and yield [6]. During the onset of stress, plants employ concerted molecular and physiological responses to tolerate and diminish the effects of the stress. A large and important part of such responses is the altered expression of stress-related proteins [7] and transcription factors [8]. Most differentially expressed proteins under abiotic stress are categorized as either metabolic, structural, or ROS-scavenging enzymes or proteins. In addition, plants respond to abiotic stresses by exploiting transcriptional changes and also translational and post-translational modifications such as glycosylation, phosphorylation, SUMOylation and methylation [9, 10].

Biotic stresses also pose a threat to worldwide cereal production. Being in close contact with the soil, roots are more prone to attacks by viruses, bacteria, and fungi. If not mitigated, this generally leads to significant decrease in yield and inevitably, economic losses [11]. In addition to microorganisms, plant-parasitic nematodes can also severely affect cereals. With their potential host range encompassing more than 3000 plant species, nematodes can also cause significant damage to worldwide agriculture. One way for plants to resist pathogen attacks is through the production of pathogen-related proteins and metabolites. Although several genes have been identified to confer resistance against the pathogens, very few proteins have been identified by using available proteomic approaches [12].

Roots are of particular interest in studying the effects of biotic and abiotic stresses in plants. They are the primary organs for water and mineral acquisition from the soil for use in plant growth and development. The plant's survival and performance greatly depend on the roots' ability to efficiently explore the soil in search of water and minerals [13]. Being entry points of water and nutrients, stable cellular integrity and optimum water uptake capacity must be maintained for proper functioning of vital processes for survival during stress. Being intrinsically complex tissues and the fact that they can be quite difficult to investigate, the roots' stress response mechanisms are not as well-studied as much as the other tissues. Roots are highly sensitive and have been shown to be the primary organs to exhibit specific cell defenses to different biotic and abiotic stresses [14].

Studying the effects of stress in plants utilizes different tools and approaches, one of which is proteomics. With recent technical improvements in the proteomics workflow, identification of plant proteins is more reproducible. With feasibility for a high-throughput analysis, the proteomics approach is becoming progressively more beneficial for studying crop plants. Furthermore, major technical advances have been made in using plant proteomics for food security [15]. Proteomics analysis is a useful tool in discovering stress-responsive and tolerance genes and pathways. Using the proteomics approach allows for identifying possibly significant changes in protein expression levels against a background of unresponsive proteins. In crop breeding, proteomics provides the advantage of detecting the stress-responsive proteins via comparison between the control and stressed plants. These proteins may then be ascertained to be consistently correlated with the expression of a particular trait [16].

Recent advancements in proteomic techniques paved ways to better explain the mechanism of stress and stress response in plants. The conventional two-dimensional electrophoresis (2-DE) coupled with mass spectrometry (MS) is still the most widely used method in resolving proteins and identifying stress-induced alterations in the plant proteome composition [17], but alternative gel-free procedures that are based on fractionation with liquid chromatography (LC) are also fast becoming popular [18]. Protein identification by MS was made easy by breakthroughs in soft ionization methods such as matrix assisted laser desorption ionization (MALDI) or electrospray ionization (ESI) and peptide fragmentation by collision-induced dissociation (CID) in tandem MS [19]. Second generation gel-free proteomic approaches which include LC-MS-based tagging techniques such as isotope-coded affinity tags (ICAT), stable isotope labeling by amino acids in cell culture (SILAC) and isobaric tags for relative and absolute quantitation (iTRAQ) allowed comparative and quantitative analysis of multiple samples and helped researchers see the effects of plant stress at a global level [18, 19].

In all these proteomics techniques, a good representation of protein population in control and stressed conditions critically relies on the basic protein extraction and preparation protocols, which pose a bit of a challenge in plants, especially in the root tissues. Aside from having relatively low protein concentration, root tissues contain large amounts of proteases, oxidative enzymes, and cellular materials such as cell wall and storage polysaccharides, lipids, and phenolic compounds that can interfere with downstream separation and analysis [20]. Plant roots have been generally considered as recalcitrant tissues for proteomic analysis because of their high degree of vacuolation and high secondary metabolite content [21]. Most proteomic studies on plant stress response utilized trichloroacetic acid (TCA)/acetone precipitation, phenol-based protocols, or their combinations and/or modifications to overcome the above challenges to a certain extent [18]. It is important to note that basic protocols might have to be modified to achieve optimal results for different plant tissues. For example, a study on the rice root proteome revealed that a high quality proteome map can be achieved from protein extraction using Mg/NP-40/TCA method [22]. Addition of a powerful nonionic surfactant, NP-40, was deemed necessary to counteract the high levels of interfering compounds such as lignins and celluloses in the rice roots. For rice root proteomics on stress response, dealing with root proteases is also a major issue due to the various functions of the stress-induced root proteases [23]. This was addressed by a simple procedure of boiling the root extract [24].

In this review, we concentrate primarily on cereals as an important group of plants for future food and feed security. We address abiotic stress tolerance in cereals primarily due to the greater combined impact of abiotic stresses on cereal yield/production compared to biotic stresses. By average, abiotic stresses can reduce yield in major crops by more than 50 % [25], while biotic stresses are responsible for 15–32 % loss [26]. Importantly, we address cereal root proteomics because a large body of information exists on the effects of abiotic stress on the aerial parts, and reviewing root responses can complement the mechanistic paradigms of stress tolerance proposed from such studies on the aerial plant parts.

In parallel, supplementing the mechanistic understanding of stress tolerance through the root response is important because it has been shown that root proteins respond more rapidly than leaf or stem proteins [27]. Also, proteins unique to roots are being rapidly recognized and need to be integrated as part of the whole plant response to growth, development and stress response. For example, in rice, proteomic analysis of all tissues during its growth course was undertaken in an effort to elucidate differential protein expression patterns. The study revealed that 36 % of proteins were unique to the roots [14]. Therefore, cereal root proteomics contributions to abiotic stress tolerance are reviewed for the particular stresses.

2.1.1 Drought

Water deficit affects numerous biological pathways and processes in plants, triggering developmental and physiological responses. Because of this, plants have evolved several adaptations to combat the effects of drought. Anatomically, plants developed spongy tissues to serve as water reservoirs, and growth is impaired to reduce leaf area and limit evaporation [28]. Responses such as leaf rolling, floral abscission, alteration of cuticle permeability [29], and floral induction [30] are also observed. The root system evidently has a critical role in response to water stress, with some plants evolving the ability to increase root growth at the very early stage of drought stress ensuring maximum water absorption in the soil. In some crops, root length, weight, volume and density are all associated with drought tolerance. Drought stress can induce changes in the dynamics of protein production and degradation as a direct response, causing either damage to the plants or trigger a cascade of physiological responses leading to tolerance [31].

Under drought stress, ROS and toxic ions produced by the plants damage essential proteins and decrease enzyme activity [32]. To counteract these deleterious effects, plants evolved protective molecular and physiological mechanisms to ensure homeostasis, detoxify the harmful molecules, and ultimately recover from the stress [33]. Proteins have important roles in stress response. They can function as protection for membranes and other proteins, and can acquire and transport water and ions in and out of the cell. Drought-inducible proteins such as chaperones, detoxification enzymes and mRNA-binding proteins play major roles in detoxification and during stress [34]. Differentially-expressed proteins under drought have been investigated in wheat, rice and *Arabidopsis* [35, 36].

Roots are the first organs to perceive the dehydration stress signal when a certain level is reached in water deficit [37]. Although the definitive mechanism of the roots' response is still unclear, recent molecular and biochemical studies revealed many abscisic acid (ABA)- and stress-responsive genes in the roots [22, 38]. In maize, high resolution 2-DE was used to identify novel proteins associated with both drought- and ABA-responsive proteins in the roots [39]. Twenty two proteins were identified using MALDI-TOF (time-of-flight) MS and were shown to be involved in energy metabolism, redox homeostasis and regulatory processes. Most of the proteins identified have regulatory and energy metabolism functions, as well as ROS scavenging and detoxifying enzymes. One protein identified is a glycine-rich RNA binding protein 2 (GRP2) believed to be involved in RNA stabilization, processing and transport. It has also been shown to possess an RNA-chaperone activity. Another protein, maize pathogenesis-related protein 10 (ZmPR10), is significantly enhanced by drought and is mainly expressed in the roots. An anionic peroxidase (APRX) is involved in the polymerization of phenolic monomers to generate the aromatic matrix suberin, along with a lignin biosynthesis enzyme, (OMT). Both of these proteins contribute to to the roots' increased drought tolerance. Enzymes for ROS detoxification were also upregulated in drought including superoxide dismutase (SOD), glutathione S-transferase (GST) and H_2O_2-decomposing antioxidant peroxiredoxins (PRXs). In rice, a novel protein RSOsPR10 was induced during drought stress, and on treatment with NaCl, jasmonic acid and probenazole almost exclusively in the roots. In the same study, it also showed upregulation during rice blast fungus infection [40]. A more recent study reported the increased expression of PR10 in the roots during drought stress and ABA-treatment [41]. Amino acid sequence of RSOsPR10 revealed homology with another protein OsPR10a/PBZ and showed similar functions during stress; however, the latter is expressed in the shoots, indicating organ-specific regulation [40].

MALDI-TOF MS was also used to identify 22 major proteins that were significantly upregulated during drought [42]. The proteins include an alcohol dehydrogenase1 (ADH1), which facilitate the interconversion between alcohols and aldehydes or ketones with the reduction of nicotinamide adenine dinucleotide NAD+ ADH1 showed significant upregulation under drought stress (in combination with heat stress) in maize roots under the promotion of ABA [42]. Increased levels of GST and cytosolic ascorbate peroxidase (APX) were also observed, both acting as regulators of ROS production during stress. Upregulation of both enzymes allows for the roots' improved tolerance to drought. Metabolic enzymes were also identified, one of which is nucleoside diphosphate kinase (NDPKs), which was significantly upregulated, and it is one of the important enzymes maintaining the balance between cellular ATP and other nucleoside triphosphates. Another enzyme, enolase 2, catalyzes the conversion of 2-phosphoglycerate (2-PGA) to phosphoenolpyruvate (PEP) during glycolysis. The enzyme was upregulated in response to drought in the roots. Fructokinase (Frk) catalyzes the transfer of a high-energy phosphate group to D-fructose to form fructose-1-phosphate. In roots, Frk was upregulated during drought. A serine/threonine-protein kinase receptor was upregulated under drought. Water channel proteins, aquaporins (a plasma membrane intrinsic protein/PIP) were also identified. PIP2-5 was upregulated under drought [42]. A similar proteomic survey was conducted in bread wheat, and found 34 root-specific differentially expressed proteins under drought similar to previous studies in maize [43].

During drought, shoot and leaf elongation is completely inhibited. In contrast, roots will often continue to grow despite the onset of stress; an important mechanism in plant adaptation [44, 45]. In maize, previous work on the mechanism of root growth adaptation during water deficit revealed that the maintenance of cell elongation is preferentially toward the apex, a response involving the modification of cell wall extension properties. This mechanism involves changes in cell wall protein (CWP) composition. Two-dimensional electrophoresis on maize root elongation zones revealed predominantly region-specific changes in cell wall protein composition in the water-stressed samples. A total of 152 drought-responsive proteins were identified and categorized according to their potential function in the cell wall. These water soluble and loosely ionically bound CWPs were either involved in ROS metabolism, defense and detoxification hydrolases, and those involved in carbohydrate metabolism. This indicates that stress-induced changes in the CWPs are involved in multiple processes that regulate the pattern of response of cell elongation within the elongation zone [46].

Transcription factors (TFs) have also been shown to confer tolerance to drought stress by regulating specific genes during stress [38]. Transcription factors are proteins that play important roles in transcriptional regulation by either activating or repressing their cognate target's promoters by binding in a sequence-specific manner to regulatory cis-elements. Transcription factors mediate their signaling input via protein-protein interaction. DNA-binding TFs are modular in structure and contain protein domains that facilitate functions such as DNA binding, multimerization, transcription activation or repression [47]. In-depth molecular studies showed a network of genes induced by drought stress with various transcription factors (TFs) regulating them, leading to specific proteins for tolerance and response [48]. Analysis of co-expression network data revealed as much as 1392 drought-responsive genes [49], most coding for TFs [50] such as bZIPs, AP2-ERF (DREB, HARDY), NAC, and WRKY [51, 52].

The NAC family is one of the largest plant-specific TFs, and has been shown to regulate a wide range of developmental processes, where several NAC proteins have been identified to interact with other proteins during defense and in response to biotic and abiotic stresses. In *Arabidopsis*, AtNAC072, AtNAC019, AtNAC055 and AtNAC102 have been identified to respond to drought, salinity, cold and submergence [53–55]. The TF OsNAC45 has a role in the development of lateral roots in rice. Transgenic lines carrying this TF showed greater drought tolerance [56]. Root-specific expression of OsNAC10 resulted to enlarged roots, which led to enhanced drought tolerance of transgenic plants and significant increase in grain yield under field drought conditions [38]. Root-specific promoters were used to overexpress OsNAC9 in transgenic rice and field performance was evaluated; with the transgenic rice showing increased yield by 13–18 %. Altered root architecture was also observed, showing increased root diameter due to enlarged xylem and augmented cortical cell size [57]. A no apical meristem (NAM) protein was identified to be an important TF, cis-regulating other drought-responsive genes in rice [52].

A shotgun proteomic approach in rice roots using nanoLC-MS/MS revealed that 38 % of proteins showed altered levels. Pathogenesis-related (PR) proteins were generally upregulated in drought-stressed plants while heat shock proteins (HSPs) were totally absent in fully watered plant roots. Proteins involved in oxidation-reduction reactions were upregulated during drought. Interestingly, two functionally contrasting protein families showed that tubulins were reduced in droughted roots while chitinases were upregulated [58]. A later study also used a shotgun proteomic approach to analyze the mechanism on how a large-effect drought QTL (*qDTY12.1*) confers drought tolerance in rice. A QTL-introgressed near isogenic line (NIL), 481-B, showing the best phenotype was compared with the recipient parent (Vandana) during proteomic analysis. Proteins for lateral root profusion, ROS detoxification enzymes (peroxidases, APX), glyceraldehyde-6-phosphate dehydrogenase, enolase and phosphoglucomutase were all upregulated in 481-B [59]. A cognate metabolite analysis study conducted by the same group on the same plant material drew high correspondence between the root proteins and metabolites generated by their activity, while also demonstrating differences in proteins in the roots and aerial parts [60]. In wheat, a comparative proteomic analysis was done to determine the response of two genotypes with contrasting drought stress responses. The drought tolerant genotype produced higher root biomass, longer roots and was able to absorb water more efficiently. Proteins belonging to defense and oxidative stress responses [Germin-like protein (GLP), GST, SOD], small heat shock proteins (HSPs), and APX families changed in abundance [61].

Another group of drought-inducible proteins are the LEA (Late Embryogenesis Abundant) proteins. Originally found in cotton (*Gossypium* sp.), these proteins accumulate late during embryogenesis and were later found in other vegetative organs under stress. The highly hydrophilic LEA protein genes have ABA response element (ABRE) and other drought response cis-elements in the promoter, and are hence inducible by ABA or drought [62]. Recently, transgenic rice containing a barley LEA protein HVA1 was shown to be highly tolerant to drought [63]. HVA1 is highly inducible by ABA, salt, cold and dehydration, and accumulates in the root apical meristem and lateral root primordia, resulting to root system expansion. Under osmotic stress, HVA1 protects the cell membrane from injury by stabilizing the proteins [64], thereby increasing water-use efficiency. It also helped promote lateral root initiation, elongation and emergence, as well as primary root elongation [63].

Aquaporins are a class of water channel proteins that are expressed in various membrane compartments of plant cells, and enhances water permeability in the vacuolar and plasma membranes (PMs) [65]. They are involved in the opening and closing of cellular gates, a process important in water balance and water use efficiency [66–68]. Aquaporins are encoded by a large multigene family, with 35 members in *Arabidopsis thaliana*, and are classified into four major subfamilies: PIPs, tonoplast intrinsic proteins (TIPs), Nod26-like intrinsic proteins (NIPs) and small and basic intrinsic proteins (SIPs). Many of these aquaporins show a cell-specific expression pattern in the root. It has been shown that aquaporins play a major role in facilitating the roots' capacity to alter their water permeability in response to stress [69]. These rapid changes are accounted for by the aquaporin-mediated changes in cell membrane permeability [70]. In rice, the PM aquaporin RWC3 was upregulated in upland rice 10 h after the water deficit, possibly by providing the cell with increased water uptake to maintain cell turgor during deficit [71]. Increased expression of four PIPs was observed in drought-stressed and ABA-treated maize plants, resulting to increased root hydraulic activity [72]. In barley, aquaporins HvPIP2;5, HvPIP2;2, and HvTIP1;1

increased water uptake capacity in the lateral roots [73]. The role of aquaporins in rice was studied by determining root hydraulic activity (Lpr) and root sap exudation rate (Sr) in well-watered and drought conditions, revealing their role in root water fluxes during drought stress and recovery [74].

Drought stress elicits major physiological and molecular responses in the roots. These responses involve the action of several important proteins that ensure the cell's integrity and proper functioning during water-limited conditions. Proteins, which protect the cells from oxidation, protect other proteins from desiccation, as well as those that change membrane permeability are often upregulated. Finally, transcription factors regulate numerous drought-responsive genes and have also been shown to confer important traits such as increased root diameter and lateral root development.

2.1.2 Salinity

Soil salinity is a major abiotic stress affecting more than 20 % of cultivated land worldwide. Excess NaCl in the soil interferes with mineral nutrition and reduces water uptake (osmotic stress). The accumulation of toxic ions (ionic stress) in plants results in cell injury [32, 75]. Plant roots are directly exposed to the saline conditions, hence are the first organs to be affected and the most sensitive to salt stress [76]. To alleviate these damaging effects, plants employ different strategies to re-establish proper cellular ion and osmotic homeostasis, as well as detoxification and repair processes. One such strategy is through upregulation of stress-related proteins [77, 78].

Roots are the primary perception and injury site during salinity stress, and it is the roots' sensitivity that limits the plant's productivity [79, 80]. In *Arabidopsis* roots, comparative proteomic analysis showed changes in the abundance of proteins in response to NaCl treatment. Using LC-MS/MS 86 proteins were identified to be stress-responsive proteins. Other proteins related to ROS scavenging, signal transduction, translation, cell wall biosynthesis, protein translation, processing and degradation, and metabolism of energy, amino acids, and hormones have also been identified [75]. Similar sets of proteins were also identified in wheat [43] and maize [81] roots.

Comparative proteomic analysis of wheat seedling roots subjected to a range of salt stress showed 198 differentially-expressed proteins with at least two-fold differences in abundance on 2-DE gels. Using MALDI-TOF MS, these proteins were identified as either involved in carbon metabolism, detoxification and defense, protein folding, and signal transduction [32]. Signal transduction proteins are important in sensing and transferring stress signals, thereby starting the cascade of signaling pathways necessary for the proper mitigation of the salt stress. A notable set of proteins identified in the study were guanine nucleotide-binding proteins (G proteins or GTPases). These act as modulators or transducers in different transmembrane signaling systems by regulating metabolic enzymes, ion channels, transporters, and controlling transcription, motility, contractility and secretion [82, 83]. Rice root PM-associated proteins were investigated following NaCl treatment. Results showed upregulation of 18 proteins by more than 1.5-fold in response to salt stress. MS analysis revealed that most of these membrane-associated proteins have roles in essential physiological processes such as membrane stabilization, signal transduction and ion homeostasis. A salt-responding leucine-rich-repeat type receptor-like kinase, OsRPK1, was identified to be also induced by cold, drought and ABA. Immunohistochemical techniques showed that the expression of OsRPK1 is localized in the PM of root cortex cells [84]. Protein ubiquitination is also an important post-translational modification that contributes to the regulation of many physiological responses. In rice roots, ubiquitination of several proteins such as pyruvate phosphate dikinase 1, HSP 81-1, probable aldehyde oxidase 3, PM ATPase, catalytic subunit 4 (UDP-forming) of cellulose synthase A and cylin-C1-1 was observed in salt-tolerant varieties [85].

Among the cereals, barley is relatively salt tolerant. Proteomic analysis of salt-stressed barley roots resulted in a relatively low number of differentially-expressed proteins [86]. Apart from the usual upregulated proteins (ROS-scavenging, protein synthesis, metabolism, disease and defense-related), the most abundant proteins observed were GST and lactoylglutathione lyase (or glyoxalase I) [86]. GST is a powerful antioxidant, catalyzing the conjugation of reduced glutathione (GSH) and is one of the major enzymes involved in the oxidative stress response, especially in salinity tolerance [87]. Lactoylglutathione lyase is involved in the glutathione-based detoxification of methylglyoxal (MG). It is a toxic by-product of carbohydrate and amino acid metabolism [86]. Elevated levels of MG indicates abiotic stress in plants. Its accumulation leads to several toxic effects in the plants, i.e. nucleic acid mutagenesis, modification, and subsequent degradation of proteins [88]. In a more recent study, significantly upregulated proteins were observed in the roots of salinity-tolerant barley lines. Proteins for signal transduction (annexin, translationally-controlled tumor protein homolog, lipoxygenases), detoxification (osmotin, vacuolar ATPase), protein folding (protein disulfide isomerase) and cell wall metabolism (UDP-glucoronic acid decarboxylase, β-D-glucan exohydrolase, UDP-glucose pyrophosphorylase) were also upregulated. This suggests that enhanced salinity tolerance is due to the increased signal transduction activity which led to the accumulation of stress-protective proteins and cell wall structural changes [89].

Annexins are ubiquitous proteins capable of binding and inserting into endomembranes and the PM [90]. Some members are capable of binding to actin, hydrolyzing ATP and GTP, acting as peroxidases or ion channels. They play central roles as regulators of stress signaling involving cytosolic free calcium and ROS [91]. Annexins are expressed throughout the plant body and many have been found in the roots [92, 93]. In *Arabidopsis*, annexins are responsible for the root epidermal PM Ca^{2+}- and K^{+} permeable conductance that is activated by extracellular hydroxyl radicals. In a study by Laohavisit et al. [90], annexins were observed to respond to high extracellular NaCl by mediating ROS-activated Ca^{2+} influx across the PM of root epidermal protoplasts.

Roots are the primary tissues affected by salinity stress. Several salt stress-responsive proteins have been identified to be involved in protecting the cells from ion toxicity. These proteins are mostly powerful antioxidants that alleviate oxidative stress. Cells are protected from ion influx by the regulation of PMs. Proteins involved in signal transduction, protein folding, carbon metabolism and post-translational modification were all implicated in enhanced salinity tolerance.

2.1.3 Cold

Cold or low temperature stress is one of the most severe abiotic stresses affecting plant growth and development in temperate regions. Cold stress causes cellular dehydration due to induced ice formation in plant tissues [94]. It also causes osmotic changes in the cell environment leading to the suppression of cellular activities. This results into reduced growth and decreased survival of the plants [95]. To cope with the adverse effects of low temperature stress, plants developed physiological strategies such as activation of primary metabolism to produce high energy, modifications of PM to maintain osmotic balance, elevating the levels of ROS-scavenging enzymes and chaperones to protect from oxidative damage, and regulation of enzymes involved in cell growth and cell wall synthesis [96–98].

Proteomic studies conducted by Lee et al. [27] on rice roots revealed that most of the proteins that were enhanced during chilling stress are involved in carbohydrate metabolism. These proteins include putative pyruvate orthophosphate dikinase precursor (PPDK), enolase, aconitate hydratase and glycine dehydrogenase. PPDK is responsible for the production of PEP which acts as a primary acceptor of CO_2. Enolase, aconitate hydratase, and glycine dehydrogenase are involved in glycolysis, Kreb's cycle and photorespiration, respectively. Enhanced levels of these enzymes suggest that plants may need to produce higher energy to cope with low temperature stress. Additionally, increased level of adenylate kinase, which catalyzes the reversible interconversion of ADP to ATP and AMP, further strengthens the idea that plants produce high energy under cold stress [99].

Modification in PM compositions and functions is one of the most important adaptation mechanisms of plants to low temperature stress. Quantitative proteomic analysis of PM of rice roots revealed that proteins which are involved in membrane permeability and signal transduction like cellular retinaldehyde-binding protein (CRALBP) and hypersensitive-induced response protein annexin are upregulated during cold or chilling stress [98]. CRALBP is a mammalian protein homologous to phosphatidylinositol/phospatidylcholine transfer proteins (PITPs) in plants [100]. PITPs are likely to regulate plant stress responses by controlling remarkable developmental pathways for polarized membrane biogenesis, which influences the symbiosis program that permits nitrogen fixation [101]. Moreover, Vincent et al. [102] proposed that Sec14p-like PITPs in *Arabidopsis thaliana* (AtSfh1p) regulate intracellular and PM phosphoinositide polarity, which directs the trafficking, Ca^{2+}-signaling and cytoskeleton functions to the growing root hair apex.

Annexins are Ca^{2+}-dependent membrane binding proteins that play important roles in membrane trafficking and organization. They are also known to regulate ion channel activity and phospholipids metabolism [103], and belong to a large structurally-related superfamily of proteins (HIR family) that includes prohibitins, stomatins and similar membrane proteins. The ion channel activity that leads to the regulation of diverse processes ranging from cell division and osmotic homeostasis to cell death is controlled by these proteins [104]. Thus, enhanced levels of CRALBP and HIR led to protection against the osmotic imbalance caused by cold stress.

Cold stress also elevates ROS, which trigger a series of harmful processes such as lipid peroxidation, degradation of proteins and DNA damage in the cell. To protect plants against them, enhanced accumulation of ROS-scavenging enzymes were observed during cold stress. Comparative proteomic analysis of cold-acclimated and non-acclimated rice revealed increased levels of SOD, catalase (CAT), APX, and glutathione reductase (GR) in rice roots when exposed in low temperature [105]. High SOD activity has been associated with stress tolerance in plants because it neutralizes the reactivity of O_2^-, which is overproduced under stress [106]. CAT, APX and GR are responsible for scavenging of H_2O_2. Moreover, enhanced levels of oxalyl-CoA dehydrogenase and glyoxalase I were observed in the roots of rice under chilling stress [27]. Oxalyl-CoA dehydrogenase is the second enzyme in the oxalate catabolism pathway which causes the decarboxylation of activated oxalate molecule that generates ROS through the Fenton reaction [19], whereas glyoxalase I plays a crucial role in methylglyoxal detoxification. Methylglyoxal is a cytotoxic compound formed as a side-product of several metabolic pathways due to several stresses including cold stress.

Under cold stress, regulatory proteins involved in signal transduction, protein biosynthesis and processing were found to be abundant in rice [27, 99] and maize [107] roots. These proteins include HSP70 putative calreticulin precursor and cysteine synthase (CysK). Heat shock proteins play important roles in keeping cellular homeostasis both under optimal growth conditions and under stress by acting as molecular chaperones that prevent the aggregation of denatured proteins and facilitates protein refolding under temperature stress [27, 107, 108]. Similarly, calreticulins were also shown to exhibit chaperone activity. Differential expression of the putative calreticulin precursor was also observed in the leaf sheath of rice under cold stress [99]. CysK, on the other hand, is the enzyme responsible for the assimilation of hydrogen sulfide to produce cysteine. Aside from its main function, CysK is found to be involved in GSH biosynthetic pathway, a reducing tripeptide that is utilized for protection against oxidative damage [109].

Low temperature stress resulted in the downregulation of phosphoglucomutase, 5-methyltetrahydropteroyltriglutamate-homocysteine S-methyl transferase, putative betaine aldehyde dehydrogenase and putative phenylalanine ammonia lyase [99]. Phosphoglucomutase is responsible for the interconversion of glucose-1-phosphate and glucose-6-phosphate. In the analysis of an *Arabidopsis* mutant, phosphoglucomutase was found to be involved in gravity perception [110]. 5-methyltetrahydropteroyltriglutamate-homocysteine S-methyl transferase, also known as cobalamin-independent methionine synthase (MetE), is the key enzyme for synthesis of methionine which is important for cell growth in plants [111]. Betaine aldehyde dehydrogenase is known as a salt-responsive protein [112] and phenylalanine ammonia lyase is known as a mechanical and bacterial infection-responsive protein [113]. Taking all these into account, downregulation of these enzymes under cold stress indicates that plants need to limit cell growth and other biochemical processes to be able to save energy to withstand the severely unfavorable environment.

Proteomic studies on rice roots reveal that plants prepare for recovery after the stress. One evidence for this is the upregulation of UDP-glucose phosphorylase, which is directly involved in synthesis of cellulose, a primary component of cell wall [99]. Enhanced level of this enzyme leads to novel synthesis of the cell wall for increased protection of the root. In addition, temperature stress-induced lipocalin is also enhanced during cold stress [27]. Expression analysis of this protein in wheat showed that it is associated with abiotic stress response and is correlated with the plant's capacity to develop freezing tolerance [114].

Under cold stress, roots develop mechanistic strategies such as production of high energy, limiting cell growth, and cell wall modification to protect the plant from the damages brought about by low temperature. These processes are facilitated by differential regulation of enzymes involved in carbohydrate and energy metabolisms, proteins that regulate signal transduction and protein turn-over, and proteins involved in cell wall synthesis and defence-responses.

2.1.4 Heat

The increasing global mean temperature seriously threatens to impact the yield and quality of major cereal crops [115]. Temperature is one of the major factors affecting plant growth and development: optimal temperatures for plant growth have been determined to range from 15–24 °C for shoots and 10–18 °C for roots of temperate plant species. While considerable effort has been made to study the proteins associated with heat stress on rice, wheat and barley, most of these studies dwell on the effects of the stress on leaves and panicles [116–118]. Upregulated proteins include those involved in photosynthesis, detoxification, energy metabolism, and protein biosynthesis.

Proteomic studies on rice anthers revealed the accumulation of stress-responsive cold shock proteins (sCSP) and sHSPs. HSPs are known to have chaperonic activities and function as reservoirs for intermediates of denatured proteins which then prevent protein aggregation due to heat, hence they play important roles in protecting metabolic activities of the cell [119]. Heat shock proteins, combined with Ca^{2+}-signaling proteins and efficient protein modification and repair mechanisms, were identified to confer heat tolerance in rice [120]. In maize, levels of HSP101 was increased in response to heat stress, more abundantly so in the developing tassel, ear, silks, endosperm and embryo; and less abundantly in vegetative and floral meristematic regions, mature pollen, leaves and roots. HSP101 belongs to a protein family whose members promote renaturation of protein aggregates and are essential for the induction of thermotolerance. Localization analysis revealed that HSP101 was highest in the root cap meristem and quiescent center of the heat-stressed roots [121]. In a later study, it was revealed that HSPs play important roles in both induced and basal thermotolerance. In addition, elongating segments of primary roots exhibited a strong ability to synthesize nucleus-localized HSPs [122].

The roots' response mechanisms to heat stress is far less investigated despite their importance in whole plant adaptation to high temperatures [115]. Roots have lower optimal growth temperatures. Hence, are more sensitive to elevated temperatures as seen in the decline in root growth and physiological function at higher soil temperature. Proteomics study was done on two species turf grass or forage grass *Agrostis stolonifera* and *A. scabra* roots subjected to heat stress. The leaves showed impaired antioxidant enzyme activities and increased lipid peroxidation in the heat-sensitive cultivar. Inducing heat injury in the roots disrupted root functions such as water and nutrient uptake and transport to the leaves, and also affected the cytokinin synthesis in the roots. Upregulation of sucrose synthase, GST, SOD, and the HSP Sti (stress-inducible protein) was observed in the heat tolerant *A. scabra* [123]. In rice, an A20/AN10 type zinc finger protein ZFP177 was found to be responsive to heat stress. It was revealed that the protein is localized in the cytoplast of both leaf and root cells. Overexpression of ZFP177 in tobacco conferred tolerance of the transgenic plants to both low and high temperatures [124]. In maize roots, combination of drought and heat stress was imposed to identify differentially-expressed proteins associated with ABA regulation. Twenty two major proteins were significantly upregulated under combined stresses. These were categorized as proteins involved in disease/defense, metabolism, cell growth/division, signal and transporters [42].

Having low optimal growth temperatures than the rest of the plant, roots are extremely sensitive to increasing temperatures. While several proteomic studies on the effects of heat stress have been done on many cereal species, these are mainly on the panicles and leaves. Some heat-responsive root proteins have been identified to be involved in metabolism, cell growth, signal transduction, transport and ROS detoxification. Other proteins play roles in modification and repair, as well as protection from heat denaturation. However, a full-scale survey of heat-responsive root proteins still needs to be undertaken.

2.1.5 Flooding

Among the most cultivated cereal crops, rice is the most flood-tolerant crop whereas maize, barley and wheat are categorized as flood-sensitive. Flooding is caused by heavy or continuous rainfall in an area with poorly drained soil and is one of the most important environmental stressors [125]. Flooding stress causes hypoxic to anoxic conditions which severely affects the roots of the plants. These conditions lead to anaerobic metabolism (glycolysis followed by alcohol fermentation) and also to programmed cell death or PCD [126]. At the developmental stage, plants can escape the low oxygen stress caused by flooding through multifaceted alterations in cellular and organ structure that promote access to and diffusion of oxygen [127].

Proteomic studies revealed that one of the common responses of plants to flooding stress is the alteration of proteins involved in primary and secondary metabolism, and energy production. For example, flood stress decreases the level of two glycolytic enzymes in wheat roots, namely fructose-1,6-bisphosphate aldolase and sucrose:fructan 6-fructosyltransferase [128]. Fructose-1,6-bisphosphate aldolase catalyzes the reversible interconversion of two triose phosphates to fructose-1,6-bisphosphate. Triose phosphates are the main carbon sources for sucrose synthesis in photosynthetic tissues, which are derived from Calvin cycle and are exported from chloroplast during the day. Sucrose:fructan 6-fructosyltransferase is involved in the synthesis of fructans. Fructans serve as the major form of carbohydrate reserves. Downregulation of these enzymes indicates that carbohydrates and energy consumption are reduced due to inhibition of photosynthesis under flooding conditions [128]. Conversely, significant induction of ADH was observed in maize roots under flood. ADH functions to enhance rates of carbohydrate breakdown, fermentation, and ATP synthesis [129].

Proteins involved in disease/defense mechanism were found to be the most upregulated in the roots of wheat seedlings under flooding stress [128]. These include two glycosylated polypeptides, α-amylase/subtilisin inhibitor, chitinase and malate dehydrogenase (MDH) or α-1,4-glucan-protein synthase. Among these proteins, α-amylase/subtilisin inhibitor functions as defense against microorganisms since it is found to inhibit both α-amylase from larvae of the red flour beetle (*Tribolium castaneum*) and subtilisin from *Bacillus subtilis* [130]. Chitinase works directly and indirectly as defense mechanism against pathogens and abiotic stresses [131]. Likewise, abundance of β-glucosidase in the roots of maize has been implicated in hormone metabolism and protection against pathogens [132].

Major proportion of downregulated proteins in wheat roots under flood are related to cell wall structure and modification [128]. These proteins include methionine synthase, β-1,3-glucanases, β-glucosidase, β-galactosidase, and (1,3;1,4)-β-glucanase precursor. Methionine synthase catalyzes the production of methionine which is important in plant cell growth. Downregulation of methionine synthase was also observed in rice roots under cold stress. β-1,3-glucanases, β-glucosidase, β-galactosidase, and (1,3;1,4)-β-glucanase belong to a family of hydrolases directly involved in the modification of cell wall polysaccharides. Downregulation of these proteins indicates that the roots of wheat seedlings respond to flooding stress by restricting cell growth through the limitation of hydrolysis of cell wall polysaccharides and assimilation of methionine. In addition, restriction of cell wall hydrolases helps to save energy as well as to preserve carbohydrate reserve, which can support the plant survival under flooding conditions [128]. Similarly, levels of ROS scavenging enzymes like APX and GR in the roots of wheat seedlings were found to be diminished since the generation of ROS is limited under these conditions [133]. Additionally, decreased levels of SOD, peroxidase and the above mentioned enzymes were observed in the roots of maize [134] and rice [135].

Proteomic analysis also revealed an enhanced level of HSPs in the roots of maize under anaerobic stress caused by flooding [136]. HSPs are small, novel set of proteins that represses the synthesis of the nonstress proteins. Furthermore, proteins with crucial roles in both cytoplasmic and organellar translation and mitochondrial elongation were found to be repressed by hypoxia which contribute to the reduction in protein synthesis during flooding stress [132]. This result further strengthens the proposition that cereal roots need to restrict cell growth to save energy to survive under severely unfavorable conditions. Proteomics study on maize roots also give an insight that roots develop ways to recover after flooding. One example is the enhanced accumulation of actin in the roots of maize after hypoxia [132]. Actin is involved in cell wall expansion for root elongation [137].

Flooding elicits hypoxic to anoxic conditions in the soil which lead to physiological and cellular changes in the roots of the plants. Proteomics analysis showed that roots improve their defense mechanism by increasing the levels of PR proteins. Downregulation of proteins involved in cell wall modification indicates that roots limit energy consumption and preserve carbohydrate reserves. Alteration in carbohydrate metabolism is also evident during flood stress, thus levels of ROS scavenging enzymes were decreased.

2.1.6 Heavy Metal Toxicity

Contamination of the soil with heavy metals (HMs) has become a major global concern as industrialization and increased dependence of agriculture on chemical fertilizers and sewage wastewater irrigation introduce considerable amounts of these toxic substances into agricultural soils. This leads to decreased crop yields as well as hazardous health issues when these metals enter the food chain [138, 139]. HMs, in general, are metals and metalloids with specific gravity of approximately 5.0 g/cm^3 or higher [140]. Some HMs are essential in plants in trace amounts as they play important roles in some physiological processes, particularly as enzyme cofactors, e.g. calcium (Ca), copper (Cu), zinc (Zn) and iron (Fe). A deficiency or excess of these HMs can be detrimental to plant growth and development. Some HMs however are nonessential and may be highly toxic in plants even at very low concentrations, e.g. cadmium (Cd), aluminum (Al), lead (Pb), mercury (Hg) and arsenic (As) [141].

The presence of specific or generic ion carriers or channels for essential nutrients allow the entry of most HMs into the plant root system. Ca^{2+}, Fe^{2+} and Zn^{2+} channels were suggested to be possible routes of uptake and transport of Cd^{2+} and Pb^{2+} [142–144]. Once inside the cells, HMs exhibit their toxicity in plants by affecting a wide range of cellular functions. They bind to specific functional groups in proteins, particularly to sulfhydryl groups of cysteine residues, and displace vital metal cations in enzyme binding sites, resulting to protein denaturation or enzyme inactivation [145]. HMs are also known to disturb redox homeostasis by stimulating the formation of free radicals and ROS, which lead to lipid peroxidation, DNA strand cleavage, and oxidative damage to proteins [146]. Ultimately, the result of heavy metal intoxication is altered plant metabolism as the plant makes strategies to combat the stress. These include the synthesis of membrane transporters and thiol-containing compounds (chelators) for vacuolar sequestration, defense proteins for ROS scavenging, and molecular chaperones for re-establishing native protein conformation. The stimulation of photosynthetic and mitochondrial respiration enzymes also aids in producing more reducing power to compensate the high-energy demand of HM-challenged cells [18].

A few proteomic studies were conducted to show the response of cereal root proteins on HM stress. In a study on rice root tissues, a dramatic decrease in the GSH levels was observed after short-term exposure of the roots to Cd, while no significant difference in the GSH levels was detected in the control and Cd-treated leaf tissues [147]. GSH is a major reservoir of non-protein thiols and plays a central role in the defense of plants against HMs and ROS [148]. About half of the number of differentially expressed proteins under Cd stress in the rice roots are also found to be oxidative stress-related proteins or antioxidant enzymes. This includes one GR, which is involved in the reduction of oxidized glutathione (GSSG) to GSH, and three GSTs, which are responsible for the direct quenching of Cd ions by forming GSH-Cd complexes [149]. These results imply that, being the primary site of exposure to Cd contamination in the soil, the rice roots allow rapid consumption of its GSH for the chelation of Cd before it can do further damage to other tissues. In maize, a similar trend in the GSH levels in the roots and leaves was observed upon Cd treatment [150]. Data from literature also suggest that Cd generally remains in the roots of maize and only small quantities are transported to the shoots of Cd-treated plants [151, 152].

Another study on the rice root proteome under Cd stress revealed the induction of different transporters such as the ABC transporters [153], which are believed to be involved in Cd sequestration in the vacuole [154, 155], and Nramp (natural resistance-associated macrophage protein), which is associated with the transport of divalent cations in plants [156, 157]. Cynnamyl alcohol dehydrogenase, an enzyme involved in the biosynthesis pathway of lignin, was also found to be upregulated in the rice roots after Cd treatment. Lignification of the root cell walls which leads to reduction in root cell wall expansion is a common defense mechanism of plants to decrease further uptake of toxic HMs from the soil [158].

A total of 27 Cu-binding proteins that are involved in antioxidant defense and detoxification, pathogenesis, regulation of gene transcription, amino acid synthesis, protein synthesis, modification, transport and degradation, cell wall synthesis, molecular signaling, and salt stress were found to be differentially expressed in a recent study on the root proteome responses of a Cu-tolerant and Cu-sensitive varieties of rice upon exposure to Cu [159]. Similar to the findings of Lee et al. [147], a GST was also found to be induced in the Cu-stressed roots of both rice varieties, which can be assumed to be involved in the direct detoxification of Cu by forming GSH-Cu complexes. A Cu/Zn-SOD and a GLP were also observed to be upregulated in both varieties. Some GLPs are known to act as SODs, which are metalloenzymes that catalyze the dismutation of the highly reactive superoxide ions ($O_2^{\cdot-}$) to H_2O_2 and O_2 [160, 161]. The induction of these proteins with SOD activity in Cu-stressed rice roots implies that they play a role in the first line of defense against ROS. It is also interesting to note that four of the identified differentially expressed proteins were PR proteins which were upregulated in the roots of the Cu-tolerant variety. PR proteins participate in a wide range of cellular functions, including cell wall rigidity, signal transduction, and antimicrobial activity [159]. The upregulation of PR proteins has also been reported in roots of other plant species exposed to heavy metal stress [162–164].

Expression patterns of maize root proteins in response to As stress were described by Requejo and Tena [109, 165]. Seven out of the eleven proteins differentially regulated by As in the maize roots were identified to be involved in cellular homeostasis for redox perturbation, including SODs, glutathione peroxidases (GPXs), and PRX. Both GPX and PRX catalyze the reduction of H_2O_2 to water coupled with the oxidation of GSH to GSSG and PRX (reduced) to PRX (oxidized). This observation suggests that oxidative stress is a major process underlying As toxicity in plants. In rice roots, 23 differentially expressed proteins were detected upon exposure to As stress, including S-adenosylmethionine synthetase (SAMS), CysK, GSTs and GR [166]. These proteins, which are closely related by their roles in sulfur metabolism, presumably work synchronously wherein GSH plays a central role in protecting cells against As stress [167].

Proteomic analyses of rice root response to As and Al suggest that the mechanism of stress response of rice to As might be similar to Al. In two independent experiments on the Al-sensitive rice cultivar, Michikogane [168], and on the Al-tolerant rice cultivar, Xiangnuo 1 [169], GSTs, SAMS, CysK, and Cu/Zn-SODs were also found to be the major enzymes upregulated in rice roots under Al stress. On the basis of their proteomic and metabolomics analyses, Fukuda et al. [168] and Yang et al. [169] suggested that SAMS and CS play functional roles in the mechanism of adaptation of rice to Al.

In barley roots, Patterson et al. [170] compared the abundance of proteins in a boron (B)-tolerant and a B-intolerant cultivar upon exposure to B stress. In addition to other stress-related proteins, the abundance of iron deficiency sensitive 2 (IDS2), IDS3, and a methylthioribose kinase, were elevated in B-tolerant plants. These three enzymes are key players in the biosynthesis of phytosiderophores, which are proteins with strong chelating activity and are known to participate in Fe uptake. These results suggest a potential link between Fe, B, and phytosiderophores under conditions of B stress, which is yet to be explored in future research.

From the above experiments, it is clear that the main mechanism explaining HM toxicity in plants is oxidative stress, and that in cereals, the first line of defense is the expression of proteins associated to HM chelation, transport and vacuolar sequestration in the roots to prevent entry and further damage of HMs to aerial organs. Direct detoxification by reduction of ROS brought about by these highly reactive substances also occurs in the roots, with the aid of enzymes such as SODs and PRXs. In all these activities, the role of GSH for stress alleviation is apparent and seemingly of immense importance, as evidenced by the induction of a vast number of enzymes related to its metabolism.

2.1.7 Nutrient Deficiency

Plants require essential mineral nutrients for normal functioning and growth. These mineral nutrients, which are required in relatively huge (macronutrients) to trace amounts (micronutrients), serve numerous functions, e.g. maintaining charge balance; acting as electron carriers, structural components, and enzyme activators; and providing osmoticum for turgor and growth. The effects of mineral nutrient deficiencies can be very subtle to very dramatic, from small changes in the pH of the cytosol and reduced export of carbohydrates to immediate termination of root growth and substantial disruption of membranes or cell walls. Taken together, each of these can result to oxidative stress, destruction of chloroplasts, and the symptoms associated to chlorosis and necrosis [171]. It is therefore understandable that the low availability of one or more of these nutrients is a key factor limiting crop yields [172]. Plants have evolved strategies to maximize nutrient availability, such as stimulation of high-affinity transporters, remobilization of nutrients from the rhizosphere or aerial portions of the plant by secreting organic acids or phosphatases, reduction of growth and photosynthesis, among others [173].

The mechanism of cereal root response to nutrient deficiency is a subject of many experiments and reviews at the genome, transcriptome and metabolome levels but remains as a scarce topic in proteomics studies. Nevertheless, root proteome responses of maize under phosphorus (P) [174] and Fe starvation [175]; rice under P deficiency [176]; and barley under nitrogen (N) deprivation [177] have already been characterized and gave important preliminary clues to the complex molecular cross-talk taking effect in the adaptation of cereal roots to nutrient deficiency.

Phosphorus is one of the key macronutrients, being an essential component of nucleic acids, phosphorylated sugars, lipids, and proteins, as well as ATP, which acts as the major energy currency of the cell. The lack of P often limits plant growth because most inorganic P are present as phosphate esters or metal ion salts, which have very low solubilities [178]. Plants have evolved adaptive strategies to cope with inadequate P supply. These include the alteration of root morphology to enhance P usage, modifications in carbon metabolism by bypassing steps that require P, and increased secretion of acid phosphatase (APase), ribonuclease (RNase) and organic acids in the roots to increase P availability in the soil [178–181].

Protein profiling of maize roots under P deprivation of a wild type (Qi-319) and low P tolerant mutant (99038) demonstrated differentially accumulated proteins under low P stress. These proteins were involved in multiple pathways, including carbon and energy metabolism, signal transduction, regulation of the cell cycle, and phytohormone metabolism [182]. Interestingly, under low P conditions, the amount of MDH, pyruvate dehydrogenase complex (PDC) E1a subunit and citrate synthase (CS) in the roots of 99038 plants significantly increased compared with Qi-319, while the amount of NAD^+-dependent isocitrate dehydrogenase (IDH) subunit 1 and aconitate hydratase (AH) significantly decreased. The accumulation of MDH, PDC and CS implies stimulation of citrate synthesis, while a decrease in IDH and AH suggests a reduction in citrate utilization. Studies have shown that of the organic acids, citrate is one of the most effective in solubilizing P in the soil [183]. Furthermore, the amount of phosphoprotein phosphatase 2A (PP2A) isoform 4 significantly increased in 99038 roots compared to Qi-319. In *Arabidopsis*, PP2A was shown to participate in hormone-mediated growth regulation, control of cell shape and plant morphology, regulation of the cell cycle, and elongation of root cortex cells [184, 185]. Taken together, the enhanced production of citrate and upregulation of PP2A in the roots of 99038 plants might explain its observed tolerance to P starvation compared to the Qi-319 wild type.

In rice roots, proteomic analysis revealed ten P starvation responsive proteins involved in metabolism and defense or stress response [176]. One of the two upregulated proteins under low P supply was identified as a member of the PR-10 family, which are generally known to have RNase activity and in most cases, are involved in defense response and plant development [186, 187]. The PR-10 proteins play yet another role in stress response by acting as RNases in the remobilization of nucleic acids and nucleotides in rice during P starvation.

Nitrogen is a macronutrient that primarily influence plant growth as it makes up many important biomolecules in plants such as proteins, nucleic acids, phytohormones, and chlorophyll. N can be taken up by plant root cells in either nitrate or ammonium form, but nitrate is the predominant form of N in agricultural soils [188]. The study of Moller et al. [177] revealed that the levels of many proteins altered in barley roots under long- and short-term N starvation are enzymes involved in nitrogen and carbon metabolism and are components of metabolic pathways implicated in stress response and regulation. Nitrite reductase, which plays a pivotal role together with nitrate reductase in the conversion of nitrate to ammonium during nitrate assimilation, was found to be upregulated in the roots, whereas it is downregulated in the shoots. This observation reflects a shift of nitrate reduction to the roots relative to the shoots under low nitrate supply [189]. Meanwhile, the downregulation of 14-3-3 proteins in roots under both long- and short-term N starvation reflects an increase in NADH:nitrite reductase activity in the roots, as 14-3-3 proteins are known to bind to a phosphorylated motif of the enzyme to effect inhibition [190].

Iron is a micronutrient that is ubiquitously present in agricultural soils. However, its bioavailability is often low due to often alkaline pH conditions of the soil [191, 192]. Fe deficiency is usually associated to oxidative stress, because most proteins acting as electron carriers in the respiratory chain are Fe-dependent. The lack of these Fe centers may result to an incomplete reduction of oxygen, ultimately leading to superoxide radical formation. The general response of maize root PM proteins exposed to low Fe conditions appears to involve proteins related to oxidative stress, growth regulation, and adaptation of nutrient uptake or translocation [175]. Two isoforms of the P-type H^+-ATPase, which are conceived to produce a proton gradient over the PM, are shown to increase with Fe deficiency. A specific member of this group, H^+-ATPase 2, is expressed in epidermal cells and was found to be involved in rhizosphere acidification for making Fe^{3+} and Cu^{2+} more soluble for plant utilization [193, 194]. One putative quinone reductase has also been identified with Fe starvation. The capacity of a naphthoquinone to transfer electrons for several NAD(P)H dependent redox enzymes purified from PM preparations has been described [195]. Vitamin K1 or phylloquinone, which is found to be a constituent of the PM, was also demonstrated as a lipid soluble electron carrier, suggesting an electron transport chain to a transmembrane constituent [196]. Considering this, the authors suggested that the increase of quinone reductase with Fe deficiency could represent a strategy to increase the transmembrane electron flow in the shortage of Fe supply. Another way of looking at these results is by considering the protective function of quinones against oxidative damage, where quinone reductases work to regenerate the quinone pool [175].

A common trend in the response of cereals under nutrient deficiency can then be seen in the plant's attempt to increase nutrient uptake by changing the expression of proteins involved in the modification of root architecture and in the production and excretion of metabolites in the roots which assist in increasing the solubility and availability of nutrients in the soil. Changes in both carbon and nitrogen metabolism also occur, as the plants utilize key enzymes that participate in the remobilization of metabolites to satisfy the nutrient needs of the growing plant under limiting nutrient supply.

2.2 Conclusions

Most of the proteomics studies done on cereal roots under different abiotic stresses showed that proteins involved in ROS detoxification, cell wall metabolism, disease and defense, and energy metabolism were upregulated. On the other hand, proteins involved in ROS detoxification and cell wall metabolism were diminished in levels under flooding stress due to limited supply of oxygen in the soil (Table 2.1; Fig. 2.1). The primary role of the roots in combating stress is also evident by the

Table 2.1 Major abiotic stress-responsive proteins in the roots of cereals

Stress	Family of proteins	Functions
Drought	CWPs	Involved in ROS metabolism, defense and detoxification hydrolases, and those involved in carbohydrate metabolism
	TFs	Transcriptional regulation
	LEA	Help promote primary and lateral root elongation; protection for protein aggregation
	mRNA-binding proteins; ROS scavengers	Detoxification
	Aquaporins	Maintenance of water balance

(continued)

Table 2.1 (continued)

Stress	Family of proteins	Functions
Salinity	ROS scavengers; detoxification enzymes	Detoxification
	PM proteins	Signal transduction; ion channels
	Cell wall metabolism proteins	Cell wall modification
	Disease and defense-related proteins	Enhance defense mechanism
	Protein synthesis, modification and turnover-related proteins	Protein homeostasis; regulation of physiological responses
Heat	Chaperones	Prevents protein aggregation; protein modification and repair mechanism
	ROS scavengers; Detoxification enzymes	Detoxification
	Disease and defense-related proteins	Enhance defense mechanism
	Ca^{2+}-signaling proteins	Protein modification and repair mechanism
Cold	Carbohydrate metabolism enzymes	Energy production
	Disease and defense-related	Enhance defense mechanism
	Regulatory proteins	Signal transduction; protein biosynthesis and processing
	ROS scavengers	Detoxification
	Cell wall metabolism proteins	Cell wall modification; enhance defense mechanism
	Cell growth-related↓	Limit energy consumption
Flood	Aerobic metabolism enzymes↓	Energy production and carbohydrate reserves preservation
	Cell wall metabolism proteins↓	Cell wall modification; carbohydrate reserves preservation
	ROS scavengers↓	Detoxification
	Disease and defense-related	Enhance defense mechanism
	Chaperones	Repression of non-stress proteins syntheses; signal transduction
Heavy metals	ROS detoxification enzymes	Detoxification
	HM chelators and transporters	Vacuolar sequestration of heavy metals; Detoxification
	Cell wall metabolism proteins	Cell wall modification; decrease uptake of HMs
Nutrient deficiency	Cell wall metabolism proteins	Cell wall modification; root elongation for increase nutrient uptake
	Nucleases	Remobilization of nucleic acids and nucleotides
	Nutrient uptake-related protein	Increase availability/solubility of nutrient in the soil

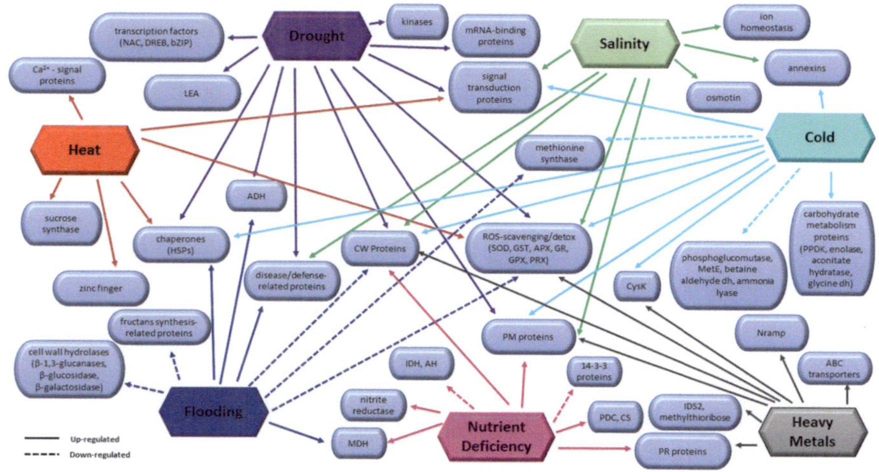

Fig. 2.1 The integrated response of cereal root proteins under abiotic stress. *NAC* NAM, ATAM and CUC, *DREB* dehydration responsive element-binding, *bZIP* basic leucine zipper, *LEA* late embryogenesis abundant, *ADH* alcohol dehydrogenase, *HSP* heat shock protein, *CW* cell wall, *ROS* reactive oxygen species, *SOD* superoxide dismutase, *GST* glutathione S-transferase, *APX* ascorbate peroxidase, *GR* glutathione reductase, *GPX* glutathione peroxidase, *PRX* peroxiredoxin, *CysK* cysteine synthase, *MetE* 5-methyltetrahydropteroyltriglutamate-homocysteine S-methyl transferase, *dh* dehydrogenase, *PPDK* pyruvate orthophosphate dikinase, *PP2A* phosphoprotein phosphatase 2A, *MDH* malate dehydrogenase, *IDH* isocitrate dehydrogenase, *AH* aconitate hydratase, *PM* plasma membrane, *PDC* pyruvate dehydrogenase complex, *CS* citrate synthase, *PR* pathogenesis-related, *Nramp* natural resistance-associated macrophage protein

Table 2.2 Drought and salinity stress-responsive root-specific proteins

Abiotic stress	Protein	Cereal	Reference
Drought	pathogenesis-related protein 10 (ZmPR10)	Maize	[39]
	O-methyltransferase (OMT),	Maize	[39]
	OsNAC10	Rice	[38]
	HVA1	Rice	[63]
	Aquaporins HvPIP2;5, HvPIP2;2, and HvTIP1;1	Barley	[73]
Salinity	leucine-rich-repeat type receptor-like kinase, OsRPK1	Rice	[84]

induction of proteins found exclusively in the root tissues (Table 2.2). Modification of the plasma membrane is also one of the adaptive mechanisms of plants in response to abiotic stresses. Under drought stress, high levels of PM-associated aquaporins enhance water permeability. Proteins that regulate signal transductions and ion channels, such as annexins, were also highly expressed in roots under salinity and temperature stresses. Additionally, cold stress induces the upregulation of PITPs which regulate phospholipid metabolism to increase membrane permeability. An induction in the abundances of PM-associated quinone reductase and H^+-ATPase in Fe deficient roots also suggests an increase in transmembrane electron flow and acidification of rhizosphere for increased iron solubility and availability, respectively. Despite all these findings, only a number of studies have been carried out for membrane proteomics due to the PM's dynamic property, which poses a challenge for isolating the full complement of membrane proteins. Advancements in PM proteome quantification have been made in both label-based (e.g. isobaric tags, isotopic labelling) and label-free forms (e.g. spectral counting, extracted ion chromatogram) [197]. Further improvement of these instruments for them to become cheaper and more efficient will lead to easy access to such tools in larger number of laboratories, which will push the frontiers of membrane and root proteomics towards a holistic understanding of plant response to stress.

References

1. FAO (2002) World agriculture: towards 2015/2030 summary report. www.fao.org
2. Who J, Consultation FE (2003) Diet, nutrition and the prevention of chronic diseases. World Health Organ Tech Rep Ser 916:I–VIII
3. Gregory PJ, George TS (2011) Feeding nine billion: the challenge to sustainable crop production. J Exp Bot 62:5233–5239
4. FAO (2009) How to feed the world in 2050. www.fao.org
5. Bita CE, Gerats T (2013) Plant tolerance to high temperature in a changing environment: scientific fundamentals and production of heat stress-tolerant crops. Front Plant Sci 4:273
6. Takehisa H, Sato Y, Antonio BA, Nagamura Y (2013) Global transcriptome profile of rice root in response to essential macronutrient deficiency. Plant Signal Behav 8:e24409

7. Aroca R, Porcel R, Ruiz-Lozano JM (2012) Regulation of root water uptake under abiotic stress conditions. J Exp Bot 63:43–57
8. Shinozaki K, Yamaguchi-Shinozaki K (2007) Gene networks involved in drought stress response and tolerance. J Exp Bot 58:221–227
9. Caruso G, Cavaliere C, Guarino C, Gubbiotti R, Foglia P, Lagana A (2008) Identification of changes in *Triticum durum L.* leaf proteome in response to salt stress by two-dimensional electrophoresis and MALDI-TOF mass spectrometry. Anal Bioanal Chem 391:381–390
10. Raorane ML, Mutte SK, Varadarajan AR, Pabuayon IM, Kohli A (2013) Protein SUMOylation and plant abiotic stress signaling: in silico case study of rice RLKs, heat-shock and Ca^{2+}-binding proteins. Plant Cell Rep 32:1053–1065
11. Balmer D, Flors V, Glauser G, Mauch-Mani B (2013) Metabolomics of cereals under biotic stress: current knowledge and techniques. Front Plant Sci 4:82
12. Mehta A, Magalhaes BS, Souza DS, Vasconcelos EA, Silva LP, Grossi-De-Sa MF et al (2008) Rooteomics: the challenge of discovering plant defense-related proteins in roots. Curr Protein Pept Sci 9:108–116
13. Gruber BD, Giehl RF, Friedel S, Von Wiren N (2013) Plasticity of the *Arabidopsis* root system under nutrient deficiencies. Plant Physiol 163:161–179
14. Nozu Y, Tsugita A, Kamijo K (2006) Proteomic analysis of rice leaf, stem and root tissues during growth course. Proteomics 6:3665–3670
15. Agrawal GK, Sarkar A, Righetti PG, Pedreschi R, Carpentier S, Wang T et al (2013) A decade of plant proteomics and mass spectrometry: translation of technical advancements to food security and safety issues. Mass Spectrom Rev 32:335–365
16. Salekdeh GH, Komatsu S (2007) Crop proteomics: aim at sustainable agriculture of tomorrow. Proteomics 7:2976–2996
17. Isaacson T, Damasceno CM, Saravanan RS, He Y, Catala C, Saladie M et al (2006) Sample extraction techniques for enhanced proteomic analysis of plant tissues. Nat Protoc 1:769–774
18. Hossain Z, Komatsu S (2013) Contribution of proteomic studies towards understanding plant heavy metal stress response. Front Plant Sci 3:310
19. Ghosh D, Xu J (2014) Abiotic stress responses in plant roots: a proteomics perspective. Front Plant Sci 5:6
20. Saravanan RS, Rose JK (2004) A critical evaluation of sample extraction techniques for enhanced proteomic analysis of recalcitrant plant tissues. Proteomics 4:2522–2532
21. Guo H, Li F, Wang S, Li S, Xiao W, Liu W (2015) Enhanced protein extraction from tobacco roots for proteomic analysis. Anal Lett 48:16–24
22. Xiang X, Ning S, Jiang X, Gong X, Zhu R, Zhu L et al (2010) Protein extraction from rice (*Oryza sativa L.*) root for two-dimensional electrophresis. Front Agric China 4:416–421
23. Kohli A, Narciso JO, Miro B, Raorane M (2012) Root proteases: reinforced links between nitrogen uptake and mobilization and drought tolerance. Physiol Plant 145:165–179
24. Raorane ML, Narciso JO, Kohli A (2016) Total soluble protein extraction for improved proteomic analysis of transgenic rice plant roots. Recomb Proteins Plants Methods Protoc pp 139–147
25. Mahajan S, Tuteja N (2005) Cold, salinity and drought stress: an overview. Arch Biochem Biophys 444:139–158
26. Oerke EC, Dehne HW (2004) Safeguarding production—losses in major crops and the role of crop protection. Crop Prot 23:275–285
27. Lee DG, Ahsan N, Lee SH, Lee JJ, Bahk JD, Kang KY et al (2009) Chilling stress-induced proteomic changes in rice roots. J Plant Physiol 166:1–11
28. Passioura JB (1996) Drought and drought tolerance. Plant Growth Regul 20:79–83
29. Taiz L, Zeiger E (2002) Plant physiology. Sinauer Associates, Inc, Sunderland
30. Lin MK, Belanger H, Lee YJ, Varkonyi-Gasic E, Taoka K, Miura E et al (2007) FLOWERING LOCUS T protein may act as the long-distance florigenic signal in the cucurbits. Plant Cell 19:1488–1506

31. Mohammadi PP, Moieni A, Komatsu S (2012) Comparative proteome analysis of drought-sensitive and drought-tolerant rapeseed roots and their hybrid F1 line under drought stress. Amino Acids 43:2137–2152
32. Guo G, Ge P, Ma C, Li X, Lv D, Wang S et al (2012) Comparative proteomic analysis of salt response proteins in seedling roots of two wheat varieties. J Proteom 75:1867–1885
33. Zadraznik T, Hollung K, Egge-Jacobsen W, Meglic V, Sustar-Vozlic J (2013) Differential proteomic analysis of drought stress response in leaves of common bean (*Phaseolus vulgaris L.*). J Proteom 78:254–272
34. Ashoub A, Beckhaus T, Berberich T, Karas M, Bruggemann W (2013) Comparative analysis of barley leaf proteome as affected by drought stress. Planta 237:771–781
35. Horn R, Chudobova I, Hansel U, Herwartz D, Koskull-Doring P, Schillberg S (2013) Simultaneous treatment with tebuconazole and abscisic acid induces drought and salinity stress tolerance in *Arabidopsis thaliana* by maintaining key plastid protein levels. J Proteom Res 12:1266–1281
36. Singh R, Jwa N-S (2013) Understanding the responses of rice to environmental stress using proteomics. J Proteome Res 12:4652–4669
37. Comstock JP (2002) Hydraulic and chemical signalling in the control of stomatal conductance and transpiration. J Exp Bot 53:195–200
38. Jeong JS, Kim YS, Baek KH, Jung H, Ha SH, Do Choi Y et al (2010) Root-specific expression of OsNAC10 improves drought tolerance and grain yield in rice under field drought conditions. Plant Physiol 153:185–197
39. Hu X, Lu M, Li C, Liu T, Wang W, Wu J et al (2011) Differential expression of proteins in maize roots in response to abscisic acid and drought. Acta Physiol Plant 33:2437–2446
40. Hashimoto M, Kisseleva L, Sawa S, Furukawa T, Komatsu S, Koshiba T (2004) A novel rice PR10 protein, RSOsPR10, specifically induced in roots by biotic and abiotic stresses, possibly via the jasmonic acid signaling pathway. Plant Cell Physiol 45:550–559
41. Liu X, Baird WV (2003) The ribosomal small-subunit protein S28 gene from *Helianthus annuus* (Asteraceae) is down-regulated in response to drought, high salinity, and abscisic acid. Am J Bot 90:526–531
42. Liu T, Zhang L, Yuan Z, Hu X, Lu M, Wang W et al (2013) Identification of proteins regulated by ABA in response to combined drought and heat stress in maize roots. Acta Physiol Plant 35:501–513
43. Peng Z, Wang M, Li F, Lv H, Li C, Xia G (2009) A proteomic study of the response to salinity and drought stress in an introgression strain of bread wheat. Mol Cell Proteom 8:2676–2686
44. Westgate M, Boyer J (1985) Osmotic adjustment and the inhibition of leaf, root, stem and silk growth at low water potentials in maize. Planta 164:540–549
45. Sharp RE, Poroyko V, Hejlek LG, Spollen WG, Springer GK, Bohnert HJ et al (2004) Root growth maintenance during water deficits: physiology to functional genomics. J Exp Bot 55:2343–2351
46. Zhu J, Alvarez S, Marsh EL, Lenoble ME, Cho IJ, Sivaguru M et al (2007) Cell wall proteome in the maize primary root elongation zone. II. Region-specific changes in water soluble and lightly ionically bound proteins under water deficit. Plant Physiol 145: 1533–1548
47. Wehner N, Hartmann L, Ehlert A, Böttner S, Oñate-Sánchez L, Dröge-Laser W (2011) High-throughput protoplast transactivation (PTA) system for the analysis of *Arabidopsis* transcription factor function. Plant J 68:560–569
48. Shinozaki K, Yamaguchi-Shinozaki K, Seki M (2003) Regulatory network of gene expression in the drought and cold stress responses. Curr Opin Plant Biol 6:410–417
49. Fukao T, Xiong L (2013) Genetic mechanisms conferring adaptation to submergence and drought in rice: simple or complex? Curr Opin Plant Biol 16:196–204
50. Yamaguchi-Shinozaki K, Shinozaki K (2005) Organization of cis-acting regulatory elements in osmotic- and cold-stress-responsive promoters. Trends Plant Sci 10:88–94

51. Todaka D, Shinozaki K, Yamaguchi-Shinozaki K (2015) Recent advances in the dissection of drought-stress regulatory networks and strategies for development of drought-tolerant transgenic rice plants. Front Plant Sci 6:84
52. Dixit S, Kumar Biswal A, Min A, Henry A, Oane RH, Raorane ML et al (2015) Action of multiple intra-QTL genes concerted around a co-localized transcription factor underpins a large effect QTL. Sci Rep 5:15183
53. Fujita M, Fujita Y, Maruyama K, Seki M, Hiratsu K, Ohme-Takagi M et al (2004) A dehydration-induced NAC protein, RD26, is involved in a novel ABA-dependent stress-signaling pathway. Plant J 39:863–876
54. Tran L-SP, Nakashima K, Sakuma Y, Simpson SD, Fujita Y, Maruyama K et al (2004) Isolation and functional analysis of *Arabidopsis* stress-inducible NAC transcription factors that bind to a drought-responsive cis-element in the early responsive to dehydration stress 1 promoter. Plant Cell 16:2481–2498
55. Christianson JA, Wilson IW, Llewellyn DJ, Dennis ES (2009) The low-oxygen-induced NAC domain transcription factor ANAC102 affects viability of *Arabidopsis* seeds following low-oxygen treatment. Plant Physiol 149:1724–1738
56. Zheng X, Chen B, Lu G, Han B (2009) Overexpression of a NAC transcription factor enhances rice drought and salt tolerance. Biochem Biophys Res Commun 379:985–989
57. Redillas MC, Jeong JS, Kim YS, Jung H, Bang SW, Choi YD et al (2012) The overexpression of OsNAC9 alters the root architecture of rice plants enhancing drought resistance and grain yield under field conditions. Plant Biotechnol J 10:792–805
58. Mirzaei M, Soltani N, Sarhadi E, Pascovici D, Keighley T, Salekdeh GH et al (2012) Shotgun proteomic analysis of long-distance drought signaling in rice roots. J Proteom Res 11:348–358
59. Raorane ML, Pabuayon IM, Varadarajan AR, Mutte SK, Kumar A, Treumann A et al (2015) Proteomic insights into the role of the large-effect QTL qDTY 12.1 for rice yield under drought. Mol Breed 35:1–14
60. Raorane ML, Pabuayon IM, Miro B, Kalladan R, Reza-Hajirezai M, Oane RH et al (2015) Variation in primary metabolites in parental and near-isogenic lines of the QTL qDTY 12.1: altered roots and flag leaves but similar spikelets of rice under drought. Mol Breed 35:1–25
61. Faghani E, Gharechahi J, Komatsu S, Mirzaei M, Khavarinejad RA, Najafi F et al (2015) Comparative physiology and proteomic analysis of two wheat genotypes contrasting in drought tolerance. J Proteom 114:1–15
62. Hundertmark M, Hincha DK (2008) LEA (late embryogenesis abundant) proteins and their encoding genes in *Arabidopsis thaliana*. BMC Genom 9:118
63. Chen YS, Lo SF, Sun PK, Lu CA, Ho TH, Yu SM (2015) A late embryogenesis abundant protein HVA1 regulated by an inducible promoter enhances root growth and abiotic stress tolerance in rice without yield penalty. Plant Biotechnol J 13:105–116
64. Chandra Babu R, Zhang J, Blum A, David Ho TH, Wu R, Nguyen HT (2004) HVA1, a LEA gene from barley confers dehydration tolerance in transgenic rice (*Oryza sativa L.*) via cell membrane protection. Plant Sci 166:855–862
65. Maurel C (1997) Aquaporins and water permeability of plant membranes. Annu Rev Plant Physiol Plant Mol Biol 48:399–429
66. Tyerman S, Niemietz C, Bramley H (2002) Plant aquaporins: multifunctional water and solute channels with expanding roles. Plant, Cell Environ 25:173–194
67. Martre P, Morillon R, Barrieu F, North GB, Nobel PS, Chrispeels MJ (2002) Plasma membrane aquaporins play a significant role during recovery from water deficit. Plant Physiol 130:2101–2110
68. Maurel C (2007) Plant aquaporins: novel functions and regulation properties. FEBS Lett 581:2227–2236
69. Johanson U, Karlsson M, Johansson I, Gustavsson S, Sjovall S, Fraysse L et al (2001) The complete set of genes encoding major intrinsic proteins in *Arabidopsis* provides a framework for a new nomenclature for major intrinsic proteins in plants. Plant Physiol 126:1358–1369
70. Javot H, Maurel C (2002) The role of aquaporins in root water uptake. Ann Bot 90:301–313

71. Lian HL, Yu X, Ye Q, Ding X, Kitagawa Y, Kwak SS et al (2004) The role of aquaporin RWC3 in drought avoidance in rice. Plant Cell Physiol 45:481–489
72. Parent B, Hachez C, Redondo E, Simonneau T, Chaumont F, Tardieu F (2009) Drought and abscisic acid effects on aquaporin content translate into changes in hydraulic conductivity and leaf growth rate: a trans-scale approach. Plant Physiol 149:2000–2012
73. Knipfer T, Besse M, Verdeil JL, Fricke W (2011) Aquaporin-facilitated water uptake in barley (*Hordeum vulgare L.*) roots. J Exp Bot 62:4115–4126
74. Grondin A, Mauleon R, Vadez V, Henry A (2016) Root aquaporins contribute to whole plant water fluxes under drought stress in rice (*Oryza sativa L.*). Plant, Cell Environ 39:347–365
75. Jiang Y, Yang B, Harris NS, Deyholos MK (2007) Comparative proteomic analysis of NaCl stress-responsive proteins in *Arabidopsis* roots. J Exp Bot 58:3591–3607
76. Rajaei S, Niknam V, Seyedi S, Ebrahimzadeh H, Razavi K (2009) Contractile roots are the most sensitive organ in *Crocus sativus* to salt stress. Biol Plant 53:523–529
77. Chinnusamy V, Jagendorf A, Zhu J-K (2005) Understanding and improving salt tolerance in plants. Crop Sci 45:437–448
78. Yan S, Tang Z, Su W, Sun W (2005) Proteomic analysis of salt stress-responsive proteins in rice root. Proteomics 5:235–244
79. Atkin RK, Barton GE, Robinson DK (1973) Effect of root-growing temperature on growth substances in xylem exudate of *Zea mays*. J Exp Bot 24:475–487
80. Steppuhn H, Raney J (2005) Emergence, height, and yield of canola and barley grown in saline root zones. Can J Plant Sci 85:815–827
81. Bernstein N, Shoresh M, Xu Y, Huang B (2010) Involvement of the plant antioxidative response in the differential growth sensitivity to salinity of leaves vs roots during cell development. Free Radic Biol Med 49:1161–1171
82. Assmann SM (2005) G proteins go green: a plant G protein signaling FAQ sheet. Science 310:71–73
83. Neves SR, Ram PT, Iyengar R (2002) G protein pathways. Science 296:1636–1639
84. Cheng Y, Qi Y, Zhu Q, Chen X, Wang N, Zhao X et al (2009) New changes in the plasma-membrane-associated proteome of rice roots under salt stress. Proteomics 9: 3100–3114
85. Liu CW, Hsu YK, Cheng YH, Yen HC, Wu YP, Wang CS et al (2012) Proteomic analysis of salt-responsive ubiquitin-related proteins in rice roots. Rapid Commun Mass Spectrom 26:1649–1660
86. Witzel K, Weidner A, Surabhi G-K, Börner A, Mock H-P (2009) Salt stress-induced alterations in the root proteome of barley genotypes with contrasting response towards salinity. J Exp Bot 60:3545–3557
87. Droog F (1997) Plant glutathione S-transferases, a tale of theta and tau. J Plant Growth Regul 16:95–107
88. Thornalley PJ (1996) Pharmacology of methylglyoxal: formation, modification of proteins and nucleic acids, and enzymatic detoxification-a role in pathogenesis and antiproliferative chemotherapy. Gen Pharmacol Vasc Syst 27:565–573
89. Mostek A, Borner A, Badowiec A, Weidner S (2015) Alterations in root proteome of salt-sensitive and tolerant barley lines under salt stress conditions. J Plant Physiol 174: 166–176
90. Laohavisit A, Shang Z, Rubio L, Cuin TA, Very AA, Wang A et al (2012) *Arabidopsis* annexin1 mediates the radical-activated plasma membrane Ca^{2+}- and K^+-permeable conductance in root cells. Plant Cell 24:1522–1533
91. Mortimer JC, Laohavisit A, Macpherson N, Webb A, Brownlee C, Battey NH et al (2008) Annexins: multifunctional components of growth and adaptation. J Exp Bot 59:533–544
92. Carroll AD, Moyen C, Van Kesteren P, Tooke F, Battey NH, Brownlee C (1998) Ca^{2+}, annexins, and GTP modulate exocytosis from maize root cap protoplasts. Plant Cell 10:1267–1276
93. Clark GB, Sessions A, Eastburn DJ, Roux SJ (2001) Differential expression of members of the annexin multigene family in *Arabidopsis*. Plant Physiol 126:1072–1084

94. Chinnusamy V, Zhu J, Zhu JK (2007) Cold stress regulation of gene expression in plants. Trends Plant Sci 12:444–451
95. Xiong L, Schumaker KS, Zhu J-K (2002) Cell signaling during cold, drought, and salt stress. Plant Cell 14:S165–S183
96. Guy CL (1990) Cold acclimation and freezing stress tolerance: role of protein metabolism. Annu Rev Plant Biol 41:187–223
97. Bohnert HJ, Sheveleva E (1998) Plant stress adaptations—making metabolism move. Curr Opin Plant Biol 1:267–274
98. Hashimoto M, Toorchi M, Matsushita K, Iwasaki Y, Komatsu S (2009) Proteome analysis of rice root plasma membrane and detection of cold stress responsive proteins. Protein Pept Lett 16:685–697
99. Hashimoto M, Komatsu S (2007) Proteomic analysis of rice seedlings during cold stress. Proteomics 7:1293–1302
100. Salama SR, Cleves A, Malehorn DE, Whitters E, Bankaitis VA (1990) Cloning and characterization of *Kluyveromyces lactis* SEC14, a gene whose product stimulates Golgi secretory function in *Saccharomyces cerevisiae*. J Bacteriol 172:4510–4521
101. Phillips SE, Vincent P, Rizzieri KE, Schaaf G, Bankaitis VA, Gaucher EA (2006) The diverse biological functions of phosphatidylinositol transfer proteins in eukaryotes. Crit Rev Biochem Mol Biol 41:21–49
102. Vincent P, Chua M, Nogue F, Fairbrother A, Mekeel H, Xu Y et al (2005) A Sec14p-nodulin domain phosphatidylinositol transfer protein polarizes membrane growth of *Arabidopsis thaliana* root hairs. J Cell Biol 168:801–812
103. Gerke V, Moss SE (2002) Annexins: from structure to function. Physiol Rev 82:331–371
104. Nadimpalli R, Yalpani N, Johal GS, Simmons CR (2000) Prohibitins, stomatins, and plant disease response genes compose a protein superfamily that controls cell proliferation, ion channel regulation, and death. J Biol Chem 275:29579–29586
105. Kuk YI, Shin JS, Burgos NR, Hwang TE, Han O, Cho BH et al (2003) Antioxidative enzymes offer protection from chilling damage in rice plants. Crop Sci 43:2109–2117
106. Bowler C, Vanmontagu M, Inze D (1992) Superoxide-dismutase and stress tolerance. Annu Rev Plant Physiol Plant Mol Biol 43:83–116
107. Kollipara KP, Saab IN, Wych RD, Lauer MJ, Singletary GW (2002) Expression profiling of reciprocal maize hybrids divergent for cold germination and desiccation tolerance. Plant Physiol 129:974–992
108. Rodziewicz P, Swarcewicz B, Chmielewska K, Wojakowska A, Stobiecki M (2014) Influence of abiotic stresses on plant proteome and metabolome changes. Acta Physiol Plant 36:1–19
109. Requejo R, Tena M (2006) Maize response to acute arsenic toxicity as revealed by proteome analysis of plant shoots. Proteomics 6(1):156–162
110. Mano E, Horiguchi G, Tsukaya H (2006) Gravitropism in leaves of *Arabidopsis thaliana* (L.) Heynh. Plant Cell Physiol 47:217–223
111. Huang G, Dong R, Allen R, Davis EL, Baum TJ, Hussey RS (2006) A root-knot nematode secretory peptide functions as a ligand for a plant transcription factor. Mol Plant Microbe Interact 19:463–470
112. Kumar S, Dhingra A, Daniell H (2004) Plastid-expressed betaine aldehyde dehydrogenase gene in carrot cultured cells, roots, and leaves confers enhanced salt tolerance. Plant Physiol 136:2843–2854
113. Zhu Q, Dabi T, Beeche A, Yamamoto R, Lawton MA, Lamb C (1995) Cloning and properties of a rice gene encoding phenylalanine ammonia-lyase. Plant Mol Biol 29:535–550
114. Charron JB, Ouellet F, Pelletier M, Danyluk J, Chauve C, Sarhan F (2005) Identification, expression, and evolutionary analyses of plant lipocalins. Plant Physiol 139:2017–2028
115. Neilson KA, Gammulla CG, Mirzaei M, Imin N, Haynes PA (2010) Proteomic analysis of temperature stress in plants. Proteomics 10:828–845
116. Skylas D, Cordwell S, Hains P, Larsen M, Basseal D, Walsh B et al (2002) Heat shock of wheat during grain filling: proteins associated with heat-tolerance. J Cereal Sci 35:175–188

117. Süle A, Vanrobaeys F, Hajós GY, Van Beeumen J, Devreese B (2004) Proteomic analysis of small heat shock protein isoforms in barley shoots. Phytochemistry 65:1853–1863
118. Rollins J, Habte E, Templer S, Colby T, Schmidt J, Von Korff M (2013) Leaf proteome alterations in the context of physiological and morphological responses to drought and heat stress in barley (*Hordeum vulgare L.*). J Exp Bot 64:3201–3212
119. Jagadish SV, Muthurajan R, Oane R, Wheeler TR, Heuer S, Bennett J et al (2010) Physiological and proteomic approaches to address heat tolerance during anthesis in rice (*Oryza sativa L.*). J Exp Bot 61:143–156
120. Shi W, Muthurajan R, Rahman H, Selvam J, Peng S, Zou Y et al (2013) Source–sink dynamics and proteomic reprogramming under elevated night temperature and their impact on rice yield and grain quality. New Phytol 197:825–837
121. Young TE, Ling J, Geisler-Lee CJ, Tanguay RL, Caldwell C, Gallie DR (2001) Developmental and thermal regulation of the maize heat shock protein, HSP101. Plant Physiol 127:777–791
122. Nieto-Sotelo J, Martinez LM, Ponce G, Cassab GI, Alagon A, Meeley RB et al (2002) Maize HSP101 plays important roles in both induced and basal thermotolerance and primary root growth. Plant Cell 14:1621–1633
123. Xu C, Huang B (2008) Root proteomic responses to heat stress in two Agrostis grass species contrasting in heat tolerance. J Exp Bot 59:4183–4194
124. Huang J, Wang MM, Jiang Y, Bao YM, Huang X, Sun H et al (2008) Expression analysis of rice A20/AN1-type zinc finger genes and characterization of ZFP177 that contributes to temperature stress tolerance. Gene 420:135–144
125. Komatsu S, Hiraga S, Yanagawa Y (2012) Proteomics techniques for the development of flood tolerant crops. J Proteome Res 11:68–78
126. Kosova K, Vitamvas P, Prasil IT, Renaut J (2011) Plant proteome changes under abiotic stress—contribution of proteomics studies to understanding plant stress response. J Proteom 74:1301–1322
127. Fukao T, Bailey-Serres J (2004) Plant responses to hypoxia—is survival a balancing act? Trends Plant Sci 9:449–456
128. Kong FJ, Oyanagi A, Komatsu S (2010) Cell wall proteome of wheat roots under flooding stress using gel-based and LC MS/MS-based proteomics approaches. Biochim Biophys Acta 1804:124–136
129. Johnson JR, Cobb BG, Drew MC (1994) Hypoxic induction of anoxia tolerance in roots of Adh1 null *Zea mays L.* Plant Physiol 105:61–67
130. Yamasaki T, Deguchi M, Fujimoto T, Masumura T, Uno T, Kanamaru K et al (2006) Rice bifunctional alpha-amylase/subtilisin inhibitor: cloning and characterization of the recombinant inhibitor expressed in *Escherichia coli*. Biosci Biotechnol Biochem 70: 1200–1209
131. Shibuya N, Minami E (2001) Oligosaccharide signalling for defence responses in plant. Physiol Mol Plant Pathol 59:223–233
132. Chang WW, Huang L, Shen M, Webster C, Burlingame AL, Roberts JK (2000) Patterns of protein synthesis and tolerance of anoxia in root tips of maize seedlings acclimated to a low-oxygen environment, and identification of proteins by mass spectrometry. Plant Physiol 122:295–318
133. Biemelt S, Keetman U, Albrecht G (1998) Re-aeration following hypoxia or anoxia leads to activation of the antioxidative defense system in roots of wheat seedlings. Plant Physiol 116:651–658
134. Ushimaru T, Kanematsu S, Shibasaka M, Tsuji H (1999) Effect of hypoxia on the antioxidative enzymes in aerobically grown rice (*Oryza sativa*) seedlings. Physiol Plant 107:181–187
135. Jackson MB, Ram PC (2003) Physiological and molecular basis of susceptibility and tolerance of rice plants to complete submergence. Ann Bot 91:227–241
136. Sachs MM, Freeling M, Okimoto R (1980) The anaerobic proteins of maize. Cell 20: 761–767

137. Wasteneys GO, Galway ME (2003) Remodeling the cytoskeleton for growth and form: an overview with some new views. Annu Rev Plant Biol 54:691–722
138. Mishra S, Dubey R (2006) Heavy metal uptake and detoxification mechanisms in plants. Int J Agric Res 1:122–141
139. Nazar R, Iqbal N, Masood A, Iqbal M, Khan R, Syeed S, Khan NA (2012) Cadmium toxicity in plant and role of mineral nutrients in its alleviation. Am J Plant Sci 3:1476–1489
140. Parker SP (1989) Heavy metals: In McGraw-Hill Dictionary of Scientific and Technical Terms, 4th edn. New York, McGraw-Hill
141. Le Gall H, Philippe F, Domon J-M, Gillet F, Pelloux J, Rayon C (2015) Cell wall metabolism in response to abiotic stress. Plants 4:112
142. Perfus-Barbeoch L, Leonhardt N, Vavasseur A, Forestier C (2002) Heavy metal toxicity: cadmium permeates through calcium channels and disturbs the plant water status. Plant J 32:539–548
143. Pourrut B, Shahid M, Dumat C, Winterton P, Pinelli E (2011) Lead uptake, toxicity, and detoxification in plants. Reviews of environmental contamination and toxicology, vol 213. Springer, New York, pp 113–136
144. Welch R, Norvell W (1999) Mechanisms of cadmium uptake, translocation and deposition in plants. Cadmium in soils and plants. Springer, New York, pp 125–150
145. Rout G, Senapti S (2013) Stress tolerance in plants: a proteomics approach. Molecular stress physiology of plants. Springer, Dordrecht
146. Sharma SS, Dietz K-J (2009) The relationship between metal toxicity and cellular redox imbalance. Trends Plant Sci 14:43–50
147. Lee K, Bae DW, Kim SH, Han HJ, Liu X, Park HC et al (2010) Comparative proteomic analysis of the short-term responses of rice roots and leaves to cadmium. J Plant Physiol 167:161–168
148. Mendoza-Cozatl D, Loza-Tavera H, Hernandez-Navarro A, Moreno-Sanchez R (2005) Sulfur assimilation and glutathione metabolism under cadmium stress in yeast, protists and plants. FEMS Microbiol Rev 29:653–671
149. Adamis PD, Gomes DS, Pinto ML, Panek AD, Eleutherio EC (2004) The role of glutathione transferases in cadmium stress. Toxicol Lett 154:81–88
150. Meuwly P, Rauser WE (1992) Alteration of thiol pools in roots and shoots of maize seedlings exposed to cadmium: adaptation and developmental cost. Plant Physiol 99:8–15
151. Rauser WE (1986) The amount of cadmium associated with Cd-binding protein in roots of *Agrostis gigantea*, maize and tomato. Plant Sci 43:85–91
152. Szalai G, Janda T, Golan-Goldhirsh A, Páldi E (2002) Effect of Cd treatment on phytochelatin synthesis in maize. Acta Biol Szeged 46:121–122
153. Aina R, Labra M, Fumagalli P, Vannini C, Marsoni M, Cucchi U et al (2007) Thiol-peptide level and proteomic changes in response to cadmium toxicity in *Oryza sativa L.* roots. Environ Exp Bot 59:381–392
154. Bovet L, Eggmann T, Meylan-Bettex M, Polier J, Kammer P, Marin E et al (2003) Transcript levels of AtMRPs after cadmium treatment: induction of AtMRP3. Plant, Cell Environ 26:371–381
155. Moons A (2003) Ospdr9, which encodes a PDR-type ABC transporter, is induced by heavy metals, hypoxic stress and redox perturbations in rice roots. FEBS Lett 553:370–376
156. Belouchi A, Kwan T, Gros P (1997) Cloning and characterization of the OsNramp family from *Oryza sativa*, a new family of membrane proteins possibly implicated in the transport of metal ions. Plant Mol Biol 33:1085–1092
157. Williams LE, Pittman JK, Hall J (2000) Emerging mechanisms for heavy metal transport in plants. Biochim Biophys Acta BBA Biomembr 1465:104–126
158. Cseh E (2002) Metal permeability, transport and efflux in plants. In: Prasad MNV, Strzałka K (eds) Physiology and biochemistry of metal toxicity and tolerance in plants. Springer, Dordrecht, pp 1–36

159. Chen C, Song Y, Zhuang K, Li L, Xia Y, Shen Z (2015) Proteomic analysis of copper-binding proteins in excess copper-stressed roots of two rice (*Oryza sativa L.*) varieties with different Cu tolerances. PLoS ONE 10:e0125367
160. Breen J, Bellgard M (2010) Germin-like proteins (GLPs) in cereal genomes: gene clustering and dynamic roles in plant defence. Funct Integr Genom 10:463–476
161. Christensen AB, Thordal-Christensen H, Zimmermann G, Gjetting T, Lyngkjaer MF, Dudler R et al (2004) The germinlike protein GLP4 exhibits superoxide dismutase activity and is an important component of quantitative resistance in wheat and barley. Mol Plant Microbe Interact 17:109–117
162. Li F, Shi J, Shen C, Chen G, Hu S, Chen Y (2009) Proteomic characterization of copper stress response in *Elsholtzia splendens* roots and leaves. Plant Mol Biol 71:251–263
163. Roth U, Von Roepenack-Lahaye E, Clemens S (2006) Proteome changes in *Arabidopsis thaliana* roots upon exposure to Cd^{2+}. J Exp Bot 57:4003–4013
164. Utriainen M, Kokko H, Auriola S, Sarrazin O, Kärenlampi S (1998) PR-10 protein is induced by copper stress in roots and leaves of a Cu/Zn tolerant clone of birch, Betula pendula. Plant, Cell Environ 21:821–828
165. Requejo R, Tena M (2005) Proteome analysis of maize roots reveals that oxidative stress is a main contributing factor to plant arsenic toxicity. Phytochemistry 66:1519–1528
166. Ahsan N, Lee DG, Alam I, Kim PJ, Lee JJ, Ahn YO et al (2008) Comparative proteomic study of arsenic-induced differentially expressed proteins in rice roots reveals glutathione plays a central role during as stress. Proteomics 8:3561–3576
167. Ahsan N, Renaut J, Komatsu S (2009) Recent developments in the application of proteomics to the analysis of plant responses to heavy metals. Proteomics 9:2602–2621
168. Fukuda T, Saito A, Wasaki J, Shinano T, Osaki M (2007) Metabolic alterations proposed by proteome in rice roots grown under low P and high Al concentration under low pH. Plant Sci 172:1157–1165
169. Yang Q, Wang Y, Zhang J, Shi W, Qian C, Peng X (2007) Identification of aluminum-responsive proteins in rice roots by a proteomic approach: cysteine synthase as a key player in Al response. Proteomics 7:737–749
170. Patterson J, Ford K, Cassin A, Natera S, Bacic A (2007) Increased abundance of proteins involved in phytosiderophore production in boron-tolerant barley. Plant Physiol 144: 1612–1631
171. Hodges SC, Constable G (2010) Plant responses to mineral deficiencies and toxicities. In: Stewart JM, Oosterhuis DM, Heitholt JJ, Mauney JR (eds) Physiology of cotton. Springer, Dordrecht, pp 142–161
172. Sanchez PA (2002) Soil fertility and hunger in Africa. Science 295:2019
173. Zhang YM, Yan YS, Wang LN, Yang K, Xiao N, Liu YF et al (2012) A novel rice gene, NRR responds to macronutrient deficiency and regulates root growth. Mol Plant 5:63–72
174. Li K, Xu C, Zhang K, Yang A, Zhang J (2007) Proteomic analysis of roots growth and metabolic changes under phosphorus deficit in maize (*Zea mays L.*) plants. Proteomics 7:1501–1512
175. Hopff D, Wienkoop S, Luthje S (2013) The plasma membrane proteome of maize roots grown under low and high iron conditions. J Proteom 91:605–618
176. Kim SG, Wang Y, Lee CH, Mun BG, Kim PJ, Lee SY et al (2011) A comparative proteomics survey of proteins responsive to phosphorous starvation in roots of hydroponically-grown rice seedlings. J Korean Soc Appl Biol Chem 54:667–677
177. Moller AL, Pedas P, Andersen B, Svensson B, Schjoerring JK, Finnie C (2011) Responses of barley root and shoot proteomes to long-term nitrogen deficiency, short-term nitrogen starvation and ammonium. Plant, Cell Environ 34:2024–2037
178. Raghothama K (1999) Phosphate acquisition. Annu Rev Plant Biol 50:665–693
179. Hernandez G, Ramirez M, Valdes-Lopez O, Tesfaye M, Graham MA, Czechowski T et al (2007) Phosphorus stress in common bean: root transcript and metabolic responses. Plant Physiol 144:752–767

180. Wasaki J, Yonetani R, Kuroda S, Shinano T, Yazaki J, Fujii F et al (2003) Transcriptomic analysis of metabolic changes by phosphorus stress in rice plant roots. Plant, Cell Environ 26:1515–1523
181. Wasaki J, Shinano T, Onishi K, Yonetani R, Yazaki J, Fujii F et al (2006) Transcriptomic analysis indicates putative metabolic changes caused by manipulation of phosphorus availability in rice leaves. J Exp Bot 57:2049–2059
182. Li K, Xu C, Li Z, Zhang K, Yang A, Zhang J (2008) Comparative proteome analyses of phosphorus responses in maize (*Zea mays L.*) roots of wild-type and a low-P-tolerant mutant reveal root characteristics associated with phosphorus efficiency. Plant J 55:927–939
183. Hinsinger P (2001) Bioavailability of soil inorganic P in the rhizosphere as affected by root-induced chemical changes: a review. Plant Soil 237:173–195
184. Rashotte AM, Delong A, Muday GK (2001) Genetic and chemical reductions in protein phosphatase activity alter auxin transport, gravity response, and lateral root growth. Plant Cell 13:1683–1697
185. Zhou H-W, Nussbaumer C, Chao Y, Delong A (2004) Disparate roles for the regulatory a subunit isoforms in *Arabidopsis* protein phosphatase 2A. Plant Cell 16:709–722
186. Dixon R, Lamb C (1990) Molecular communication in interactions between plants and microbial pathogens. Annu Rev Plant Physiol Plant Mol Biol 41:339–367
187. Green PJ (1994) The ribonucleases of higher plants. Annu Rev Plant Physiol Plant Mol Biol 45:421–445
188. Hagedorn F, Bucher JB, Schleppi P (2001) Contrasting dynamics of dissolved inorganic and organic nitrogen in soil and surface waters of forested catchments with Gleysols. Geoderma 100:173–192
189. Marschner H (2011) Marschner's mineral nutrition of higher plants. Academic press, Cambridge
190. Bachmann M, Huber JL, Liao PC, Gage DA, Huber SC (1996) The inhibitor protein of phosphorylated nitrate reductase from spinach (*Spinacia oleracea*) leaves is a 14-3-3 protein. FEBS Lett 387:127–131
191. Schmidt W (1999) Mechanisms and regulation of reduction-based iron uptake in plants. New Phytol 141:1–26
192. Grusak MA (1995) Whole-root iron(III)-reductase activity throughout the life cycle of iron-grown *Pisum sativum L.* (Fabaceae): relevance to the iron nutrition of developing seeds. Planta 197:111–117
193. Sondergaard TE, Schulz A, Palmgren MG (2004) Energization of transport processes in plants. Roles of the plasma membrane H^+-ATPase. Plant Physiol 136:2475–2482
194. Buch-Pedersen MJ, Pedersen BP, Veierskov B, Nissen P, Palmgren MG (2009) Protons and how they are transported by proton pumps. Pflugers Arch 457:573–579
195. Luthje S, Hopff D, Schmitt A, Meisrimler CN, Menckhoff L (2009) Hunting for low abundant redox proteins in plant plasma membranes. J Proteom 72:475–483
196. Lüthje S, Van Gestelen P, Córdoba-Pedregosa MC, González-Reyes JA, Asard H, Villalba JM et al (1998) Quinones in plant plasma membranes—a missing link? Protoplasma 205:43–51
197. Yadeta KA, Elmore JM, Coaker G (2013) Advancements in the analysis of the *Arabidopsis* plasma membrane proteome. Front Plant Sci 4:86

Chapter 3
A Proteomic View of the Cereal and Vegetable Crop Response to Salinity Stress

Katja Witzel and Hans-Peter Mock

Abstract Salt stress is a major factor with worldwide negative impact on agricultural productivity. Crop plants with higher tolerance towards salinity would allow sustainable production in less favorable environments resulting from soil or irrigation conditions. Future breeding strategies will depend on novel insights into molecular mechanisms requested to adopt current elite varieties. Extant genetic variation for tolerance within some crop germplasm provides a sound basis for elucidating the mechanisms underlying naturally evolved tolerance, so elaborating reliable and scalable phenotyping platforms to permit the efficient evaluation of extensive collections of plant germplasm is a necessary development. The genetic basis of salinity tolerance is complex, but the advent of "omics" technologies has expanded the informativeness of contrasts between accessions which show a differential response to the stress. Characterization at the level of the proteome is a relatively recent development, but one which has had already demonstrated a measure of success. After briefly outlining physiological aspects and adaptation strategies in general, we will summarize the proteomic studies directed to cereals and vegetable crops separately to better identify general responses, but also to pinpoint strategies specific for either monocotyledonous or dicotyledonous crops. While the most frequently used analytical proteomics platform remains two-dimensional gel electrophoresis (2DE), there is growing use made of liquid chromatography (LC)-based separation technology, which provides a higher degree of sensitivity and has a lower requirement with respect to sample amount.

Keywords LC-based proteomics · Plant natural variation · Two-dimensional gel electrophoresis · Cereals · Legumes

---

K. Witzel
Leibniz Institute of Vegetable and Ornamental Crops (IGZ),
Theodor-Echtermeyer-Weg 1, 14979 Grossbeeren, Germany

H.-P. Mock (✉)
Leibniz Institute of Plant Genetics and Crop Plant Research (IPK),
Corrensstraße 3, 06466 Stadt Seeland, Gatersleben, Germany
e-mail: mock@ipk-gatersleben.de

G.H. Salekdeh (ed.), *Agricultural Proteomics Volume 2*,
DOI 10.1007/978-3-319-43278-6_3

3.1 The Physiological Response to Salinity Stress

Plants are exposed to a variety of both biotic and abiotic stress factors, which together are responsible for a major loss in production [1]. In many parts of the world, soil salinity, whether occurring naturally or induced by irrigation practice, represents one of the most severe abiotic stress agents [2]. As a result, improving the level of salinity tolerance expressed by crop varieties is becoming an ever more important breeding priority.

Saline soils are those which contain excessive levels of the cations Na^+, Ca^{2+} and Mg^{2+}, and the anions SO_4^{2-} and Cl^-; those in which Na^+ is the dominant cation are considered to be sodic rather than saline. Saline soils suffer from low porosity, and the flow of water into the plant root is further hindered by their high osmotic potential. The effect of salinity on plant growth is due to the combination of osmotic stress and the toxicity of the cations [2]. The former inhibits the uptake of water, generating a syndrome which has much in common with drought stress. Cell expansion and cell wall synthesis, protein synthesis, stomatal conductance and photosynthetic activity are all compromised. A certain length of exposure to salinity stress is required for the cellular content of Na^+ and other cations to reach a toxic level, and it has even been suggested that the presence of these cations during the early phase of a stress episode can be beneficial, since their export to the vacuole aids the plant in combatting osmotic stress [3]. Na^+ enters the root passively along an electrochemical potential gradient, while Cl^- entry is restricted by a negative plasma membrane potential. Once inside the plant, the ions are dispersed throughout the plant via the xylem; following their arrival in the leaf, their concentration tends to rise as the plant loses water through transpiration. When the ion sequestration capacity of the vacuole becomes exhausted, the cytosolic ion concentration inevitably starts to rise. While most plants respond to this stress by pumping cytosolic Na^+ into the cell wall, the consequence of this activity is to force the cell wall to shrink and dehydrate. High cellular concentrations of ions tend to induce the production of reactive oxygen species (ROS), which damage cell membranes, many proteins and ultimately DNA.

3.2 Adaptive Mechanisms for Salt Tolerance

The first line of defence against salinity stress is a barrier to the entry of cations into the plant [4]. This comprises a combination of reducing the ion uptake by the root, inhibiting the loading of ions into or their unloading out of the xylem and preventing their export via the phloem to growing tissue [5, 6]. At the cellular level, Na^+/H^+ antiporters effect the removal of Na^+ from the cytosol and sequester it within the vacuole achieved by Na^+/H^+ antiporters driven by the pH gradient across the tonoplast [reviewed in 7]. The genes identified to date as important for bolstering salinity tolerance have been classified into three main groups.

The products of the first group are concerned with the control of ionic uptake and transport [8]. Among these is the *Arabidopsis thaliana* Na^+/H^+ antiporter AtNHX1: AtNHX1 over-expressors are able to maintain plant growth and set seed when they are challenged with 200 mM NaCl, whereas wild type plants cannot withstand more than 100 mM NaCl [9]. The over-expression of the plasma membrane Na^+/H^+ antiporter AtSOS1 also boosts the plant's salinity tolerance: the roots of these transgenic plants tolerate saline soil, significantly improving plant survival [10]. The high affinity K^+ transporter AtHKT regulates the root-to-shoot transport of Na^+ by removing Na^+ from the xylem sap [11]: AtHKT1 knock-down plants are hyper-sensitive to salinity stress [12], while the introduction of the durum wheat genes TmHKT1; 4-A2 and TmHKT1; 5-A into bread wheat results in a pronounced enhancement in salinity tolerance [13]. The electrochemical gradient across the plasma membrane required for H^+ antiporter activity is created by a combination of P-type H^+-ATPase-mediated pumping of protons into the apoplast, and V-type H^+-ATPase- and pyrophosphatase-mediated pumping of protons across the tonoplast into the vacuole [2]. The over-expression of AtAVP1 (which encodes a vacuolar pyrophosphatase) allows the plant to withstand the level of stress imposed by the presence of 250 mM NaCl in the growing medium [14].

Products of the second group act as protectants against osmotic stress. Osmotic adjustment under stress conditions can be achieved by the accumulation of various osmolytes, notably sugars, organic acids, polyols and nitrogenous compounds such as glycinebetaine and proline. These compounds accumulate in the cytosol to balance the osmotic pressure exerted by ions sequestered in the vacuole, and thereby succeed in maintaining the turgor needed for continued cell growth [15]. They are often referred to as 'compatible solutes' because they do not interfere with enzyme function, even when present at a high concentration. Proline is universally accumulated in response to salinity stress. Tobacco plants expressing a mutated form of the gene encoding δ(1)-pyrroline-5-carboxylate synthetase (the rate-limiting enzyme in proline synthesis), in which the normal feedback inhibition exerted by proline has been abolished, accumulated much more proline than is normal, and this effect was magnified when the plants were exposed to 200 mM NaCl [16]. Similarly, the over-expression of genes involved in either mannitol and/or trehalose synthesis enhanced the level of the plants' salinity tolerance [17, 18]. Since the reactive oxygen species (ROS) induced by osmotic stress need to be neutralized to prevent their causing secondary damage to the cell, genes encoding the various ROS-scavenging enzymes (superoxide dismutase, catalase, ascorbate peroxidase and others) also belong to this class [19].

Finally, the third group of gene products act to promote plant growth in saline soil. These proteins are intimately involved in signalling, and include a number of transcription factors and protein kinases. For example, the over-expression of a Ca^{2+}-dependent protein kinase in rice has a beneficial effect on salinity tolerance [20], while the knock-down of a similar gene in *A. thaliana* has the opposite effect [21]. Various abscisic acid-responsive transcription factors have been shown to modulate plant performance under salinity stress conditions [22].

3.3 Salinity Stress Limits the Production of Cereal and Vegetable Crops

Species vary considerably with respect to their salinity tolerance, and their physiological response to the stress differs markedly [7]. In many cases, the level of sensitivity is dependent on the developmental stage of the plant. Germination is relatively insensitive to salinity, although more sensitive species can suffer from delayed germination. The seedling stage is typically highly susceptible to damage, particularly under field conditions, where salinity is often at its highest near the soil surface as a result of evaporation. Once the plant is able to extend its root deeper into the soil profile, it tends to become increasingly tolerant [23, 24].

Grains of crops, such as wheat, rice, maize or barley, are highly valuable for the nutrition of humans and livestock as they contain a high percentage of sugars, starch, storage proteins and fatty acids, and they provide roughly half of the calories consumed worldwide. According to their growth habitat and salt tolerance, with maize being a sensitive plant and barley being one of the most tolerant, yields may be drastically reduced by soil salinization [25]. According to FAOSTAT, 2.5 billion tons of cereals and 1.1 billion tons of vegetables were produced in 2013. Vegetables are grown to a lesser extent in production volume when compared to cereals. But in terms of economic yield, plants from Solanaceae, Cucurbitaceae or other families are important players in modern agriculture. Vegetables are indispensable in the human diet and provide vitamins, minerals and health-promoting plant secondary metabolites. Vegetables are in general more sensitive towards salinity as compared to cereals, with bean being more sensitive as compared to cowpea [25].

Complete genome sequences and large-scale EST sequencing projects from various cereal and vegetable crops facilitated the use of large-scale gene expression analysis on the genome and transcriptome level to study salt stress responses [26, 27]. But in order to elucidate gene function, the investigation of the gene product, the protein, is inevitable. Proteomics can also play a role in molecular marker-based breeding programs. There are some reviews highlighting the potential of proteomics-based dissection of salt tolerance mechanisms in crops [28, 29]. This review will focus on current developments and achievements in this field with respect to the years 2011–2015.

3.3.1 The Proteomic Response of Cereals to Salinity Stress

Rice (*Oryza sativa* L.) is generally considered as salt sensitive but due to its ability to grow well under flooded conditions, facilitating leaching of salts, it is also planted on salt-affected soils. Rice does harbour genetic variation with respect to salinity sensitivity/tolerance. When Lee et al. [30] contrasted the 2DE profiles of leaf proteins extracted from a pair of cultivars contrasting with respect to sensitivity, 23 features increased in intensity under either stress conditions, while 7 features were common to both cultivars. Candidate proteins were involved in energy pathway (RuBisCo activase, triose phosphate isomerase, ATP synthase), disease and defense-related proteins (beta-1,3-glucanase, class III peroxidase) and ROS-related proteins (dehydroascorbate reductase, 2-cys peroxiredoxin). In a comparison of the leaf proteomes of derivatives of a mutagenesis breeding programme, Ghaffari et al. [31] were able to show that exposure to salinity stress led to the up-regulation of 34 proteins, most of which were concerned with photosynthesis, carbohydrate metabolism and oxidative stress. An experiment involving a shorter, but more intensive stress treatment identified 65 differentially abundant root and 38 leaf proteins; in the more tolerant accession, the function of the up-regulated proteins included protein modification, nucleoside synthesis, lipid synthesis and energy metabolism [32]. A similar analysis compared the effect of a 16 days exposure to salinity stress on both the leaf and root proteome of the salt sensitive IR29 cultivar with the salt tolerant FL478 cultivar, a recombinant inbred line from a cross of IR29 with the high tolerant Pokkali [33]: here, 39 leaf and 59 root proteins were identified as being responsive to salinity stress in at least one of the two test accessions. Antioxidative proteins were up-regulated in both accessions, so were not considered as explanatory for the observed differential level of salinity sensitivity/tolerance. However, proteins related to polyamine and protein synthesis accumulated in the roots of the sensitive cultivar, reflecting its attempt to adapt to the stress. A similar comparison was followed using leaf proteins extracted from a salinity-challenged sensitive Thai jasmine rice KDML105 cultivar and Pokkali [34]; of the 2DE features up-regulated in Pokkali, 12 were shown to be involved in photosynthesis and ROS detoxification. A short duration, high salinity treatment was applied by Li et al. [35] to identify responsive gene products in the rice shoot. The experiment successfully identified 52 2DE features; those representing proteins involved in photosynthesis and carbon assimilation were down-regulated by the stress, while proteins associated with either metabolism or antioxidation were increased in abundance. A characterization of the short-term response of the leaf proteome to stress by LC-MS for peptide separation and iTRAQ labeling for peptide quantification identified 56 differentially abundant proteins, which were grouped into 20 functional categories [36]; the most prominent of these groups were photosynthesis, antioxidation and oxidative phosphorylation. A leaf proteome comparison between a sensitive and a tolerant cultivar identified the cyclophilin OsCYP2 as being associated with tolerance [37]. When OsCYP2 was subsequently over-expressed, the transgenic proved to be more tolerant than the wild type. The function of OsCYP2 appeared to be related to signalling in a pathway also involved in the plants' response to other abiotic stresses. In cells maintained in a suspension culture, [38] were able to identify 106 proteins using a differential gel electrophoresis (DIGE) platform and 521 using iTRAQ, of which only 58 were in common. Of these, 111 proteins were revealed to be responsive to salinity stress. In combination with a parallel set of metabolomic data, the indications were that the response of the suspension cells shared some similarities with that observed in planta: in particular, carbohydrate and energy metabolism pathways, redox signalling pathways, auxin/indole-3-acetic acid pathways and the synthetic pathways leading to osmoprotectants were all stress-responsive. The proteomic response of the reproductive organs of rice has also been investigated. When anther proteins of a sensitive and a tolerant cultivar were compared, 18 differentially expressed proteins were identified by Sarhadi et al. [39], most of which are concerned with carbohydrate metabolism and anther/pollen wall remodelling. A phosphoproteomic analysis of the leaf microsomal fraction was carried out by Chang et al. [40] in an attempt to characterize the phosphorylation status of proteins embedded in the cellular membrane; the observation of several phosphorylation sites in aquaporins was taken to imply a regulatory role of water flux under stress conditions. When the proteome of an over-expressor of a gene encoding SnRK2 kinase was compared to that of the wild type in plants exposed to salinity stress Nam et al. [41] observed a considerable reduction in the number of differentially expressed proteins. Since the basal expression level of proteins up-regulated in the wild type by salinity stress was changed in the transgenic plants before any stress was applied, the interpretation was that rice possesses a constitutively activated stress responsive pathway.

In contrast to rice, wheat is more sensitive to salinity stress during its vegetative phase than during its reproductive phase. In general, wheat can exceed higher salt levels than rice, while durum wheat (*Triticum durum* Desf.) is more tolerant as compared to bread wheat (*Triticum aestivum* L.). When ascorbic acid treatment was applied to promote germination in durum wheat, Fercha et al. [42] noted that 83 proteins changed their level of expression level (as opposed to 72 upon imbibition in water), many of which were involved in protein metabolism, antioxidant protection, repair processes and methionine-related metabolism. In a follow-up study, the effect of salinity on the proteome of the germinating embryo and its surrounding tissues was described [43]. Exogenous application of salicylic acid is able to reduce salinity-induced growth inhibition. Inspection of 2DE-acquired bread wheat leaf proteomes of plants exposed to salinity stress and treated with the phytohormone salicylic acid (SA) identified 38 proteins as being differentially regulated by both salinity and SA; those for which a function could be assigned were involved in signal transduction, defence, energy, metabolism and photosynthesis antioxidant activity and indirect effects by activating rather unspecific stress-related pathways that allow for a better germination rate [44]. A comparison of the effect of a short-term exposure to salinity stress on the root proteome of a tolerant and a sensitive bread wheat cultivar identified a set of 144 up-regulated proteins, whose functions spanned energy metabolism, protein metabolism, signal transduction and antioxidant activity [45]. A similar comparison of the leaf proteome identified the salinity-induced up-regulation of various ROS scavenging and photosynthesis-related proteins in the more tolerant cultivar [46]. DIGE 2DE characterized the short-term response to salinity stress of an elite Chinese bread wheat cultivar; a functional analysis of 52 proteins showing altered expression showed that the major cellular impact of salinity stress was on carbon metabolism [47]. In durum wheat, Capriotti et al. [48] were able to show that proteins involved in photosynthesis, transcription and signal transduction were all suppressed in stressed plants. Two reports were available within the period 2011–2015 on subcellular proteomics approaches in response to salinity. Kamal et al. [49] subjected chloroplasts to a proteomic analysis which had been extracted from salinity stressed bread wheat seedlings; the analysis revealed that 65 proteins were stress-responsive, most of which were related to photosynthesis and ROS detoxification. A contrast of the mitochondrial proteomes of bread wheat and a salinity tolerant wheat × *Lophopyrum elongatum* hybrid performed by Jacoby et al. [50] showed that salinity stress enhanced the presence in the hybrid of manganese superoxide dismutase, aconitase and serine hydroxymethyl transferase.

Barley is more salinity tolerant as a species than wheat. A number of combinations of contrasting pairs of cultivars have been compared on the proteome level with the intention of identifying candidate proteins associated with tolerance. An example is represented by the comparison of the seedling leaf proteomes of the cultivar pair Afzal (tolerant) and L-527 (sensitive) following a short duration stress episode [51]. In all, 22 proteins were identified as being responsive to the stress, most of which mainly involved in photosynthesis, ROS detoxification and energy metabolism. The effect of a more prolonged stress episode was described by Fatehi et al. [52]; here, 20 proteins were identified as being salinity stress responsive, with only few overlapping those responsive to the short duration stress treatment. When the root proteomes of cvs. Steptoe and Morex were inspected by Witzel et al. [53], the proteins induced earliest proved to be associated with either the oxidative stress response or the methylerythritol 4-phosphate pathway, while those induced later involved energy and primary metabolism, protein synthesis and transport. To evaluate the similarity in tolerance mechanisms, a tolerant cultivar XZ16 of the wild ancestor of barley (*Hordeum spontaneum* K. Koch) was compared with a tolerant barley cultivar CM72 [54]. Here, 20 differential 2DE features were identified in the root proteome and 21 in the leaf proteome. There were apparent resemblances in protein abundance patterns between the tested cultivars when challenged with salinity. However, also genotype-exclusive changes were observed reflecting a higher tissue tolerance in XZ16 as compared to CM72. A 3 weeks exposure of *H. spontaneum* to salinity stress led to the identification of 16 proteins which were up-regulated in the leaf [55]: these included superoxide dismutase and thioredoxin, as well as nucleoside diphosphate kinase and an oxygen-evolving enhancer protein. A relatively novel approach to increase salt tolerance in crops is the application of mutualistic endophytes that colonize host plants and thereby enhancing growth. The presence of the root fungal endophyte *Piriformospora indica* is known to afford barley plants some protection from salinity damage. An analysis of the leaf proteome of inoculated plants has been provided by Alikhani et al. [56]; the outcome was that the abundance of several stress-responsive proteins was unaffected by the presence of the endophyte, but there was evidence for the up-regulation of a Myb transcription factor and a papain-like cysteine protease.

The least tolerant of the cereals is maize (*Zea mays* L.). A comparison of the root proteomes of a pair of contrasting cultivars has revealed that the more tolerant one was able to make some adjustments to its carbohydrate metabolism, while the more sensitive one modified its protein metabolism, redox homeostasis and carbohydrate metabolism [57]. An iTRAQ approach was selected to compare the proteome adjustments to salinity in salt tolerant cultivar F63 with the salt sensitive cultivar

F35 [58]. Here, the more tolerant cultivar accumulated cysteine protease, a lichenase-2 precursor and a xyloglucan endotransglycosylase homolog in its roots. Germination under saline conditions is critical for plant development. The proteomic consequences of salinity during germination suggested delays to storage protein degradation and accumulation of ATP accumulation [59].

By far the most salinity tolerant cereal is rye, but very little effort has been invested in identifying the molecular basis of its tolerance. A single published paper covers the effect of salinity on the leaf proteome: here, some 17 differentially abundant 2DE features were identified and a number of ROS scavenging enzymes were shown to be up-regulated [60].

3.3.2 The Proteomic Response of Vegetables to Salinity Stress

Tomato (*Solanum lycopersicum* L.) is the most widely grown fruit vegetable in the world with 163 million tons produced for fresh consumption and 5.4 million tons produced for processing (FAOSTAT, 2013). While greenhouse production of tomato is seldom affected by salinity, most of the crop worldwide is field-grown and thus is exposed to soil salinized by poor irrigation practice. The species is considered to be moderately sensitive to salinity [61]. The root proteomes of a relatively tolerant var. *cerasiforme* and a sensitive conventional cultivar have been compared by iTRAQ LC-MS/MS [62]. An increased abundance of proteins associated with root growth, carbohydrate metabolism, antioxidant activity and stress signal transduction were observed in the more tolerant cultivar. A similar comparison involving four diverse cultivars was conducted by Manaa et al. [63]. A number of cultivar-specific and salinity stress—responsive proteins were detected by 2DE; among the potentially interesting candidates were a transcription factor and several proteins involved in cell wall reinforcement. With respect to the tomato leaf proteome, Manaa et al. [64] contrasted a relatively tolerant cultivar with a sensitive one; the higher abundance of antioxidant proteins in the former was taken to afford a measure of protection for the plant's photosynthetic machinery. Unlike the domesticated tomato, its wild relative S. *chilense* is adapted to saline soils. When its leaf and root proteome were investigated, a number of proteins involved in antioxidation, detoxification and ion uptake/transport were found to be up-regulated by the imposition of salinity stress [65]. Fruit yield and sugar composition is heavily affected by salinity. A prominent feature of the fruit pericarp proteome was the enrichment for ethylene synthesis-related proteins induced by salinity stress, associated with accelerated fruit ripening [66]. Although silicon is non-essential for plant growth, its supply has a positive effect on the salinity tolerance of tomato, for reasons which are not yet fully resolved. The silicon-aided growth enhancement has been correlated with a greater abundance of potassium transporters and an altered transcriptional regulation [67]. The thylakoid protein complexes present in the chloroplasts developed in silicon-fortified plants appeared to be more stable than those developed in the absence of fortification [68]. Plant growth under saline conditions can also be supported by the exogenous supply of polyamides; the effect of this treatment on the 2DE-acquired leaf proteome included 39 features, some of which represented proteins associated with the defence response or antioxidative activity [69]. In the root, salinity has been noted to up-regulate proteins involved in abscisic acid signalling and ROS scavenging [70].

Potato (*Solanum tuberosum* L.) is considered as a moderately salinity sensitive species, especially during tuber bud initiation [25]. A combined transcriptomic and proteomic investigation of the salinity stress response of cv. Désirée showed that the leaf content of several photosynthesis-related proteins was reduced, while protein metabolism was up-regulated [71].

Cucumber is highly sensitive to salinity in irrigation water, but some protection is given by the application of polyamines. An analysis of its leaves following exposure to salinity confirmed that the provision of polyamine enhanced the production of proteins involved in protein synthesis, antioxidation and S-adenosylmethionine synthase [72]. Cucurbitaceae also represent a model plant system for phloem translocation processes. When the phloem sap of salinity-stressed plants of a tolerant and a sensitive cultivar was compared, 745 proteins were identified, of which 111 proved to be responsive to salinity treatment [73]; carbon fixation pathway proteins were suppressed by the stress in the sensitive cultivar but not in the tolerant one.

Lettuce (*Lactuca sativa* L.) is salinity sensitive, especially at the young seedling and bolting stages. An analysis of its leaves harvested from plants subjected to salinity stress showed that the content of superoxide dismutase was enhanced, along with that of proteins involved in ethylene metabolism [74].

The grain amaranth (*Amaranthus cruentus* L.) is regarded one of the most salinity tolerant non-cereal crop species. When its root proteome was investigated in plants subjected to salinity stress, 77 responsive 2DE features were identified [75]. These included enzymes involved in ROS scavenging, nucleotide metabolism and fatty acid and vitamin synthesis. An assessment of the mesophyll and bundle sheath chloroplast proteome demonstrated the accumulation of ATP synthase subunits and electron cycling proteins as being a characteristic feature of salinity tolerance [76].

Proteomic analyses of the salinity stress response have been conducted for both cowpea (*Vigna unguiculata*) and grasspea (*Lathyrus sativus* L.). An analysis of the leaf proteome in cowpea plants subjected to salinity stress identified 22 proteins which were altered in abundance by the stress: these included proteins associated with photosynthesis and energy metabolism [77]. In grasspea, a short duration salinity stress episode induced various proteins in the 14-3-3 signalling pathway and ROS scavenging enzymes [78].

3.4 Conclusion and Future Perspectives

Improving the level of salinity tolerance expressed by both cereal and vegetable crops is a critical plant breeding challenge. The molecular basis of the response to salinity stress in the model plant *A. thaliana* has been exhaustively investigated and increasingly this knowledge is being applied to understanding the response in non-model species. Current technologies related to DNA sequencing are sufficiently generic and cost-effective to be applied to almost any species, and are being used widely to generate markers to support conventional breeding [26, 79]. To date, however, proteomics platforms have not been much deployed to reveal the mechanistic basis of plant salinity tolerance, mainly because the methodology is challenging with respect of both sample isolation and the subsequent separation, quantification and identification procedures. The gap between differential gene transcription, which can be captured readily by high throughput analysis of mRNA, and the differential accumulation of specific proteins remains very wide, and will remain so until protein separation methods can be substantially refined and streamlined [80]. Nevertheless, state-of-the-art LC-based separation technology is capable of detecting several thousand proteins, with a level of sensitivity much greater than is possible using 2DE. The majority of the proteomics literature to date (summarized in Table 3.1) has relied on a gel-based separation platform, but it is likely that gel-free alternatives will soon replace them, at least for non-targeted analyses. 2DE will likely retain utility in characterizing post-translational modified forms, protein isoforms and protein complexes. A second basis for the slow uptake of proteomics to determine the mechanistic basis of salinity tolerance is that plant responses to the stress are typically organ- or even tissue-specific, and unlike DNA, there is no means presently available for amplifying material extracted from very small biological samples; thus leaf extracts, for example, tend to be swamped by the highly abundant protein RuBisCo. Some attempts have been made nonetheless to analyse the proteome of chloroplasts, mitochondria, cellular membranes and the fruit pericarp (Table 3.1), and it can be anticipated that spatially-resolved proteome responses will make a contribution in future to our understanding of the complexity of the plant salinity stress response.

A widely-used approach for the identification of functional proteins is to compare the proteomes of two or more cultivars/accessions which differ in their response to salinity stress (Fig. 3.1). Using parent accessions of segregating populations provide the possibility to assess the functionality of candidates in offspring lines with similar responses as compared to the parent lines, prior to time-consuming genetic manipulation such as transgene expression or knock-out generation.

Salt tolerance is a complex agronomic trait. Combining omics technologies to a systems biology approach is not a novel direction in salt stress investigations but it has not been applied to crop research in full extension. The step forward from collecting proteomics data to modeling and functional prediction will open new

Table 3.1 An overview of published salinity stress-related proteomic analyses in cereal and vegetable crops over the period 2011–2015

Species	Salt treatment and duration	Tissue	Analytical platform	Reference
Rice				
	45 and 90 mM, 4 days	Leaf	2DE	Lee et al. [30]
	250 mM, 2 days	Leaf	2DE	Li et al. [35]
	120 mM, 7 days	Leaf	2DE	Jankangram et al. [34]
	100 mM, 10 days	Leaf	2DE	Ruan et al. [37]
	200 mM, 2 days	Leaf microsome fraction	LC-MS/MS	Chang et al. [40]
	150 mM, 3 and 7 h	Root	2DE	Nam et al. [41]
	100 mM, reproductive phase	Anther	2DE	Sarhadi et al. [39]
	100 mM, 1, 5 and 24 h	Suspension cultured cells	2DE DIGE and iTRAQ LC-MS/MS	Liu et al. [38]
	120 mM, 6 days	Leaf	2DE	Ghaffari et al. [31]
	250 mM, 30 min	Leaf and root	2DE	Liu et al. [32]
	120 mM, 16 days	Leaf and root	2DE	Hosseini et al. [33]
	150 mM, 1 day	Leaf	iTRAQ and LC-MS/MS	Xu et al. [36]
Wheat				
	170, 250, 340, 425 mM, 2 days	Leaf	2DE DIGE	Gao et al. [47]
	85, 250, 425 mM, 2 days	Root	2DE	Guo et al. [45]
	150 mM, 1, 2 and 3 days	Chloroplast	2DE	Kamal et al. [49]
	250 mM, 3 days	Leaf	2DE	Kang et al. [44]
	250 mM, germination	Grain	LC-MS/MS	Fercha et al. [42]
	200 mM, 7 weeks	Mitochondria	2DE DIGE	Jacoby et al. [50]
	100 and 200 mM, 10 days	Leaf	LC-MS/MS	Capriotti et al. [48]
	250 mM, germination	Grain tissues	LC-MS/MS	Fercha et al. [43]
	200 mM, 17 days	Leaf	2DE	Maleki et al. [46]
Barley				
	300 mM, 1 day	Leaf	2DE	Rasoulnia et al. [51]
	300 mM, 3 weeks	Leaf	2DE	Fatehi et al. [52]
	100 and 300 mM, 14 days	Leaf	2DE	Alikhani et al. [56]
	300 mM, 3 weeks	Leaf	2DE	Fatehi et al. [55]
	100 and 150 mM, 1, 4, 7, 10 days	Root	2DE	Witzel et al. [53]
	200 mM, 2 days	Leaf and root	2DE	Wu et al. [54]

(continued)

Table 3.1 (continued)

Species	Salt treatment and duration	Tissue	Analytical platform	Reference
Maize				
	150 mM, 3 days	Root	2DE	Cheng et al. [57]
	100 and 200 mM, 0, 24, 42, 48, 60 h	Grain embryo	2DE	Meng et al. [59]
	160 mM, 2 days	Root	iTRAQ and LC-MS/MS	Cui et al. [58]
Rye				
	200 mM, 4 days	Leaf	2DE	Lee et al. [60]
Tomato				
	100 mM, 14 days	Root	2DE	Manaa et al. [63]
	200 mM, 25 days	Leaf and root	2DE DIGE	Zhou et al. [65]
	40 mM, fruit ripening	Fruit pericarp	2DE	Manaa et al. [66]
	100 mM, 14 days	Leaf	2DE	Manaa et al. [64]
	200 mM, 2 days	Root	iTRAQ and LC-MS/MS	Nveawiah-Yoho et al. [62]
	50 mM, 3 days	Root	iTRAQ and LC-MS/MS	Gong et al. [70]
	25 and 50 mM, 5 days	Chloroplast	BlueNative PAGE	Muneer et al. [68]
	25 and 50 mM, 5 days	Root	2DE	Muneer et al. [67]
	75 mM, 4 days	Leaf	2DE	Zhang et al. [69]
Potato				
	150 mM, 3, 8 days	Leaf	2DE DIGE	Evers et al. [71]
Cucumber				
	75 mM, 3 days	Leaf	2DE	Li et al. [72]
	75 mM, 3 days	Phloem sap	iTRAQ and LC-MS/MS	Fan et al. [73]
Lettuce				
	100 mM, 15 and 30 days	Leaf	LC-MS/MS	Lucini et al. [74]
Amaranth				
	150 mM, 1 h, 1 and 7 days	Root	2DE	Huerta-Ocampo et al. [75]
	300 mM, 5 days	Chloroplast	BlueNative PAGE	Joaquin-Ramos et al. [76]
Cowpea				
	75 mM, 17 days	Leaf	2DE	Abreu et al. [77]
Grasspea				
	500 mM, 12, 24 and 36 h	Leaf	2DE	Chattopadhyay et al. [78]

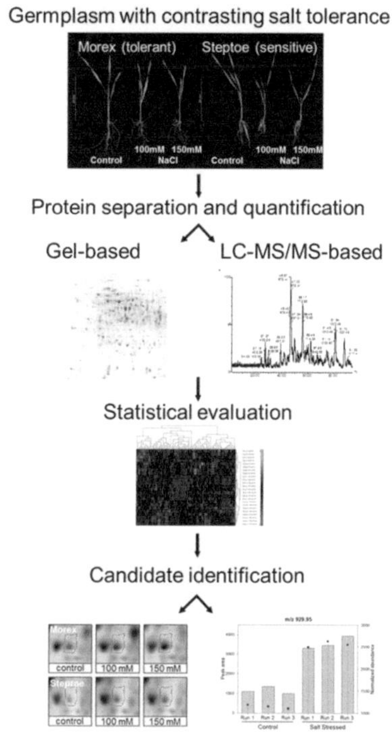

Fig. 3.1 A workflow used to identify candidate proteins contributing salt tolerance in barley

avenues for the sustainable production of plants that are adjusted to unfavorable environmental conditions.

Acknowledgments Financial support of proteomic research on plant abiotic stress defense to H.-P. Mock by the BMBF (BASALT; FKZ 01DH13018) and by the FNR (PROKAR; FKZ 22023411) are gratefully acknowledged.

References

1. Vij S, Tyagi AK (2007) Emerging trends in the functional genomics of the abiotic stress response in crop plants. Plant Biotechnol J 5:361–380
2. Munns R (2005) Genes and salt tolerance: bringing them together. New Phytol 167:645–663
3. Fricke W (2004) Rapid and tissue-specific accumulation of solutes in the growth zone of barley leaves in response to salinity. Planta 219:515–525
4. Deinlein U, Stephan AB, Horie T, Luo W, Xu G, Schroeder JI (2014) Plant salt-tolerance mechanisms. Trends Plant Sci 19:371–379
5. Munns R, James RA, Lauchli A (2006) Approaches to increasing the salt tolerance of wheat and other cereals. J Exp Bot 57:1025–1043
6. Munns R (2002) Comparative physiology of salt and water stress. Plant, Cell Environ 25:239–250

7. Munns R, Husain S, Rivelli AR, James RA, Condon AG, Lindsay MP et al (2002) Avenues for increasing salt tolerance of crops, and the role of physiologically based selection traits. Plant Soil 247:93–105
8. Kronzucker HJ, Britto DT (2011) Sodium transport in plants: a critical review. New Phytol 189:54–81
9. Apse MP, Aharon GS, Snedden WA, Blumwald E (1999) Salt tolerance conferred by overexpression of a vacuolar Na^+/H^+ antiport in *Arabidopsis*. Science 285:1256–1258
10. Shi H, Lee BH, Wu SJ, Zhu JK (2003) Overexpression of a plasma membrane Na^+/H^+ antiporter gene improves salt tolerance in *Arabidopsis thaliana*. Nat Biotechnol 21:81–85
11. Huang S, Spielmeyer W, Lagudah ES, Munns R (2008) Comparative mapping of HKT genes in wheat, barley, and rice, key determinants of Na^+ transport, and salt tolerance. J Exp Bot 59:927–937
12. Rus A, Lee BH, Munoz-Mayor A, Sharkhuu A, Miura K, Zhu JK et al (2004) AtHKT1 facilitates Na^+ homeostasis and K^+ nutrition in planta. Plant Physiol 136:2500–2511
13. James RA, Blake C, Byrt CS, Munns R (2011) Major genes for Na^+ exclusion, Nax1 and Nax2 (wheat HKT1; 4 and HKT1; 5), decrease Na^+ accumulation in bread wheat leaves under saline and waterlogged conditions. J Exp Bot 62:2939–2947
14. Gaxiola RA, Li J, Undurraga S, Dang LM, Allen GJ, Alper SL et al (2001) Drought- and salt-tolerant plants result from overexpression of the AVP1 H^+-pump. Proc Natl Acad Sci USA 98:11444–11449
15. Ashraf M, Foolad MR (2007) Roles of glycine betaine and proline in improving plant abiotic stress resistance. Environ Exp Bot 59:206–216
16. Hong Z, Lakkineni K, Zhang Z, Verma DP (2000) Removal of feedback inhibition of delta(1)-pyrroline-5-carboxylate synthetase results in increased proline accumulation and protection of plants from osmotic stress. Plant Physiol 122:1129–1136
17. Abebe T, Guenzi AC, Martin B, Cushman JC (2003) Tolerance of mannitol-accumulating transgenic wheat to water stress and salinity. Plant Physiol 131:1748–1755
18. Garg AK, Kim JK, Owens TG, Ranwala AP, Choi YD, Kochian LV et al (2002) Trehalose accumulation in rice plants confers high tolerance levels to different abiotic stresses. Proc Natl Acad Sci USA 99:15898–15903
19. Mittler R, Vanderauwera S, Gollery M, Van Breusegem F (2004) Reactive oxygen gene network of plants. Trends Plant Sci 9:490–498
20. Saijo Y, Kinoshita N, Ishiyama K, Hata S, Kyozuka J, Hayakawa T et al (2001) A Ca^{2+}-dependent protein kinase that endows rice plants with cold- and salt-stress tolerance functions in vascular bundles. Plant Cell Physiol 42:1228–1233
21. Zhao R, Sun H, Zhao N, Jing X, Shen X, Chen S (2015) The *Arabidopsis* Ca^{2+}-dependent protein kinase CPK27 is required for plant response to salt-stress. Gene 563:203–214
22. Lata C, Prasad M (2011) Role of DREBs in regulation of abiotic stress responses in plants. J Exp Bot 62:4731–4748
23. Maas EV, Poss JA (1989) Salt sensitivity of wheat at various growth-stages. Irrig Sci 10:29–40
24. Maas EV, Hoffman GJ, Chaba GD, Poss JA, Shannon MC (1983) Salt sensitivity of corn at various growth-stages. Irrig Sci 4:45–57
25. Maas EV (1993) Testing crops for salinity tolerance. Proceedings of the workshop on adaptation of plants to soil stresses. University of Nebraska, Lincoln, pp 234–247
26. Zhuang J, Zhang J, Hou XL, Wang F, Xiong AS (2014) Transcriptomic, proteomic, metabolomic and functional genomic approaches for the study of abiotic stress in vegetable crops. Crit Rev Plant Sci 33:225–237
27. Shelden MC, Roessner U (2013) Advances in functional genomics for investigating salinity stress tolerance mechanisms in cereals. Front Plant Sci 4:123
28. Barkla BJ, Castellanos-Cervantes T, De Leon JL, Matros A, Mock HP, Perez-Alfocea F et al (2013) Elucidation of salt stress defense and tolerance mechanisms of crop plants using proteomics: current achievements and perspectives. Proteomics 13:1885–1900

29. Abreu IA, Farinha AP, Negrao S, Goncalves N, Fonseca C, Rodrigues M et al (2013) Coping with abiotic stress: proteome changes for crop improvement. J Proteom 93:145–168
30. Lee DG, Park KW, An JY, Sohn YG, Ha JK, Kim HY et al (2011) Proteomics analysis of salt-induced leaf proteins in two rice germplasms with different salt sensitivity. Can J Plant Sci 91:337–349
31. Ghaffari A, Gharechahi J, Nakhoda B, Salekdeh GH (2014) Physiology and proteome responses of two contrasting rice mutants and their wild type parent under salt stress conditions at the vegetative stage. J Plant Physiol 171:31–44
32. Liu CW, Chang TS, Hsu YK, Wang AZ, Yen HC, Wu YP et al (2014) Comparative proteomic analysis of early salt stress responsive proteins in roots and leaves of rice. Proteomics 14:1759–1775
33. Hosseini SA, Gharechahi J, Heidari M, Koobaz P, Abdollahi S, Mirzaei M et al (2015) Comparative proteomic and physiological characterisation of two closely related rice genotypes with contrasting responses to salt stress. Funct Plant Biol 42:527–542
34. Jankangram W, Thammasirirak S, Jones MG, Hartwell J, Theerakulpisut P (2011) Proteomic and transcriptomic analysis reveals evidence for the basis of salt sensitivity in Thai jasmine rice (*Oryza sativa* L. cv. KDML 105). Afr J Biotechnol 10:16157–16166
35. Li XJ, Yang MF, Zhu Y, Liang Y, Shen SH (2011) Proteomic analysis of salt stress responses in rice shoot. J Plant Biol 54:384–395
36. Xu J, Lan H, Fang H, Huang X, Zhang H, Huang J (2015) Quantitative proteomic analysis of the rice (*Oryza sativa* L.) salt response. PLoS ONE 10:e0120978
37. Ruan SL, Ma HS, Wang SH, Fu YP, Xin Y, Liu WZ et al (2011) Proteomic identification of OsCYP2, a rice cyclophilin that confers salt tolerance in rice (*Oryza sativa* L.) seedlings when overexpressed. BMC Plant Biol 11:1
38. Liu DW, Ford KL, Roessner U, Natera S, Cassin AM, Patterson JH et al (2013) Rice suspension cultured cells are evaluated as a model system to study salt responsive networks in plants using a combined proteomic and metabolomic profiling approach. Proteomics 13:2046–2062
39. Sarhadi E, Bazargani MM, Sajise AG, Abdolahi S, Vispo NA, Arceta M et al (2012) Proteomic analysis of rice anthers under salt stress. Plant Physiol Biochem 58:280–287
40. Chang IF, Hsu JL, Hsu PH, Sheng WA, Lai SJ, Lee C et al (2012) Comparative phosphoproteomic analysis of microsomal fractions of *Arabidopsis thaliana* and *Oryza sativa* subjected to high salinity. Plant Sci 185–186:131–142
41. Nam MH, Huh SM, Kim KM, Park WJ, Seo JB, Cho K et al (2012) Comparative proteomic analysis of early salt stress-responsive proteins in roots of SnRK2 transgenic rice. Proteome Sci 10:25
42. Fercha A, Capriotti AL, Caruso G, Cavaliere C, Gherroucha H, Samperi R et al (2013) Gel-free proteomics reveal potential biomarkers of priming-induced salt tolerance in durum wheat. J Proteom 91:486–499
43. Fercha A, Capriotti AL, Caruso G, Cavaliere C, Samperi R, Stampachiacchiere S et al (2014) Comparative analysis of metabolic proteome variation in ascorbate-primed and unprimed wheat seeds during germination under salt stress. J Proteom 108:238–257
44. Kang G, Li G, Zheng B, Han Q, Wang C, Zhu Y et al (2012) Proteomic analysis on salicylic acid-induced salt tolerance in common wheat seedlings (*Triticum aestivum* L.). Biochim Biophys Acta 1824:1324–1333
45. Guo G, Ge P, Ma C, Li X, Lv D, Wang S et al (2012) Comparative proteomic analysis of salt response proteins in seedling roots of two wheat varieties. J Proteom 75:1867–1885
46. Maleki M, Naghavi MR, Alizadeh H, Poostini K, Mishani CA (2014) Comparison of protein changes in the leaves of two bread wheat cultivars with different sensitivity under salt stress. Annu Res Rev Biol 4:1784–1797
47. Gao L, Yan X, Li X, Guo G, Hu Y, Ma W et al (2011) Proteome analysis of wheat leaf under salt stress by two-dimensional difference gel electrophoresis (2D-DIGE). Phytochemistry 72:1180–1191

48. Capriotti AL, Borrelli GM, Colapicchioni V, Papa R, Piovesana S, Samperi R et al (2014) Proteomic study of a tolerant genotype of durum wheat under salt-stress conditions. Anal Bioanal Chem 406:1423–1435
49. Kamal AH, Cho K, Kim DE, Uozumi N, Chung KY, Lee SY et al (2012) Changes in physiology and protein abundance in salt-stressed wheat chloroplasts. Mol Biol Rep 39:9059–9074
50. Jacoby RP, Millar AH, Taylor NL (2013) Investigating the role of respiration in plant salinity tolerance by analyzing mitochondrial proteomes from wheat and a salinity-tolerant amphiploid (wheat × Lophopyrum elongatum). J Proteome Res 12:4807–4829
51. Rasoulnia A, Bihamta MR, Peyghambari SA, Alizadeh H, Rahnama A (2011) Proteomic response of barley leaves to salinity. Mol Biol Rep 38:5055–5063
52. Fatehi F, Hosseinzadeh A, Alizadeh H, Brimavandi T, Struik PC (2012) The proteome response of salt-resistant and salt-sensitive barley genotypes to long-term salinity stress. Mol Biol Rep 39:6387–6397
53. Witzel K, Matros A, Strickert M, Kaspar S, Peukert M, Muhling KH et al (2014) Salinity stress in roots of contrasting barley genotypes reveals time-distinct and genotype-specific patterns for defined proteins. Mol Plant 7:336–355
54. Wu D, Shen Q, Qiu L, Han Y, Ye L, Jabeen Z et al (2014) Identification of proteins associated with ion homeostasis and salt tolerance in barley. Proteomics 14:1381–1392
55. Fatehi F, Hosseinzadeh A, Alizadeh H, Brimavandi T (2013) The proteome response of *Hordeum spontaneum* to salinity stress. Cereal Res Commun 41:78–87
56. Alikhani M, Khatabi B, Sepehri M, Nekouei MK, Mardi M, Salekdeh GH (2013) A proteomics approach to study the molecular basis of enhanced salt tolerance in barley (*Hordeum vulgare* L.) conferred by the root mutualistic fungus *Piriformospora indica*. Mol BioSyst 9:1498–1510
57. Cheng YJ, Chen GQ, Hao DR, Lu HH, Shi ML, Mao YX et al (2014) Salt-induced root protein profile changes in seedlings of maize inbred lines with differing salt tolerances. Chil J Agric Res 74:468–476
58. Cui D, Wu D, Liu J, Li D, Xu C, Li S et al (2015) Proteomic analysis of seedling roots of two maize inbred lines that differ significantly in the salt stress response. PLoS ONE 10:e0116697
59. Meng LB, Chen YB, Lu TC, Wang YF, Qian CR, Yu Y et al (2014) A systematic proteomic analysis of NaCl-stressed germinating maize seeds. Mol Biol Rep 41:3431–3443
60. Lee KW, Choi GJ, Kim KY, Ji HC, Park HS, Lee SH (2013) A proteomic approach to identify salt-responsive proteins in rye. Pak J Bot 45:1489–1496
61. Shahbaz M, Ashraf M, Al-Qurainy F, Harris PJC (2012) Salt tolerance in selected vegetable crops. Crit Rev Plant Sci 31:303–320
62. Nveawiah-Yoho P, Zhou J, Palmer M, Sauve R, Zhou SP, Howe KJ et al (2013) Identification of proteins for salt tolerance using a comparative proteomics analysis of tomato accessions with contrasting salt tolerance. J Am Soc Hortic Sci 138:382–394
63. Manaa A, Ben Ahmed H, Valot B, Bouchet JP, Aschi-Smiti S, Causse M et al (2011) Salt and genotype impact on plant physiology and root proteome variations in tomato. J Exp Bot 62:2797–2813
64. Manaa A, Mimouni H, Wasti S, Gharbi E, Aschi-Smiti S, Faurobert M et al (2013) Comparative proteomic analysis of tomato (*Solanum lycopersicum*) leaves under salinity stress. Plant Omics 6:268–277
65. Zhou SP, Sauve RJ, Liu Z, Reddy S, Bhatti S, Hucko SD et al (2011) Identification of salt-induced changes in leaf and root proteomes of the wild tomato, *Solanum chilense*. J Am Soc Hortic Sci 136:288–302
66. Manaa A, Faurobert M, Valot B, Bouchet JP, Grasselly D, Causse M et al (2013) Effect of salinity and calcium on tomato fruit proteome. OMICS 17:338–352
67. Muneer S, Jeong BR (2015) Proteomic analysis of salt-stress responsive proteins in roots of tomato (*Lycopersicon esculentum* L.) plants towards silicon efficiency. Plant Growth Regul 77:133–146

68. Muneer S, Park YG, Manivannan A, Soundararajan P, Jeong BR (2014) Physiological and proteomic analysis in chloroplasts of *Solanum lycopersicum* L. under silicon efficiency and salinity stress. Int J Mol Sci 15:21803–21824
69. Zhang Y, Zhang H, Zou ZR, Liu Y, Hu XH (2015) Deciphering the protective role of spermidine against saline-alkaline stress at physiological and proteomic levels in tomato. Phytochemistry 110:13–21
70. Gong B, Zhang C, Li X, Wen D, Wang S, Shi Q et al (2014) Identification of NaCl and $NaHCO_3$ stress responsive proteins in tomato roots using iTRAQ-based analysis. Biochem Biophys Res Commun 446:417–422
71. Evers D, Legay S, Lamoureux D, Hausman JF, Hoffmann L, Renaut J (2012) Towards a synthetic view of potato cold and salt stress response by transcriptomic and proteomic analyses. Plant Mol Biol 78:503–514
72. Li B, He L, Guo S, Li J, Yang Y, Yan B et al (2013) Proteomics reveal cucumber Spd-responses under normal condition and salt stress. Plant Physiol Biochem 67:7–14
73. Fan H, Xu Y, Du C, Wu X (2015) Phloem sap proteome studied by iTRAQ provides integrated insight into salinity response mechanisms in cucumber plants. J Proteom 125:54–67
74. Lucini L, Bernardo L (2015) Comparison of proteome response to saline and zinc stress in lettuce. Front Plant Sci 6:240
75. Huerta-Ocampo JA, Barrera-Pacheco A, Mendoza-Hernandez CS, Espitia-Rangel E, Mock HP, Barba De La Rosa AP (2014) Salt stress-induced alterations in the root proteome of *Amaranthus cruentus* L. J Proteome Res 13:3607–3627
76. Joaquin-Ramos A, Huerta-Ocampo JA, Barrera-Pacheco A, De Leon-Rodriguez A, Baginsky S, Barba De La Rosa AP (2014) Comparative proteomic analysis of amaranth mesophyll and bundle sheath chloroplasts and their adaptation to salt stress. J Plant Physiol 171:1423–1435
77. De Abreu CE, Araujo Gdos S, Monteiro-Moreira AC, Costa JH, Leite Hde B, Moreno FB et al (2014) Proteomic analysis of salt stress and recovery in leaves of *Vigna unguiculata* cultivars differing in salt tolerance. Plant Cell Rep 33:1289–1306
78. Chattopadhyay A, Subba P, Pandey A, Bhushan D, Kumar R, Datta A et al (2011) Analysis of the grasspea proteome and identification of stress-responsive proteins upon exposure to high salinity, low temperature, and abscisic acid treatment. Phytochemistry 72:1293–1307
79. Witzel K, Neugart S, Ruppel S, Schreiner M, Wiesner M, Baldermann S (2015) Recent progress in the use of ‘omics technologies in brassicaceous vegetables. Front Plant Sci 6:244
80. Matros A, Kaspar S, Witzel K, Mock HP (2011) Recent progress in liquid chromatography-based separation and label-free quantitative plant proteomics. Phytochemistry 72:963–974

Chapter 4
Proteomics of Flooding-Stressed Plants

Mudassar Nawaz Khan and Setsuko Komatsu

Abstract Climate change is a growing worldwide concern with respect to food security. Abiotic stresses are responsible for huge annual losses in agricultural productivity. In particular, flooding is a serious threat for many crops, including wheat and soybean, which exhibit dramatic reductions in growth and yield that result in the annual loss of billions of dollars. Flooding induces various adverse morphological and physiological effects, and forces plants to shift from aerobic to anaerobic metabolism through modifications at the molecular level. Proteomic analyses have greatly contributed to unraveling the flooding stress-response mechanisms that are adopted by different plant species, particularly soybean. The proteomic study of post-flooding recovery mechanisms has contributed to the search for flooding-responsive proteins and those that play essential roles in the transition from stress to post-stress conditions. This review summarizes the major findings from proteomic studies that have examined flooding stress-response mechanisms in important crop species. Furthermore, protein abundance changes and their significance during post-flooding recovery are discussed.

Keywords Proteomics · Flooding stress · Plants

4.1 Introduction

Meeting the food needs of the growing worldwide population has become more challenging in the twenty-first century due to global climate changes. The effects of climatic changes include extreme weather conditions, such as increasing temperature

M.N. Khan · S. Komatsu
Graduate School of Life and Environmental Sciences, University of Tsukuba, Tsukuba 305-8572, Japan

M.N. Khan · S. Komatsu (✉)
National Institute of Crop Science, National Agriculture and Food Research Organization, Kannondai 2-1-18, Tsukuba 305-8518, Japan
e-mail: skomatsu@affrc.go.jp

G.H. Salekdeh (ed.), *Agricultural Proteomics Volume 2*,
DOI 10.1007/978-3-319-43278-6_4

and precipitation, severe drought, and frequent flooding events. It is predicted that the frequency and intensity of heat stress, drought, and flooding events will continue to increase [1]. Extremes in precipitation have increasingly limited food and forest production worldwide, and have adversely affected natural water cycles [2]. For these reasons, climate change, which is attributed to both natural and human causes [3], is an issue of major concern with respect to agriculture, and multidisciplinary efforts are needed to cope with these changes.

Severe climatic changes have caused increased flooding events over the past six decades [4], and the flooding frequency is predicted to continue to increase in this century in Asia, North and South America, and Africa [5]. Flooding due to heavy rainfall in poorly drained areas is a major abiotic stressor for many important agricultural crops [6], as gaseous exchange between plant tissues and the atmosphere [7] and the availability of light are markedly reduced under flooding conditions [8]. Flooding also alters the chemical characteristics of soil, including increasing the pH and decreasing the redox potential [9], which increases the uptake of toxic metals by plants [10]. The reduced gas exchange that is induced by flooding results in oxidative damage and shifts plant metabolism to anaerobic pathways [11]. Flooding-induced damage to agricultural crops is a major limiting factor in meeting the ever-growing food needs of the global population.

Plants exhibit diverse responses to flooding that include changes at the morphological, metabolic, and molecular levels. Flooding-tolerant plants, such as rice, have developed two main strategies to cope with submergence: low-oxygen escape and low-oxygen quiescence [4, 12]. In the escape strategy, the rate of gas exchange between the plant and environment is increased in plant tissues located above the water level, leaves bend upward (hyponasty), and shoot elongation is enhanced [13]. The less energy-consuming quiescence strategy involves the restriction of growth through changes in metabolism [13]. In particular, high-energy consuming processes, such as DNA replication, protein synthesis, and cell wall synthesis, are reduced and metabolism shifts from aerobic respiration to anaerobic glycolysis. In contrast to these strategies, flooding-intolerant plants, such as soybean, primarily respond to flooding stress by increasing aerenchyma formation in roots, shoots, and secondary tissues, thereby enabling sustained oxidative phosphorylation [14]. In addition, the development of adventitious roots and formation of leaf gas films are also synchronized to enable gas diffusion [15, 16]. Plants responses to flooding are diverse and depend on the plant species and severity of the stress.

Advances in high-throughput proteomics have helped unravel the complicated biological processes associated with plant stress responses. Gel-based and gel-free proteomic approaches with label-based and label-free protein quantification have been used extensively to identify stress-responsive proteins at both the organ and subcellular level [17]. In plant proteomics, obtaining high-quality proteins from plant organs and subcellular organelles is extremely challenging due to the large abundance of proteases, oxidative enzymes, and secondary metabolites in plant cells and tissues [18]. However, the development of trichloroacetic acid precipitation and phenol extraction method has markedly improved the efficiency of plant protein extraction [19]. Proteomic techniques are useful tools for elucidating cellular responses to flooding stress, including the role of target enzymes [17]. In the present review, the findings from proteomic analyses of different plant species exposed to flooding/waterlogging stress, including the organ-specific proteins that are regulated in response to flooding conditions, are described in detail. In addition, the findings from proteomic studies examining the mechanisms underlying post-flooding recovery are also discussed.

4.2 Proteomic Analyses of Plants Under Flooding Stress

Plants respond to flooding in various ways depending on the plant species and specific conditions. Most notably, plants adopt escape or quiescence strategies, or develop aerenchyma. Proteomic analysis allows gene expression changes to be studied at the protein level, and can be further extended to post-translation modifications. Various proteomic approaches have been applied to the study of plant responses to flooding. The results from these proteomic studies in different plant species are summarized in Table 4.1.

4.2.1 Soybean

Soybean (*Glycine max*) is a protein and oil-rich legume crop that is grown in many parts of the world. The genome of soybean cultivar *Williams 82* consists of 950 Mbp [43] and that of cultivar *Enrei* consists of approximately 947 Mbp [44]. The sequenced genome information of soybean has greatly contributed to our understanding of plant interactions with the environment. Soybean is flooding intolerant and exhibits dramatically reduced growth and yields under flooding conditions [45]. In particular, exposure to flooding reduces hypocotyl pigmentation and length, and decreases both root and shoot growth, although roots are primarily affected in the initial stages of flooding stress [16, 46]. Flooding also damages soybean seeds without radicle protrusion by the physical disruption of cells caused by the rapid uptake of water [47]. Soybean appears to attempt to reduce flooding injury through extensive adventitious root development, which is reported to enhance oxygen transport from the stem to the roots [48, 49]. Aerenchyma development primarily occurs in new adventitious roots, whereas the primary roots of flooded plants exhibited tightly packed cortical cells [14, 50].

Proteomic techniques have been used to study the physiological and molecular responses of soybean to flooding stress [17, 51]. Hashiguchi et al. [46] and Nanjo et al. [52] analyzed protein changes that occur during the first 24 h of flooding stress and found that the abundance of proteins related to glycolysis, fermentation, and the cell wall were increased, whereas reactive oxygen species (ROS)-scavenging

Table 4.1 Proteomic analyses of flooding-stressed plants

Plant	Stress duration (additional stress)	Functional category of proteins	Protein abundance	Reference
Soybean	3 h	Cell wall, protein metabolism	Decreased	[20]
	3 h	Transport, RNA regulation	Increased	[21]
	12 h	Glycolysis, fermentation	Increased	[22]
	1 day	Glycolysis, fermentation, cell wall	Increased	[23]
		ROS scavenging, cell structure, amino acid metabolism	Decreased	
	1–3 days (Al nanoparticles)	Glycolysis, fermentation, tricarboxylic acid, amino acid metabolism, nucleotide metabolism	Increased	[24]
	1–4 days	Metal handling	Decreased	[25]
		Stress, protein	Increased	
	2 days (gibberellic acid) (calcium)	Glycolysis, fermentation	Increased	[26]
		Jasmonate synthesis, defense, redox	Decreased	[27]
		Tricarboxylic acid, electron transport chain	Increased	[28]
		stress	Decreased	[29]
		Secondary metabolism, cell cycle, protein degradation/synthesis	Decreased	[30]
		Cell wall, hormone metabolism, protein metabolism, DNA synthesis	Decreased	[31]
	2 and 4 days (abscisic acid) (Ag nanoparticles)	Cell wall, tricarboxylic acid, secondary metabolism	Increased	[32]
		Stress	Decreased	
		Fermentation	Increased	[33]
		Defense	Decreased	
		Energy	Increased	[34]
	3 days	Cell wall	Decreased	[23]
Wheat	2 days	Defense, redox homeostasis, energy, cell wall	Increased	[35]
	7 days	Stress, defense	Increased	[36]
		Cell wall, glycolysis	Decreased	
Rice	3 days	Glycolysis	Increased	[37]
	3 to 6 days	Stress, fermentation	Increased	[38]
Tomato	1 to 3 days	Photosynthesis, energy	Decreased	[39]
	1 and 3 days	Stress, defense, fermentation, hormone metabolism, secondary metabolism, programmed cell death	Increased	[40]
	3 days	Fermentation	Increased	[41]
	5 days	Stress, disease	Increased	[42]

enzymes, and proteins related to cell organization and amino acid metabolism were decreased, indicating growth suppression. Nanjo et al. [53] also analyzed 12 h flooding stress-induced changes in soybean and concluded that glycolysis and fermentation enzymes and inducers of heat shock proteins are key elements in the early responses to flooding stress. Glycolysis and the detoxification-linked methylglyoxal pathway were activated and sucrose degradation was reduced. In addition, Yin et al. [20] examined changes in protein phosphorylation in the early stages of flooding stress and provided evidence that this post-translational modification is linked with mechanisms of soybean tolerance in root tips via the ethylene signaling pathway within 3 h of flooding exposure. Further, proteomic analysis of nuclear-localized phosphoproteins in flooded soybean root tips indicated that zinc finger domain-containing protein, glycine-rich protein, and rRNA processing protein, which are related to the abscisic acid response, are phosphorylated in response to flooding stress [21].

In addition to analyses of the soybean proteins affected in the initial stages flooding stress, various proteomic studies have examined 2-days flooding-stress responses. Proteomic analyses of subcellular organelles, including mitochondria [54], endoplasmic reticulum [55] cell wall [26], and nucleus [56], have been reported. Several cell wall-related proteins were suppressed [26], whereas tricarboxylic acid cycle-related proteins and proteins involved in the electron transport chain were increased in abundance as ATP production decreased [27]. Protein folding, translocation, and degradation-related heat shock proteins were also increased in the roots and cotyledons of soybean under flooding stress [57]. Flooding exposure for 2 days also reduced the N-glycosylation of stress-related and protein degradation-related proteins, whereas glycoproteins involved in glycolysis were activated [29]. Kamal et al. [25] reported that ferritin functions as a protective agent against the oxidative damage caused by flooding. Taken together, the findings from these reports suggest that flooding induces marked changes in the levels of numerous organelle-specific and organ-specific proteins, leading to growth suppression.

The roles of phytohormones, calcium, and nanoparticles in flooding stress responses of soybean have also been investigated. Komatsu et al. [32] reported that abscisic acid enhances the flooding tolerance of soybean through regulation of zinc finger proteins and energy conservation via the glycolytic system. The treatment of flooding-stressed soybean with gibberellic acid increased the abundance of secondary metabolism, cell, and protein synthesis/degradation-related proteins [58]. Proteins involved in protein metabolism and modifications, hormone metabolism, cell wall metabolism, and DNA synthesis are also decreased by flooding stress; however, the levels of these proteins are restored in soybean upon calcium treatment [30]. Silver nanoparticles were shown to reduce oxygen deprivation under flooding conditions by increasing the abundance of fermentation-related and detoxification-linked glyoxalase II 3 proteins [33]. The treatment of flooding-stress soybean with aluminium oxide nanoparticles also promoted growth by altering the regulation of energy metabolism and cell death [24]. The findings from these proteomic studies have revealed that flooding retards the growth of soybean and that glycolysis and various fermentation pathways are activated to partially compensate for flooding-induced energy deficiency.

4.2.2 Wheat

Wheat (*Triticum aestivum*) is one of the main food crops worldwide as wheat gluten and storage proteins provide one-fifth of the total calories of the world population [59, 60]. The large 17 Gb size of the common wheat genome [61] likely contributes to the extensive capacity of wheat plants to adopt to various ecological conditions [62]. Information from the genome sequencing of wheat is providing insight into the understanding and approaches for developing wheat varieties with higher tolerance to environmental stresses. Wheat is sensitive to flooding and displays decreases in yield of up to 65 % under severe flooding conditions [63]. Morphological analysis of waterlogged wheat showed the reduced growth as root length and dry mass were decreased. Lysigenous aerenchyma formation also occurs in wheat in response to waterlogging [35].

Few proteomic studies concerning wheat responses to flooding have been reported. Kong et al. [36] analyzed the cell wall proteome of wheat roots under flooding stress using gel-based and gel-free proteomics and found that proteins involved in glycolysis and cell wall structure and modification were predominantly decreased, whereas defense and disease response proteins were increased. The findings from this study indicate that wheat seedlings restrict cell growth and thus reduce energy consumption through coordinating methionine incorporation and cell wall hydrolysis. Haque et al. [35] reported that wheat proteins related to energy changes, redox homeostasis, defense responses, and the cell wall are increased, whereas respiration and energy metabolism-related proteins are decreased under waterlogging stress. The authors concluded that wheat seedlings adopt alternative forms of respiration and promote cell degeneration as simultaneous metabolic and anatomic responses in roots under hypoxic conditions. The aforementioned studies provide valuable insight into alterations of the protein profiles that occur in response to waterlogging/flooding stress in wheat. Primarily, wheat responds to waterlogging stress by increasing the abundance of disease-related proteins and altering energy metabolism, although overall growth and grain yields are suppressed.

4.2.3 Rice

Rice (*Oryza sativa*) is a staple food in many parts of the world, particularly Southeast Asia [64]. Rice is a model monocot plant whose genome has been fully sequenced and annotated [65, 66]. The genome size of rice cultivar *Oryza glaberrima* is 357 Mbp [67]. Rice is considered to be a flooding tolerant crop and is able to grow under fully anoxic conditions [68]; however, flooding/submergence is associated with reduced rice yields. In tropical regions, rice sowing is typically performed by distributing seeds within paddy fields that are submerged in water [69]. During germination, the coleoptile grows much faster when submerged, thus enabling the seedling to more rapidly reach the water surface and escape from the unfavorable hypoxic/anoxic conditions [68].

The availability of the fully sequenced and annotated rice genome has facilitated proteomic studies, particularly for rice plants exposed to hypoxia/anoxia, which frequently occurs under flooding conditions. Sadiq et al. [38] performed a proteomic analysis of rice coleoptiles under anoxic conditions and revealed that proteins related to stress responses and fermentation were increased in abundance. A proteomic analysis focusing on the anoxic-to-oxic transition in rice revealed that total heme content, cytochrome absorbance spectra, and electron carrier cytochrome C increased markedly on air adaptation [70]. The findings from this report indicate that heme synthesis is decreased in the absence of oxygen and that the blockage of mitochondrial biogenesis is fully reversible in this anoxia-tolerant species. Huang et al. [37] detected enhanced rates of glycolysis and ATP formation in rice coleoptiles under prolonged anoxia. The observed response mechanisms in rice to submergence stress indicate that oxygen deficiency reduces mitochondrial respiration, although glycolysis and fermentation pathways are stimulated. Notably, ethylene-related signaling pathways have not been identified in rice by proteomic analysis. Further proteomic studies are expected to help identify the possible rice-specific mechanisms that promote tolerance to submergence/flooding.

4.2.4 *Tomato*

Tomato (*Lycopersicon esculentum*) is a versatile vegetable that is consumed fresh as well as in the form of processed products [71]. The genome of the inbred tomato cultivar *Heinz 1706* is approximately 900 Mbp [72, 73]. Tomato is an excellent source of dietary antioxidants, as it is rich in vitamins, carotenoids, and phenolic compounds [74]. Tomato plants are sensitive to waterlogging stress, which leads to reduced photosynthesis due to stem closure, decreased chlorophyll content, increased hydrogen peroxide levels, leaf chlorosis and senescence, reduced stem elongation, and adventitious root formation [75, 40]. Vidoz et al. [76] reported that ethylene stimulates auxin accumulation in the base of rice plants and induces the growth of pre-formed root initials that lead to the formation of a new root system to replace the roots damaged by submergence. Else et al. [77] reported that stomatal closure depresses internal CO_2 concentrations and is linked to subsequent changes in chlorophyll fluorescence. These authors also demonstrated that stomatal opening is promoted by the aeration of adventitious roots.

Only a limited number of studies have examined changes in protein abundance in tomato under waterlogging/flooding conditions. However, one such study revealed that proteins related to primary metabolism and various cellular processes are affected by waterlogging stress [40]. Among the increased proteins, a number of proteins belonging to hormone and secondary metabolism, fermentation, including alcohol dehydrogenase and enolase, programmed cell death, stress and defense mechanisms were identified. A proteomic study on tomato leaves from 5-week-old plants under waterlogging stress also detected large-scale changes in protein abundance [39]. The differentially changed proteins were predominantly related to diverse functional categories such as energy metabolism, photosynthesis, defense/disease resistance, and protein biosynthesis. The degradation of photosynthesis-related proteins was associated with leaf senescence and decreased leaf chlorophyll content. These findings from limited proteomic studies in tomato indicate that in response to waterlogging stress, defense, hormone, and secondary metabolism-related proteins are accumulated, whereas photosynthesis-related proteins are degraded.

4.2.5 Maize

Maize (*Zea mays*) is an important crop that is used for human food, livestock feed, industrial processing, and various other purposes, and grows over a wide range of temperatures. Maize has a large genome of approximately 2.4 Gbp [78] and is used as a model monocot for the study of plant genetics. Chen et al. [42] reported that the flooding treatment of maize damaged the photosynthetic systems of the first and second leaves, and also affected the third and fourth leaves. Maize adapts to flooding stress by altering the chlorophyll a/b ratios and increasing basal shoot diameter [79]. The roots of flooded maize become negatively gravitropic and shoot length tends to decrease. However, flooding treatment increased the shoot stem diameter by 24 % and raised the hydrogen peroxide content [79].

Chen et al. [42] analyzed alterations in protein abundance in maize leaves under flooding stress and revealed that flooding damaged the leaf photosynthetic systems, thereby reducing energy production, and led to the accumulation of ROS. The increase in hydrogen peroxide levels induced the accumulation of translationally controlled tumor protein, which may regulate programmed cell death. Moreover, polyamine synthesis was enhanced under flooding stress and the abundance of disease-resistance proteins was increased. In contrast to neutral pH conditions, under which the levels of peroxidase were decreased, alkaline peroxidases were increased in maize exposed to flooding stress. Notably, the observed increase in ROS levels was attributed to a decrease in ROS-scavenging enzymes and resulted in damage to the photosystems [79]. Chang et al. [41] analyzed protein synthesis patterns in maize under low-oxygen stress and reported that a number of metabolic enzymes, including alcohol dehydrogenase and enolase, were preferentially synthesized during hypoxic acclimation. The findings from this study suggested that multiple suites of gene products may combine to provide tolerance to flooding.

Flooding reduces photosynthetic activity in maize as ROS accumulate; however, the levels of several fermentation-related enzymes are increased in an attempt to acclimate to flooding stress conditions.

4.3 Organ-Specific Proteomics of Flooding-Stressed Plants

Plant responses to flooding vary depending on the organ and may lead to different physiological and molecular modifications for coping with the stress. Organ-specific proteomic analysis allows the identification of proteins involved in stress-response mechanisms in specific plant organs. Proteins associated with the primary function of an organ are uniquely activated in that organ/tissue [80]. The main limitation of organ-specific analyses is detecting changes in low-abundance proteins, which are often difficult to quantify due to physical or chemical interference from high-abundance proteins [81]. The main approach for overcoming this problem is the elimination of highly abundant proteins using specific chemical methods.

4.3.1 Leaf

The leaf is an important plant organ as it fixes the carbon required for energy generation through photosynthesis. Leaves also play a vital role in the transport of essential elements and water from the roots to aerial parts of the plant. However, the normal physiology of the leaf is adversely affected by flooding, which induces growth inhibition, reduced stomatal transpiration, and decrease in chlorophyll a/b [82]. The decrease in chlorophyll content is more severe in older leaves that are closer to flooded roots. Flooding has been linked with reductions in plant biomass due to decreased stomatal conductance [83], as well as biochemical changes, such as altered RuBisCO levels [84]. In pea and maize, waterlogging has been reported to lead to leaf chlorosis [85].

A number of methods have been used to study the leaf proteome and increase the identification of low-abundance proteins, which are often difficult to detect due to the presence of numerous high-abundance proteins such as RuBisCo [81]. High-abundance proteins in leaves can be eliminated from samples using a polyethylene glycol fractionation method [39, 86], which involves extracting proteins from leaves using Mg/Nonidet P-40 buffer and then fractionating the obtained samples using 15 % polyethylene glycol [39]. Using this method, RuBisCO was successfully eliminated from other tomato leaf proteins prior to analysis [39]. In another method described by Hashimoto et al. [87], an anti-RuBisCO LSU antibody-affinity column loaded with protein A-Sepharose as a resin was used to prepare leaf extracts for analysis. Ion-exchange chromatography fractionation [88],

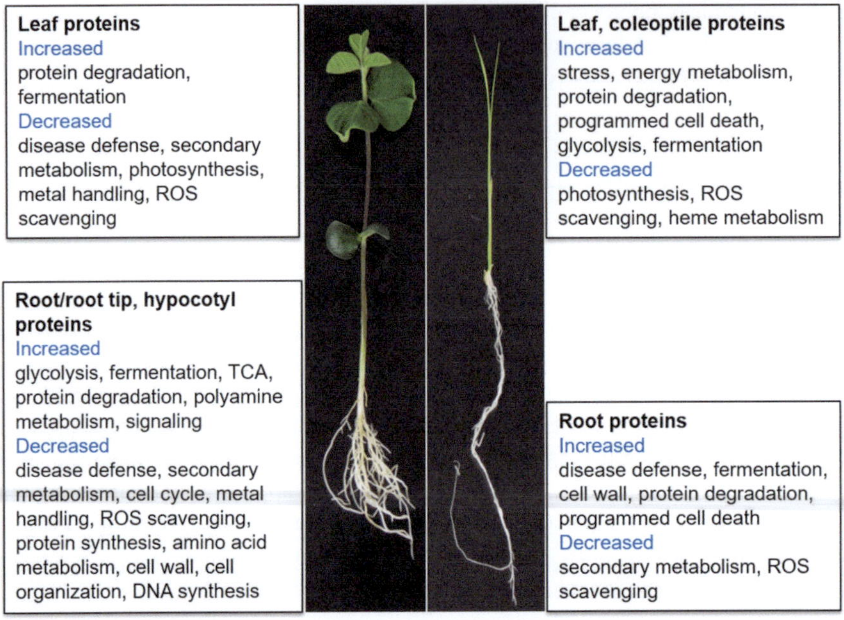

Fig. 4.1 Overview of plant organ-specific responses to flooding stress. Soybean and rice were selected as representative dicot and monocot species, respectively. Proteins that changed in response to flooding stress are categorized based on function and are shown in organs as either increased or decreased proteins. *TCA* tricarboxylic acid

immune-affinity chromatography [89], and affinity-based solid-phase techniques [90] have also been used to either enrich for low-abundance proteins or remove high-abundance proteins in leaves.

Protein abundance changes in leaves, cotyledons, and coleoptiles of plants exposed to flooding stress have been detected (Fig. 4.1). Khatoon et al. [91] reported that isoflavone reductase and proteins related to disease defense are decreased in the leaves of flooding-stressed soybean seedlings. The reduced levels of isoflavone reductase indicate that the efficiency of the antioxidant system is decreased in flooded soybean. Proteomic analysis of soybean cotyledons under flooding stress indicated a decrease in calcium oxalate crystals changing the physiological calcium levels in the soybean tissues as calcium ion was accumulated in the cotyledon [32]. However, the heat shock protein HSP70, which is involved in protein folding, translocation, and degradation, was increased in the cotyledons of flooding-stressed soybean. Kamal et al. [25] conducted proteomic analysis of soybean cotyledon under flooding conditions at an early growth stage and reported that sucrose metabolism-related proteins were decreased, whereas fermentation-related proteins were increased. In addition, ferritin levels were also found to be decreased, suggesting that iron may accumulate in soybean under flooding conditions.

Proteomic analysis of tomato leaf under waterlogging stress [39] revealed that stress and energy metabolism-related proteins were increased in abundance, whereas photosynthesis and protein biosynthesis-related proteins, including RuBisCO and RuBisCO activase, were decreased. The decrease of both RuBisCO and RuBisCO activase was attributed to increased amounts of ROS in tomato leaves. The waterlogging-induced decrease in protein biosynthesis and the activation of proteases led to the injury of leaves. However, to combat waterlogging stress and promote cellular survival, heat shock proteins were increased in abundance. Although the functional categorization of wheat leaf proteins revealed that the majority of proteins are involved in energy production, and primary and secondary metabolism [92], leaf proteomic analysis of wheat plants under flooding stress has yet not been reported. In maize leaf, the photosynthetic machinery was damaged due to flooding-induced ROS production and resulted in reduced energy metabolism. The increased levels of hydrogen peroxide under flooding stress also induced plant cell death [42].

A few studies have examined the role of the coleoptile in flooding responses in monocot plants. During rice germination, the coleoptile grows much faster when submerged compared to aerobic conditions, enabling the seedling to more rapidly reach the water surface and escape from the oxygen-deficient environment [68]. Primary leaf growth is stopped in rice seedlings germinated under anoxia and is speculated to minimize energy expenditure before the coleoptile tip reaches the water surface. Sadiq et al. [38] analyzed coleoptile proteins in rice under anoxic conditions and revealed that a number of fermentation-related proteins were increased. Cytoskeleton-related proteins are also increased in rice coleoptiles under anoxic conditions, were found to be involved in the anoxia-stimulated growth of the coleoptile. In addition, ROS-scavenging ascorbate peroxidase (APX) and glutathione S-transferase were decreased, and the rates of glycolysis and ATP formation were enhanced in anoxic coleoptiles [38]. Under anoxic condition, orthophosphate dikinase, alcohol dehydrogenase, and pyruvate kinase were induced in rice coleoptile for the generation of pyrophosphate for sucrose hydrolysis and continuation of glycolysis [37]. Together, the findings from these leaf proteomic analyses demonstrate that flooding stress results in damage to the leaf photosynthetic machinery and leads to ROS accumulation and reduced ROS scavenging. However, in the coleoptile, shoot elongation is enhanced as a mechanism to escape the anoxic conditions that are experienced in waterlogged soil.

4.3.2 Root

Roots are critical for the survival of plants and maintaining cellular homeostasis as they absorb water and nutrients from the soil and supply them throughout the plant body. Under flooding stress, roots undergo several structural and functional modifications at the molecular, cellular, and phenotypic level [93]. Flooding weakens the hydraulic conductivity of roots, leading to reduced root permeability [94]. This, in turn, results in lower water absorption and mineral uptake, leading to decreased rates of photosynthesis, altered hormonal balance, and development of aerenchyma and adventitious roots [51].

Proteomic techniques have served as an important tool for analyzing flooding-response mechanisms in the roots of various plant species. Similar to leaves, the analysis of the root proteome requires additional treatment procedures to enrich for low-abundance proteins in root extracts [81]. Among the enrichment methods developed for root proteomic studies, the most widely used protein extraction method is trichloroacetic acid/acetone precipitation [52]. However, Ahsan et al. [95] reported that treatment of root with Mg/Nonidet P-40 buffer followed by extraction with alkaline phenol and methanol/ammonium acetate produced high-quality proteome maps with well-separated and high-intensity spots on two-dimensional polyacrylamide gels. In addition, the preparation of combinatorial peptide ligand libraries has also been used for the enrichment of low-abundance plant proteins [96].

The soybean root proteome has been analyzed in several studies that have examined temporal changes in protein abundance during flooding (Fig. 4.1). Investigations analyzing changes in total root proteins and sub-cellular proteins revealed the suppression of lignification and energy production, despite the up-regulation tricarboxylic acid cycle proteins [26, 23]. In addition, an imbalance in the post-translational regulation of proteins involved in carbohydrate metabolism was detected in flooding-stressed roots [22]. The degradation of ubiquitinated proteins is also increased in roots exposed to flooding [97], and the N-glycosylation of proteins related to stress and protein degradation is reduced [29]. Several phytohormones, including gibberellic acid [58], abscisic acid [32], ethylene [20], and calcium [30], have been linked with flooding stress responses in soybean roots. The treatment of flooding-stressed roots with gibberellic acid restored the levels of proteins involved in secondary metabolism, cell cycle, and protein synthesis/degradation. Abscisic acid also contributes to flooding tolerance by regulating nuclear-localized proteins [21], and ethylene significantly promotes soybean growth under flooding stress conditions [20]. Calcium treatment recovered the levels of proteins involved in protein metabolism and modification, cell wall metabolism, DNA synthesis, development, and cell signaling, which were all decreased in response to flooding stress. Taken together, the results from the proteomic analyses of root proteins indicate that flooding suppresses energy metabolism and cell wall lignification in roots, although glycolysis, fermentation, TCA, and defense-related proteins are increased in abundance.

The root tip contains actively dividing meristematic cells and is vital for seedling establishment [98]. Proteomic investigations of soybean root tips have revealed that proteins related to stress, glycolysis, redox, and protein processing are present at higher levels than those found in the other root areas [99]. Nanjo et al. [52] analyzed the protein profiles in root tips of soybean exposed to flooding stress and concluded that proteins involved in glycolysis, fermentation, nucleotide metabolism, and cell wall metabolism were increased compared to untreated plants. However, numerous proteins related to cell organization and amino acid metabolism were decreased, indicating that root tip growth is suppressed under flooding conditions. The proteomic analysis of nuclear proteins in the root tips of soybean has also indicated that RACK1 protein has a vital role in plant flooding responses [56].

Won Oh et al. [31] analyzed the nuclear proteins in the root tips of flooded soybean and detected changes in the levels of proteins related to nucleic acid metabolism. Specifically, proteins involved in protein synthesis, post-translational modification, and protein degradation were increased in abundance, whereas proteins involved in transcription, post-transcriptional processing, DNA synthesis, and chromatin structure were decreased. Phosphoenol pyruvate synthesis by way of oxaloacetate produced in the TCA cycle is stimulated in response to flooding in soybean root tips [100]. A recent phosphoproteomic study suggested that the ethylene signaling pathway plays a key role in mediating stress tolerance in the initial stages of flooding via the modulation of protein phosphorylation [20]. Treatment with abscisic acid affects the flooding responses of early stage soybean by regulating nuclear-localized proteins [21]. These findings indicate that soybean root tips, which are actively growing regions of the root, perceive stress signals, but exhibit suppressed growth under flooding conditions, leading to root tip death.

Proteomics has also been applied to the study of flooding response mechanisms in the roots of plants other than soybean. Among the few studies that have been conducted in wheat, Kong et al. [36] revealed that carbohydrate metabolism-related proteins were decreased in abundance in roots under flooding conditions, suggesting that energy consumption is reduced as a stress-survival mechanism. In contrast, a number of defense-related proteins were increased in roots to resist against the environmental stress. Haque et al. [35] reported that metabolic adjustments occur in wheat roots to cope with different degrees of stress. In particular, proteins related to energy, redox homeostasis, defense, and cell wall metabolism were increased in abundance. In tomato roots exposed to waterlogging stress, secondary metabolism, defense, and programmed cell death-related proteins were increased [40]. Thus, in response to flooding stress, wheat and tomato undergo metabolic adjustments that involve decreased energy production, leading to growth inhibition.

4.4 Proteins Regulated Under Flooding Stress

4.4.1 Proteins Related to Glycolysis and Fermentation

Soil oxygen deprivation is the most inevitable consequence of flooding and forces submerged plants to shift from aerobic to anaerobic respiration [101]. This shift in respiration allows plants to regenerate NAD^+ through ethanol fermentation by selectively synthesizing flooding-inducible proteins involved in sucrose breakdown, glycolysis, and fermentation [4]. The results of proteomic analyses have revealed that the levels of several glycolysis-related proteins, including fructose-bisphosphate aldolase, phosphoglycerate kinase [46, 27], glyceraldehyde-3-phosphate dehydrogenase [26], enolase [22, 102], sugar isomerase, phosphofructo-kinase [102], and pyruvate kinase [52, 102, 103], are increased in soybean under flooding stress (Table 4.2). These findings indicate that activation of glycolysis and fermentation pathways is an important initial response to protecting plants from flooding-induced damage [104]. In wheat, fructose-1,6-bisphosphate aldolase and sucrose-fructan 6-fructosyl transferase were decreased under flooding stress.

Fermentation is stimulated under anaerobic conditions and leads to the accumulation of fermentation-related proteins, such as alcohol dehydrogenase (ADH) and pyruvate carboxylase. The flooding-induced accumulation of ADH in soybean roots/root tips [20, 25, 102, 104, 28, 105], rice coleoptiles [38, 37], and tomato roots [40] indicates that activation of the alcohol fermentation pathway is one method that plants attempt to cope with hypoxic conditions. The reduction of acetaldehyde to ethanol by ADH with concurrent reoxidation of NAD^+ was observed to be essential for the continuation of glycolysis. The fermentation-related enzyme pyruvate decarboxylase was increased in soybean roots [81, 102, 104] and rice coleoptiles [38] in response to flooding/waterlogging stress. Similarly, aldehyde dehydrogenase, which functions in fermentative metabolism, is also increased in soybean under flooding stress [54, 34]. The proteins related to glycolysis and fermentation are increased under flooding stress in different plant species to accelerate energy production via non-oxidative pathways, even though overall growth is suppressed.

4.4.2 *Energy-Related Proteins*

As flooding causes oxygen deprivation that shifts aerobic metabolism to anaerobic metabolism, net energy production is decreased. Flooding stress induces impairment of the electron transport chain in the roots and hypocotyls of soybean seedlings [54]. In particular, inner membrane carrier proteins and proteins related to complexes III, IV, and V of the electron transport chain are decreased in abundance, whereas succinate-semialdehyde dehydrogenase, 2-oxoglutarate dehydrogenase, and gamma-amino butyrate are significantly increased, suggesting that the gamma-aminobutyrate shunt replenishes the intermediates that have been depleted by flooding stress and that are needed for energy production via non-oxidative pathways [54]. Oxaloacetate produced in the TCA cycle stimulates phosphoenol pyruvate synthesis in response to flooding in soybean root tips [100]. This pathway provides indirect stimulation for the continuation of glycolysis.

A number of energy metabolism-related proteins, including citrate synthase, glutamate dehydrogenase, and adenosine kinase, are decreased in wheat roots under waterlogging stress [35]. In addition, energy-related proteins such as beta-amylase,

Table 4.2 Proteins regulated in plants in response to flooding stress

Functional category of proteins	Proteins description	Protein abundance	Plant	References
Glycolysis/fermentation	Fructose-bisphosphate aldolase	Increased	Soybean	[46]
			Soybean	[26]
		Decreased	Wheat	[36]
			Tomato	[39]
	Phosphoglycerate kinase	Increased	Soybean	[46, 26]
			Soybean	[42]
		Decreased	Maize	
	Glyceraldehyde-3-phosphate dehydrogenase	Increased	Soybean	[26]
			Soybean	
	Enolase	Increased	Soybean	[22, 102]
	Suger isomerase	Increased	Soybean	[102]
	Phosphofructo-kinase 3	Increased	Soybean	[102]
	UDP-glucose pyrophosphorylase	Increased	Soybean	[46]
			Soybean	[26, 22]
	UDP-glucose 6 dehydrogenase	Increased	Soybean	[22]
	Pyruvate kinase	Increased	Soybean	[102, 103]
	ADH	Increased	Soybean	[54, 22, 102, 28, 105]
			Tomato	[40]
			Rice	[38]
	Pyruvate decarboxylase	Increased	Soybean, Rice	[38, 102]
	Aldehyde dehydrogenase	Increased	Soybean	[28]
			Soybean	[34]
Energy	Cytochrome c reductase	Decreased	Soybean	[54]
	Cytochrome c oxidase	Decreased		
	ATP synthase	Decreased		
	Isocitrate dehydrogenase	Increased		
	Malate dehydrogenase	Increased		
	Succinate-semialdehyde dehydrogenase	Increased		
	2-Oxoglutarate dehydrogenase	Increased		
	Glutamate dehydrogenase	Decreased	Wheat	[35]
	Citrate synthase	Decreased		
	Adenosine kinase	Decreased	Maize	[42]
	Beta-amylase	Decreased		
	Carboxykinase	Decreased		
	RuBisCO binding alpha-subunit	Decreased	Tomato	[39]
	RuBisCO activase	Decreased		

(continued)

Table 4.2 (continued)

Functional category of proteins	Proteins description	Protein abundance	Plant	References
ROS scavenging	Peroxidase, SOD,CAT, APX	Decreased	Soybean	[27, 42, 34, 106, 107]
			Maize	
Cell wall loosening	Polygalactouronase inhibitor,	Increased	Soybean	[23]
	Expansin-like-B1-like protein	Increased		
	Cinnamyl alcohol dehydrogenase	Decreased		[27]
	Cellulose synthase-interactive protein	Decreased		
	Germin	Decreased		
	Lipoxygenase	Decreased		
	Methionine synthase	Decreased	Wheat	[36]
	Beta-1,3-glucanases	Decreased		
	Beta-1,3-glucosidase	Decreased		
Protein degradation	20S proteasome	Increased	Soybean	[22]
	26S proteasome	Increased		
	COP9 signalosome	Increased		[97]
	Ubiquitin-conjugating enzyme spm2	Increased	Wheat	[35]

malate dehydrogenase, fructose-1,6-bisphosphatase, and phosphoenol pyruvate carboxykinase are decreased in response to flooding stress, indicating that gluconeogenesis is suppressed in wheat under these conditions [42]. RuBisCO was reported to be degraded under the high ROS conditions found in tomato leaves under waterlogging stress. RuBisCO subunit binding-protein alpha subunit and RuBisCO activase are decreased in maize leaves under flooding stress [39], suggesting that decreased chlorophyll content is associated with RuBisCO degradation and leaf senescence. Flooding stress decreased the net energy production in plants, as photosystem and its components were largely damaged by ROS.

4.4.3 Reactive Oxygen Species Scavenging-Related Proteins

ROS are primarily recognized as toxic byproducts of aerobic metabolism and are controlled by various types of antioxidants and antioxidative enzymes. Studies of ROS in plants have clearly demonstrated that these molecules play important roles in signaling related to growth, development, and biotic and abiotic stress responses [108]. The development of well-organized scavenging mechanisms to overcome ROS toxicity likely led to the use of reactive molecules as signal transducers in plant cells. Plant cells can initiate and enhance ROS production related to cell signaling by increasing the levels of enzymes such as respiratory-burst NADPH oxidases. ROS production in cellular organelles, such as plastids, mitochondria, and peroxisomes, is involved in the initiation of various signaling cascades [109]. ROS-mediated signaling appears to be controlled by the balance between the production and scavenging of ROS intermediates [110].

ROS scavengers, such as peroxidase, APX, cytosolic APX, and superoxide dismutase (SOD), are decreased in abundance in soybean under flooding conditions [55, 26, 111]. Kausar et al. [106] confirmed that the level of peroxide-scavenging APX is decreased in soybean exposed to flooding stress. In an organ-specific proteomic study, the ROS scavengers SOD and catalase (CAT) were also decreased in the roots and leaves of soybean under flooding stress. The levels of SOD were particularly low in the cell wall [26]. APX, CAT, peroxidase, and SOD have recently been linked to increased biophoton emissions under flooding stress [34]. In maize, the accumulation of hydrogen peroxide and other ROS in leaves results in decreased photosynthesis [42]. APX and glutathione S-transferase are also decreased in rice coleoptiles under anoxia [38]. When present at optimal levels, ROS are beneficial for normal metabolism and cell signaling; however, flooding results in the accumulation of ROS to toxic levels. The findings from these studies indicate that the decreased abundance of ROS scavengers in plants under flooding stress is associated with growth suppression.

4.4.4 Cell Wall Loosening-Related Proteins

Cell walls serve as the outer protective boundary and also function in the sensing and transduction of stress signals between the apoplast and symplast. Cell wall metabolism and structure are adversely affected by flooding stress [26]. Investigation of the function of the cell wall of flooded soybean seedlings revealed that lipoxygenases, germin-like protein precursors, stem glycoprotein precursors, and SOD are decreased in abundance [26]. The findings from this study suggest that flooding suppresses the lignification of roots through a decrease in ROS scavenging enzymes and jasmonate biosynthetic activity. In response to flooding stress, proteins related to cell wall synthesis are decreased in soybean [22]. It has also been reported that the synthesis of rhamnose, which is a vital component of plant cell walls, is also decreased in soybean under flooding stress [23].

Proteins involved in cell wall modification, such as polygalactouronase inhibitor-like and expansin-like B1-like proteins, are increased in abundance in the roots, root tips, and hypocotyls of soybean seedlings under flooding stress [23]. Additionally, several cell wall synthesis-related proteins, such as cinnamyl-alcohol dehydrogenase and cellulose synthase-interactive protein-like protein, were decreased in abundance in soybean hypocotyl. Proteomic analysis of cell wall proteins in flooding-stressed wheat revealed that the levels of many cell wall-specific proteins, including methionine synthase, β-1,3-glucanases, and β-1,3-glucosidase, were reduced compared to those found in non-treated plants [36]. The findings from this study suggest that flooding stress induces the assimilation of methionine and promotes cell wall hydrolysis, thereby restricting growth. Thus, it can be concluded that under flooding stress, cell wall synthesis-related proteins decrease, cell wall loosening-related proteins increase, and cell wall lignification is suppressed.

4.4.5 Ubiquitination Proteasome-Related Proteins

Proteins involved in proteolysis, protein folding, and storage were found to be changed in abundance in response to flooding, indicating that these proteins are involved in removing flooding-damage-induced non-active proteins [55, 25, 58]. Heat shock proteins act as molecular chaperones in preventing protein aggregation, translocation of nascent chains across membranes, assembly or disassembly of multimeric protein complexes, and targeting proteins for lysosomal or proteasomal degradation [55]. The ubiquitin/proteasome-mediated proteolysis of enzymes involved in glycolysis and fermentation pathways may be negatively controlled under the hypoxic condition caused by flooding stress [55]. Flooding alters the abundance of the 20S proteasome subunits in soybean and affects the activity of the 26S proteasome [22]. Ubiquitin-mediated proteolytic processes are active in roots under flooding stress and lead to the degradation of root tip cells and death of root cap cells [97]. The abundance of ubiquitinated proteins in soybean roots was also found to decrease under flooding, but increased to levels similar to controls after de-submergence. Among the identified proteins, ubiquitin-conjugating enzyme spm2 was increased in wheat roots and might be important for protein degradation during cell degeneration in the process of aerenchyma formation under waterlogging stress [35]. The ubiquitination process is affected by flooding and functions to alter plant responses to stress.

4.4.6 Proteins Regulated During Recovery from Flooding Stress

The post-flooding recovery period has been relatively poorly studied in plants. However, elucidating the mechanisms involved in post-flooding recovery may provide valuable insight towards the development of flooding-tolerant plants. A few proteomic studies have examined the protein profiles of soybean plants following the removal of flooding stress [103, 107, 112]. Salavati et al. [107] reported that soybean plants recovered following removal of flooding stress, although growth recovery was delayed with increasing duration of stress exposure. In particular, root elongation, hypocotyl elongation, and the development of first leaves were delayed in the flooding-exposed seedlings as compared to untreated control seedlings. Similar morphological changes in soybean roots were observed by Khan et al. [103]

during the post-flooding recovery period. Soybean seedlings flooded for 2 and 4 days exhibited recovery rates that were directly proportion to the stress duration. A rapid increase in root and hypocotyl lengths and fresh weights was noted during the recovery period. In addition, the pigmentation of the roots, hypocotyls, and cotyledons was reduced under flooding conditions, but was restored to the level of control seedlings during the recovery period. However, for plants exposed to 6 days of flooding stress, recovery was not observed following stress removal and all seedlings died.

Salavati et al. [107] analyzed the proteomic profiles of soybean during post-flooding recovery and reported that Gro-EL-like chaperone ATPase, 26 S proteasome regulatory subunit 7, 26 S regulatory subunit S 10B, and cyclophilin were decreased in seedlings recovering from flooding stress, whereas globulin-like protein, Kunitz trypsin protease inhibitor, and peptidyl-prolyl cis-trans isomerase 1 were increased. Khan et al. [103] also investigated post-flooding recovery mechanisms in soybean roots and found that several proteins, including disease resistance dirigent-like proteins, aldolase-type TIM barrel family protein, O-methyl transferase 1, leucine-rich repeat family protein, quinone reductase, isoflavone reductases, and peroxidases were increased, whereas protein synthesis-related proteins, acid phosphatase, and lipoxygenase were markedly decreased during post-flooding recovery. Notably, cytoskeletal organization, cell expansion, and programmed cell death-related proteins were only detected in gel-based proteomic analysis [107], whereas ROS scavenging peroxidases were only found to increase by gel-free proteomic analysis [103]. Taken together, these findings indicate that soybean root recovers from flooding by altering cell structure, strengthening cell wall lignification, and scavenging toxic ROS.

In a study analyzing post-flooding recovery in soybean hypocotyl, it was reported that enzymes involved in glucose and secondary metabolism were changed in response to flooding [112]. Pyruvate kinase was increased in abundance under flooding conditions as a response to the increased glycolytic activity. However, during the post-flooding period, the abundance of pyruvate kinase decreased to normal levels, indicating that the cellular glycolytic activity was restored to normal levels. Nucleotidylyl transferase and beta-ketoacyl reductase were also increased in abundance under flooding, but decreased to normal levels the during recovery stage. However, the enzyme activities of these two enzymes gradually increased during the recovery stage, indicating the occurrence of a significant metabolic shift in secondary metabolism. The findings from these reports suggest that proteins involved in ROS scavenging, cell wall metabolism, cell structure, and primary metabolism are significantly changed during post-flooding recovery in roots, whereas proteins involved in secondary metabolism are altered during the post-flooding recovery in hypocotyl. To date, as very few proteomic studies have examined plant responses following the removal of flooding stress, further exploration of proteins involved in post-flooding recovery is expected to identify marker proteins and contribute to efforts for developing flooding-tolerant crops.

4.5 Conclusion and Future Prospective

Abiotic stresses are major limiting factors for plant growth and yields. Flooding stress is a widely occurring abiotic stress for many important agricultural crops. Every year, flooding accounts for billions of dollars in losses worldwide, and a substantial proportion of this amount is directly attributed to crop damage. Plants exhibit both species-specific and common responses to flooding at the morphological, physiological, metabolic, and molecular level. Flooding stress predominantly results in oxidative damage due to oxygen deficiency and forces plants to adopt an anaerobic mode of metabolism. Flooding stress induces disease-related proteins and limits the activity of energy synthesizing machinery, which decreases the net energy production in plants. In addition, exposure to flooding leads to cell wall loosening, which affects intracellular homeostasis. All of these events involve complex crosstalk signaling mechanisms that result in changes at the molecular level. Understanding the response mechanisms and identification of marker proteins are expected to contribute to the development of flooding-tolerant plants.

To date, only a few proteomic studies have examined flooding stress-response mechanisms in plants other than soybean, and knowledge regarding the molecular mechanisms involved in post-flooding recovery is also limited. Analyzing plant responses during post-flooding recovery in soybean and other flooding-intolerant crops will lead to a better understanding of the mechanisms involved in stress recovery. The proteomic analysis of plants during post-flooding recovery can identify proteins that undergo changes in abundance during the transition from stress to post-stress periods. Overexpression of the genes of identified indicator proteins may increase plant tolerance to flooding stress and thereby increase crop yields. Extensive proteomic studies are also needed to unravel the stress- and post-stress-response mechanisms in less studied plants.

Acknowledgments This work was supported by JSPS KAKENHI Grant Number 15H04445.

References

1. Mittler R, Blumwald E (2010) Genetic engineering for modern agriculture: challenges and perspectives. Annu Rev Plant Biol 61:443–462
2. Fukao T, Yeung E, Bailey-Serres J (2011) The submergence tolerance regulator SUB1A mediates crosstalk between submergence and drought tolerance in rice. Plant Cell 23:412–427
3. Hashiguchi A, Ahsan N, Komatsu S (2010) Proteomics application of crops in the context of climatic changes. Food Res Int 43:1803–1813
4. Bailey-Serres J, Voesenek L (2008) Flooding stress: acclimations and genetic diversity. Annu Rev Plant Biol 59:313–339
5. Hirabayashi Y, Mahendran R, Koirala S, Konoshima L, Yamazaki D, Watanabe S et al (2013) Global flood risk under climate change. Nat Clim Change 3:816–821

6. Jackson M, Colmer T (2005) Response and adaptation by plants to flooding stress. Ann Bot 96:501–505
7. Armstrong W, Drew M (2002) Root growth and metabolism under oxygen deficiency. Plant Roots Hidden Half 3:729–761
8. Vervuren P, Blom C, De Kroon H (2003) Extreme flooding events on the Rhine and the survival and distribution of riparian plant species. J Ecol 91:135–146
9. Sun L, Chen S, Chao L, Sun T (2007) Effects of flooding on changes in Eh, pH and speciation of cadmium and lead in contaminated soil. Bull Environ Contam Toxicol 79:514–518
10. Michalcová D, Gilbert JC, Lawson CS, Gowing DJ, Marrs RH (2011) The combined effect of waterlogging, extractable P and soil pH on α-diversity: a case study on mesotrophic grasslands in the UK. Plant Ecol 212:879–888
11. Gibbs J, Greenway H (2003) Review: mechanisms of anoxia tolerance in plants. I. Growth, survival and anaerobic catabolism. Funct Plant Biol 30:353
12. Voesenek L, Colmer T, Pierik R, Millenaar F, Peeters A (2006) How plants cope with complete submergence. New Phytol 170:213–226
13. Fukao T, Xu K, Ronald PC, Bailey-Serres J (2006) A variable cluster of ethylene response factor-like genes regulates metabolic and developmental acclimation responses to submergence in rice. Plant Cell 18:2021–2034
14. Shimamura S, Yamamoto R, Nakamura T, Shimada S, Komatsu S (2010) Stem hypertrophic lenticels and secondary aerenchyma enable oxygen transport to roots of soybean in flooded soil. Ann Bot 106:277–284
15. Jackson M, Armstrong W (1999) Formation of aerenchyma and the processes of plant ventilation in relation to soil flooding and submergence. Plant Biol 1:274–287
16. Sauter M (2013) Root responses to flooding. Curr Opin Plant Biol 16:282–286
17. Komatsu S, Sakata K, Nanjo Y (2015) 'Omics' techniques and their use to identify how soybean responds to flooding. J Anal Sci Technol 6:1–8
18. Isaacson T, Damasceno CM, Saravanan RS, He Y, Catalá C, Saladié M et al (2006) Sample extraction techniques for enhanced proteomic analysis of plant tissues. Nat Protoc 1:769–774
19. Ghosh D, Xu J (2014) Abiotic stress responses in plant roots: a proteomics perspective. Root Syst Biol 5:73–86
20. Yin X, Sakata K, Komatsu S (2014) Phosphoproteomics reveals the effect of ethylene in soybean root under flooding stress. J Proteome Res 13:5618–5634
21. Yin X, Komatsu S (2015) Quantitative proteomics of nuclear phosphoproteins in the root tip of soybean during the initial stages of flooding stress. J Proteomics 119:183–195
22. Nanjo Y, Skultety L, Ashraf Y, Komatsu S (2010) Comparative proteomic analysis of early-stage soybean seedlings responses to flooding by using gel and gel-free techniques. J Proteome Res 9:3989–4002
23. Nanjo Y, Nakamura T, Komatsu S (2013) Identification of indicator proteins associated with flooding injury in soybean seedlings using label-free quantitative proteomics. J Proteome Res 12:4785–4798
24. Mustafa G, Sakata K, Komatsu S (2015) Proteomic analysis of flooded soybean root exposed to aluminum oxide nanoparticles. J Proteomics 128:280–297
25. Kamal AHM, Rashid H, Sakata K, Komatsu S (2015) Gel-free quantitative proteomic approach to identify cotyledon proteins in soybean under flooding stress. J Proteomics 112:1–13
26. Komatsu S, Kobayashi Y, Nishizawa K, Nanjo Y, Furukawa K (2010) Comparative proteomics analysis of differentially expressed proteins in soybean cell wall during flooding stress. Amino Acids 39:1435–1449
27. Komatsu S, Sugimoto T, Hoshino T, Nanjo Y, Furukawa K (2010) Identification of flooding stress responsible cascades in root and hypocotyl of soybean using proteome analysis. Amino Acids 38:729–738
28. Komatsu S, Thibaut D, Hiraga S, Kato M, Chiba M, Hashiguchi A et al (2011) Characterization of a novel flooding stress-responsive alcohol dehydrogenase expressed in soybean roots. Plant Mol Biol 77:309–322

29. Mustafa G, Komatsu S (2014) Quantitative proteomics reveals the effect of protein glycosylation in soybean root under flooding stress. Front Plant Sci 5:627
30. Oh M, Komatsu S, Nanjo Y (2014) Gel-free proteomic analysis of soybean root proteins affected by calcium under flooding stress. Front Plant Sci 5:559
31. Won OhM, Nanjo Y, Komatsu S (2014) Identification of nuclear proteins in soybean under flooding stress using proteomic technique. Protein Pept Lett 21:458–467
32. Komatsu S, Han C, Nanjo Y, Altaf-Un-Nahar M, Wang K, He D et al (2013) Label-free quantitative proteomic analysis of abscisic acid effect in early-stage soybean under flooding. J Proteome Res 12:4769–4784
33. Mustafa G, Sakata K, Hossain Z, Komatsu S (2015) Proteomic study on the effects of silver nanoparticles on soybean under flooding stress. J Proteomics 122:100–118
34. Kamal AHM, Komatsu S (2015) Involvement of reactive oxygen species and mitochondrial proteins in biophoton emission in roots of soybean plants under flooding stress. J Proteome Res 14:2219–2236
35. Haque E, Kawaguchi K, Komatsu S (2011) Analysis of proteins in aerenchymatous seminal roots of wheat grown in hypoxic soils under waterlogged conditions (supplementary material). Protein Pept Lett 18:912–924
36. Kong FJ, Oyanagi A, Komatsu S (2010) Cell wall proteome of wheat roots under flooding stress using gel-based and LC MS/MS-based proteomics approaches. Biochim Biophys Acta 1804:124–136
37. Huang S, Greenway H, Colmer TD, Millar AH (2005) Protein synthesis by rice coleoptiles during prolonged anoxia: implications for glycolysis, growth and energy utilization. Ann Bot 96:703–715
38. Sadiq I, Fanucchi F, Paparelli E, Alpi E, Bachi A, Alpi A et al (2011) Proteomic identification of differentially expressed proteins in the anoxic rice coleoptile. J Plant Physiol 168:2234–2243
39. Ahsan N, Lee DG, Lee SH, Kang KY, Bahk JD, Choi MS et al (2007) A comparative proteomic analysis of tomato leaves in response to waterlogging stress. Physiol Plant 131:555–570
40. Ahsan N, Lee D-G, Lee S-H, Lee K-W, Bahk JD, Lee B-H (2007) A proteomic screen and identification of waterlogging-regulated proteins in tomato roots. Plant Soil 295:37–51
41. Chang WW, Huang L, Shen M, Webster C, Burlingame AL, Roberts JK (2000) Patterns of protein synthesis and tolerance of anoxia in root tips of maize seedlings acclimated to a low-oxygen environment, and identification of proteins by mass spectrometry. Plant Physiol 122:295–318
42. Chen Y, Chen X, Wang H, Bao Y, Zhang W (2014) Examination of the leaf proteome during flooding stress and the induction of programmed cell death in maize. Proteome Sci 12:1
43. Arumuganathan K, Earle E (1991) Nuclear DNA content of some important plant species. Plant Mol Biol Report 9:208–218
44. Shimomura M, Kanamori H, Komatsu S, Namiki N, Mukai Y, Kurita K et al (2015) The *Glycine max* cv. Enrei genome for improvement of Japanese soybean cultivars. Int J Genomics 1–8
45. Githiri S, Watanabe S, Harada K, Takahashi R (2006) QTL analysis of flooding tolerance in soybean at an early vegetative growth stage. Plant Breed 125:613–618
46. Hashiguchi A, Sakata K, Komatsu S (2009) Proteome analysis of early-stage soybean seedlings under flooding stress. J Proteome Res 8:2058–2069
47. Nakayama N, Hashimoto S, Shimada S, Takahashi M, Kim Y, Oya T et al (2004) The effect of flooding stress at the germination stage on the growth of soybean [*Glycine max*] in relation to initial seed moisture content. Jpn J Crop Sci (Jpn) 74:325–329
48. Shimamura S, Mochizuki T, Nada Y, Fukuyama M (2003) Formation and function of secondary aerenchyma in hypocotyl, roots and nodules of soybean (*Glycine max*) under flooded conditions. Plant Soil 251:351–359
49. Lee K-H, Park SW, Kwon YW (2003) Enforced early development of adventitious roots increases flooding tolerance in soybean. Jpn J Crop Sci 72:82–88

50. Bacanamwo M, Purcell LC (1999) Soybean root morphological and anatomical traits associated with acclimation to flooding. Crop Sci 39:143–149
51. Hossain Z, Komatsu S (2014) Potentiality of soybean proteomics in untying the mechanism of flood and drought stress tolerance. Proteomes 2:107–127
52. Nanjo Y, Skultety L, UváČKová LU, Klubicová KN, Hajduch M, Komatsu S (2011) Mass spectrometry-based analysis of proteomic changes in the root tips of flooded soybean seedlings. J Proteome Res 11:372–385
53. Nanjo Y, Maruyama K, Yasue H, Yamaguchi-Shinozaki K, Shinozaki K, Komatsu S (2011) Transcriptional responses to flooding stress in roots including hypocotyl of soybean seedlings. Plant Mol Biol 77:129–144
54. Komatsu S, Yamamoto A, Nakamura T, Nouri M-Z, Nanjo Y, Nishizawa K et al (2011) Comprehensive analysis of mitochondria in roots and hypocotyls of soybean under flooding stress using proteomics and metabolomics techniques. J Proteome Res 10:3993–4004
55. Komatsu S, Kuji R, Nanjo Y, Hiraga S, Furukawa K (2012) Comprehensive analysis of endoplasmic reticulum-enriched fraction in root tips of soybean under flooding stress using proteomics techniques. J Proteomics 77:531–560
56. Komatsu S, Hiraga S, Nouri MZ (2014) Analysis of flooding-responsive proteins localized in the nucleus of soybean root tips. Mol Biol Rep 41:1127–1139
57. Komatsu S, Makino T, Yasue H (2013) Proteomic and biochemical analyses of the cotyledon and root of flooding-stressed soybean plants. PLoS ONE 8:e65301
58. Won OhM, Nanjo Y, Komatsu S (2014) Analysis of soybean root proteins affected by gibberellic acid treatment under flooding stress. Protein Pept Lett 21:911–947
59. Gill BS, Appels R, Botha-Oberholster A-M, Buell CR, Bennetzen JL, Chalhoub B et al (2004) A workshop report on wheat genome sequencing international genome research on wheat consortium. Genetics 168:1087–1096
60. Reynolds M, Bonnett D, Chapman SC, Furbank RT, Manès Y, Mather DE et al (2010) Raising yield potential of wheat. I. Overview of a consortium approach and breeding strategies. J Exp Bot 62:439–452
61. Safar J, Simkova H, Kubalakova M, Cihalikova J, Suchankova P, Bartos J et al (2010) Development of chromosome-specific BAC resources for genomics of bread wheat. Cytogenet Genome Res 129:211–223
62. Brenchley R, Spannagl M, Pfeifer M, Barker GL, D'amore R, Allen AM et al (2012) Analysis of the bread wheat genome using whole-genome shotgun sequencing. Nature 491:705–710
63. McDonald G, Setter TL, Waters I, Tugwell R (2006) Screening for waterlogging tolerance of wheat in the field in Western Australia. In: Proceedings of the 13th Australian society of agronomy conference, Perth, Western Australia, pp 10–14
64. Sasaki T, Burr B (2000) International Rice Genome Sequencing Project: the effort to completely sequence the rice genome. Curr Opin Plant Biol 3:138–142
65. Goff SA, Ricke D, Lan T-H, Presting G, Wang R, Dunn M et al (2002) A draft sequence of the rice genome (*Oryza sativa* L. ssp. japonica). Science 296:92–100
66. Yu J, Hu S, Wang J, Wong GK-S, Li S, Liu B et al (2002) A draft sequence of the rice genome (*Oryza sativa* L. ssp. indica). Science 296:79–92
67. Kim H, Hurwitz B, Yu Y, Collura K, Gill N, Sanmiguel P et al (2008) Construction, alignment and analysis of twelve framework physical maps that represent the ten genome types of the genus *Oryza*. Genome Biol 9:R45
68. Magneschi L, Perata P (2009) Rice germination and seedling growth in the absence of oxygen. Ann Bot 103:181–196
69. Yamauchi M, Aragones DV, Casayuran PR, Cruz PC, Asis CA, Cruz RT (2000) Seedling establishment and grain yield of tropical rice sown in puddled soil. Agron J 92:275–282
70. Millar AH, Trend AE, Heazlewood JL (2004) Changes in the mitochondrial proteome during the anoxia to air transition in rice focus around cytochrome-containing respiratory complexes. J Biol Chem 279:39471–39478

71. Toor R, Lister C, Savage G (2005) Antioxidant activities of New Zealand-grown tomatoes. Int J Food Sci Nutr 56:597–605
72. Michaelson MJ, Price HJ, Ellison JR, Johnston JS (1991) Comparison of plant DNA contents determined by Feulgen microspectrophotometry and laser flow cytometry. Am J Bot 78:183–188
73. Tomato Genome Consortium (2012) The tomato genome sequence provides insights into fleshy fruit evolution. Nature 485:635–641
74. Lenucci MS, Cadinu D, Taurino M, Piro G, Dalessandro G (2006) Antioxidant composition in cherry and high-pigment tomato cultivars. J Agric Food Chem 54:2606–2613
75. Jackson MB, Davies WJ, Else MA (1996) Pressure–flow relationships, xylem solutes and root hydraulic conductance in flooded tomato plants. Ann Bot 77:17–24
76. Vidoz ML, Loreti E, Mensuali A, Alpi A, Perata P (2010) Hormonal interplay during adventitious root formation in flooded tomato plants. Plant J 63:551–562
77. Else MA, Janowiak F, Atkinson CJ, Jackson MB (2009) Root signals and stomatal closure in relation to photosynthesis, chlorophyll a fluorescence and adventitious rooting of flooded tomato plants. Ann Bot 103:313–323
78. Schnable PS, Ware D, Fulton RS, Stein JC, Wei F, Pasternak S et al (2009) The B73 maize genome: complexity, diversity, and dynamics. Science 326:1112–1115
79. Meisrimler C-N, Buck F, Lüthje S (2014) Alterations in soluble Class III peroxidases of maize shoots by flooding stress. Proteomes 2:303–322
80. Watson BS, Asirvatham VS, Wang L, Sumner LW (2003) Mapping the proteome of barrel medic (*Medicago truncatula*). Plant Physiol 131:1104–1123
81. Komatsu S, Hossain Z (2013) Organ-specific proteome analysis for identification of abiotic stress response mechanism in crop. Front Plant Sci 4:71
82. Gomes AS, Kozlowski T (1980) Growth responses and adaptations of *Fraxinus pennsylvanica* seedlings to flooding. Plant Physiol 66:267–271
83. Mielke MS, A-aF De Almeida, Gomes FP, MaG Aguilar, PaO Mangabeira (2003) Leaf gas exchange, chlorophyll fluorescence and growth responses of Genipa americana seedlings to soil flooding. Environ Exp Bot 50:221–231
84. Maayan I, Shaya F, Ratner K, Mani Y, Lavee S, Avidan B et al (2008) Photosynthetic activity during olive (*Olea europaea*) leaf development correlates with plastid biogenesis and Rubisco levels. Physiol Plant 134:547–558
85. Przywara G, Stępniewski W (1999) The influence of waterlogging at different temperatures on penetration depth and porosity of roots and on stomatal diffusive resistance of pea and maize seedlings. Acta Physiol Plant 21:405–411
86. Zhu W, Hu J, Wang X, Tian J, Komatsu S (2015) Organ-specific analysis of mahonia using gel-free/label-free proteomic technique. J Proteome Res 14:2669–2685
87. Hashimoto M, Komatsu S (2007) Proteomic analysis of rice seedlings during cold stress. Proteomics 7:1293–1302
88. Widjaja I, Naumann K, Roth U, Wolf N, Mackey D, Dangl JL et al (2009) Combining subproteome enrichment and Rubisco depletion enables identification of low abundance proteins differentially regulated during plant defense. Proteomics 9:138–147
89. Cellar NA, Kuppannan K, Langhorst ML, Ni W, Xu P, Young SA (2008) Cross species applicability of abundant protein depletion columns for ribulose-1, 5-bisphosphate carboxylase/oxygenase. J Chromatogr B 861:29–39
90. Xu Y, Wang B-C, Zhu Y-X (2007) Identification of proteins expressed at extremely low level in Arabidopsis leaves. Biochem Biophys Res Commun 358:808–812
91. Khatoon A, Rehman S, Hiraga S, Makino T, Komatsu S (2012) Organ-specific proteomics analysis for identification of response mechanism in soybean seedlings under flooding stress. J Proteomics 75:5706–5723
92. Donnelly BE, Madden RD, Ayoubi P, Porter DR, Dillwith JW (2005) The wheat (*Triticum aestivum* L.) leaf proteome. Proteomics 5:1624–1633
93. Atkinson NJ, Urwin PE (2012) The interaction of plant biotic and abiotic stresses: from genes to the field. J Exp Bot 63:3523–3543

94. Clarkson DT, Carvajal M, Henzler T, Waterhouse RN, Smyth AJ, Cooke DT et al (2000) Root hydraulic conductance: diurnal aquaporin expression and the effects of nutrient stress. J Exp Bot 51:61–70
95. Ahsan N, Komatsu S (2009) Comparative analyses of the proteomes of leaves and flowers at various stages of development reveal organ-specific functional differentiation of proteins in soybean. Proteomics 9:4889–4907
96. Boschetti E, Bindschedler LV, Tang C, Fasoli E, Righetti PG (2009) Combinatorial peptide ligand libraries and plant proteomics: a winning strategy at a price. J Chromatogr A 1216:1215–1222
97. Yanagawa Y, Komatsu S (2012) Ubiquitin/proteasome-mediated proteolysis is involved in the response to flooding stress in soybean roots, independent of oxygen limitation. Plant Sci 185:250–258
98. Drew MC, Cobb BG, Johnson JR, Andrews D, Morgan PW, Jordan W et al (1994) Metabolic acclimation of root tips to oxygen deficiency. Ann Bot 74:281–286
99. Mathesius U, Djordjevic MA, Oakes M, Goffard N, Haerizadeh F, Weiller GF et al (2011) Comparative proteomic profiles of the soybean (*Glycine max*) root apex and differentiated root zone. Proteomics 11:1707–1719
100. Komatsu S, Nakamura T, Sugimoto Y, Sakamoto K (2014) Proteomic and metabolomic analyses of soybean root tips under flooding stress. Protein Pept Lett 21:865–884
101. Hossain Z, López-Climent MF, Arbona V, Pérez-Clemente RM, Gómez-Cadenas A (2009) Modulation of the antioxidant system in citrus under waterlogging and subsequent drainage. J Plant Physiol 166:1391–1404
102. Oh M, Komatsu S (2015) Characterization of proteins in soybean roots under flooding and drought stresses. J Proteomics 114:161–181
103. Khan MN, Sakata K, Hiraga S, Komatsu S (2014) Quantitative proteomics reveals that peroxidases play key roles in post-flooding recovery in soybean roots. J Proteome Res 13:5812–5828
104. Komatsu S, Yamamoto R, Nanjo Y, Mikami Y, Yunokawa H, Sakata K (2009) A comprehensive analysis of the soybean genes and proteins expressed under flooding stress using transcriptome and proteome techniques. J Proteome Res 8:4766–4778
105. Alam I, Sharmin SA, Kim K-H, Yang JK, Choi MS, Lee B-H (2010) Proteome analysis of soybean roots subjected to short-term drought stress. Plant Soil 333:491–505
106. Kausar R, Hossain Z, Makino T, Komatsu S (2012) Characterization of ascorbate peroxidase in soybean under flooding and drought stresses. Mol Biol Rep 39:10573–10579
107. Salavati A, Khatoon A, Nanjo Y, Komatsu S (2012) Analysis of proteomic changes in roots of soybean seedlings during recovery after flooding. J Proteomics 75:878–893
108. Baxter A, Mittler R, Suzuki N (2014) ROS as key players in plant stress signalling. J Exp Bot 65:1229–1240
109. Gill SS, Tuteja N (2010) Reactive oxygen species and antioxidant machinery in abiotic stress tolerance in crop plants. Plant Physiol Bioch 48:909–930
110. Komatsu S, Kamal AHM, Makino T, Hossain Z (2014) Ultraweak photon emission and proteomics analyses in soybean under abiotic stress. Biochim Biophys Acta 1844:1208–1218
111. Shi F, Yamamoto R, Shimamura S, Hiraga S, Nakayama N, Nakamura T et al (2008) Cytosolic ascorbate peroxidase 2 (cAPX 2) is involved in the soybean response to flooding. Phytochemistry 69:1295–1303
112. Khan MN, Sakata K, Komatsu S (2015) Proteomic analysis of soybean hypocotyl during recovery after flooding stress. J Proteomics 121:15–27

Chapter 5
Proteomic Analysis of Crop Plants Under Low Temperature: A Review of Cold Responsive Proteins

Javad Gharechahi, Golandam Sharifi, Setsuko Komatsu and Ghasem Hosseini Salekdeh

Abstract Cold stress is one of the major environmental limiting factors affecting growth, development, and geographical distribution of many plant species especially in temperate regions where plants are seasonally exposed to low or freezing temperatures. During evolution, plants have evolved physiological, morphological, and molecular mechanisms to tolerate low temperatures. Understanding the molecular basis of plant cold stress tolerance is of great importance since it provides valuable information required for improving cold stress tolerance of existing crop cultivars with high yield and limited tolerance to low temperatures. At molecular level, the response of plants to low temperatures include changes in expression, abundance, post translational modifications, and subcellular localization of proteins which allow plants to adapt to low temperatures and to withstand freezing. Comparative proteomic analyses have revealed a list of proteins that are newly synthesized, induced or repressed in response to low temperature in different crop species. These include proteins whose functions are directly involved in cold and freezing tolerance including dehydrins, late embryogenesis abundant (LEA) proteins, cold regulated (COR) proteins, anti-freezing proteins, and pathogenesis related (PR) proteins. Cold stress is always associated with changes in cellular homeostasis which leads to an increased generation of reactive oxygen species (ROS) and therefore, elicits profound changes in abundance and activities

J. Gharechahi
Chemical Injuries Research Center,
Baqiyatallah University of Medical Sciences, Tehran, Iran

G. Sharifi
Department of Basic Sciences, Iranian Institute for Encyclopedia Research,
Tehran, Iran

S. Komatsu
National Institute of Crop Science, National Agriculture
and Food Research Organization, Tsukuba 305-8518, Japan

G.H. Salekdeh (✉)
Department of Systems Biology, Agricultural Biotechnology Research Institute of Iran,
Agricultural Research, Education, and Extension Organization, Karaj, Iran
e-mail: hsalekdeh@yahoo.com

G.H. Salekdeh (ed.), *Agricultural Proteomics Volume 2*,
DOI 10.1007/978-3-319-43278-6_5

of ROS scavenging enzymes. In addition, during cold stress, the abundance of proteins associated with carbohydrate metabolism, photosynthesis, protein synthesis and processing including those involved in folding and repairs of damaged proteins changes significantly to alleviate the damaging effects of low temperature. In this chapter, proteomics tools and techniques are introduced and the challenges faced with plant materials are discussed. Subsequently, the function of cold responsive proteins is reviewed in the context of cold stress tolerance in different crop species.

Keywords Low temperature · Cold stress · Cold acclimation · Proteomics · 2-DE · Mass spectrometry

5.1 Introduction

Plants are continuously exposed to diverse environmental stressors and are unable to escape from stress conditions because they are sessile organisms. This has caused plants to evolve adaptive physiological and molecular mechanisms to sense environmental stimuli and to activate defense responses that allow them to tolerate stress conditions. Cold stress is one of the major environmental constrains that largely affects agricultural production as well as geographical distribution of many plant species [1]. Cold stress impedes plant growth and development and consequently reduces its yield by imposing osmotic and oxidative stresses, inhibiting cellular metabolisms, perturbing gene transcription, and finally by inducing cell death through cellular dehydration and extracellular ice formation [2, 3]. When plants are faced with low temperatures, they either experience chilling stress (temperatures below 20 °C) or freezing stress (temperatures below 0 °C) [2, 4].

In some 42 % of total land area on the earth, plants may periodically experience low or sub-optimal temperatures even temperatures below −20 °C [5]. In such areas, plants have to either tolerate cold stress through activating molecular mechanisms that allow them to survive low temperatures or escape cold or freezing stresses through avoiding tissue freezing by supercooling and/or lowering the freezing point by synthesizing and accumulating antifreeze substances such as antifreeze proteins, amino acids, and sugars [5, 6]. In addition, plants have also evolved mechanisms to avoid exposure to cold or freezing stress through overwintering as dormant seeds with little freezable water and by delaying the transition from vegetative to reproductive stage of growth [4, 7, 8].

In temperate plants, cold and freezing stress tolerance is induced and achieved through exposure to a period of low or non-freezing temperatures, a process known as cold acclimation [1]. During cold acclimation, plants undergo a programmed remodeling of cell and tissue structures that largely depends on changes in gene expression and metabolism [1, 2]. During this process, plant metabolism is redirected toward synthesis of osmolytes and anti-freeze molecules such as sugars (saccharose, raffinose, stachyose, and trehalose), sugar alcohols (sorbitol, ribitol, and inositol), amines (proline and glycine betaine), and anti-oxidative molecules (glutathione, ascorbic acid, and α-tocopherol) [5, 9]. Accumulation of osmolytes in cold-acclimated plants increases osmotic potential and consequently decreases the cytoplasmic freezing point and promotes the stabilization of cell membrane and proteins [10]. Low temperatures induce changes in gene expression that are responsible for aforementioned metabolic reprograming and directs synthesis of new gene products such as anti-freeze proteins, dehydrins, reactive oxygen species (ROS) scavenging enzymes, late embryogenesis abundant (LEA) proteins, pathogenesis related (PR) proteins, and chaperones [4, 5, 7, 11]. In addition, during cold acclimation, the lipid and protein compositions of plasma membrane are changed to maintain its integrity and functionality under subsequent low or freezing temperatures. Interestingly, the proportion of unsaturated fatty acids increases to maintain membrane fluidity in cold-acclimated plants [9].

In order to avoid flowering during cold seasons, most temperate plants (especially winter-habit plants) have acquired the ability to delay their flowering time through a phenomenon referred to as vernalization [2, 12]. During vernalization, plants are exposed to a long period of low temperature to get competence for flowering in spring when temperature and photoperiod become suitable. Indeed, vernalization is an adaptation process that allows plants to pass cold seasons (autumn and winter) as seedlings to ensure that flowering occurs under temperature-favorable conditions in spring [12, 13]. Although both cold acclimation and vernalization require exposure to low temperatures, they are quite different in terms of the duration of cold treatment. Full vernalization always requires an extended period of low temperature while cold acclimation is achieved within only 1 or 2 days of cold treatment [14]. In contrast to vernalization, cold acclimation is a reversible process that is rapidly lost upon exposure to high temperatures making plants susceptible to subsequent cold and freezing stresses.

Resistance to cold stress is a complex trait that largely depends on the species and the genetic potential as well as the developmental stage of plants. Within crop plants, species such as wheat, barley, oat, and rye are capable of cold acclimation and vernalization and show variable degrees of resistance to cold and freezing but species such as maize and rice are incapable of cold acclimation and therefore, are cold-sensitive [2]. Traditional breeding programs have significantly improved cold and freezing tolerance in certain plant species; however, the applicability of these approaches largely depends on the existence of genetic variation. Since most variations in cold stress tolerance in economically-important species have been explored, further improvements using traditional breeding approaches are largely limited and labor-intensive [10]. On the other hand, advances in recombinant DNA technologies have revolutionized breeding for cold tolerance especially by crossing species barriers and allowing to transfer cold tolerance genes between distantly-related organisms. Identification of cold tolerance genes and their functional characterization under cold stress needs in-depth exploration of biochemical, physiological, and molecular responses of plant to cold stress. As noted above, response to cold stress is a complex phenomenon that involves the interplay of several structural, molecular, metabolic, and regulatory pathways, which needs to be inspected at global systems biology level by integrating high-throughput genomic, transcriptomic, proteomic, and metabolomic data. Analyses of genomic and transcriptomic data in response to low temperatures provide a snapshot of genetic and gene expression potential of plant, respectively, for cold hardening. However, the limitations associated with these approaches as well as the existence of a broad spectrum of post-transcriptional/post-translational regulation mechanisms that are involved in fine-tuning the abundance and the activity of proteins as final gene products have encouraged molecular biologists to focus their research on proteins. The main objective of proteomics is the detailed characterization of proteins expressed in a given cell, tissue, organ, and/or organism under particular environmental or developmental conditions. Proteomic analyses could provide valuable information about the abundance, sub-cellular localization, physical or functional interactions, and post-translational modifications (PTMs) of proteins in their cellular context [15]. Compared with other omics-based approaches, data generated by proteomics are of great importance because proteins are the key players in nearly all cellular processes.

In this chapter, proteomics tools and techniques are introduced and their challenges with plant materials are discussed. Subsequently, the function of cold responsive proteins is reviewed in the context of cold stress tolerance in different crop species.

5.2 Proteomics Tools and Techniques

The success of a comparative proteomic analysis largely depends on the ability to extract, fractionate, quantify, and identify proteins contained in a given sample. During a plant proteomics study, each of these steps may impose some technical challenges that need to be dealt with in more details. Protein extraction from plant materials is usually challenging because plant tissues have relatively low protein contents and are rich in proteases, polysaccharides, polyphenols, lipids, secondary metabolites, and other interfering compounds that could affect the quality of the extracted proteins and interfere with subsequent fractionations and downstream analyses [16, 17]. In addition, in green tissues of plants, the high abundance of photosynthetic proteins such as small and large subunits of ribulose-1,5-bisphosphate carboxylase/oxygenase (RuBisCO), hiders the detection of low abundant proteins and further complicates the analysis of plant proteins. Hence, to achieve a complete coverage of all proteins expressed in a sample, the protein extraction procedure must be able to capture diverse proteins with varying abundance, molecular weight, charge, hydrophobicity, and modifications [17]. In most common protocols that are used for protein extraction from plant tissues, proteins are either precipitated using trichloroacetic acid (TCA) in acetone [18] or solubilized in phenol [19]. The applicability of these protein extraction methods largely depends on the nature and the type of plant tissues to be analyzed. The TCA/acetone precipitation is known to be suitable for protein extraction from young plants [17]. However, the phenol extraction procedure has successfully been utilized for the extraction of proteins from recalcitrant plant tissues such as woody tissues [20], olive leaves [21], saffron corm and callus [22], banana and avocado fruits [23].

In order to reduce sample complexity, protein extraction is usually followed by protein separation using one or two-dimensional gel electrophoresis (2-DE) and/or column chromatography based approaches (Fig. 5.1). Since plant tissues are usually complex and heterogeneous in nature, a single analytical technique could not provide a detailed picture of their proteome [16]. For complete separation and greater resolution of the proteome, it is always necessary to combine different separation techniques to achieve full coverage of the proteome. Traditional approach used for the separation of proteins is 2-DE, which combines the separation of proteins based on the net charge (isoelectric focusing, IEF) and the molecular weight (sodium dodecyl sulfate-polyacrylamide gel electrophoresis, SDS-PAGE). Even thought this approach has several advantages including high reproducibility and resolution, ease of use, wide molecular weight coverage, and the ability to detect PTMs but the limitations associated with this method such as difficulty in resolving membrane and highly acidic or basic proteins, inability to detect low abundant proteins, and the relatively low throughput have restricted its applicability for high throughput proteomic analysis [16, 24]. Gel-free separation methods often involve a combination of different liquid chromatography-based (LC) approaches which separate proteins or peptides based on their hydrophobicity (reverse phase, RP), charge (ion exchange), and/or affinity [15]. The main advantage of LC-based separations is the fact that they can be coupled with mass spectrometry for inline protein identification. In an exemplified LC-MS/MS approach, shotgun proteomics, or multi-dimensional protein identification technology (MudPIT), the protein mixture is enzymatically digested and the digested peptides are subjected to a combination of cation exchange (CEX) and RP chromatography separation coupled inline with tandem MS (MS/MS) for protein identification (Fig. 5.1) [25].

Protein identification using MS is achieved through either peptide mass fingerprinting (PMF) or by obtaining the amino acid sequence of protein using MS/MS analysis (de novo sequencing) [26, 27]. In the former case, the MS, usually a matrix-assisted laser desorption/ionization-time of flight (MALDI-TOF) instrument measures the accurate masses of peptides derived from the enzymatic digestion of the target protein and then the measured masses are queried against the theoretical masses of peptides derived from in silico digestion of proteins in the database [28]. This method is only applicable for proteins that their sequences are present in the databases [29]. In addition, it is not suited for the identification of proteins present in a complex mixture of unseparated proteins. The PMF method is therefore best suited for the identification of proteins from species for which complete genome sequence is available and also for identification of protein in 2-DE separated spots [29]. MS/MS instruments such as MALDI-quadrupole/TOF (MALDI-Q/TOF) and electrospray ionization-triple quadrupole (ESI-TQ) consist of two mass analyzers that are separated by a collision induced dissociation (CID) cell which is fed by a single ionization source (ESI or MALDI) [29]. During MS/MS

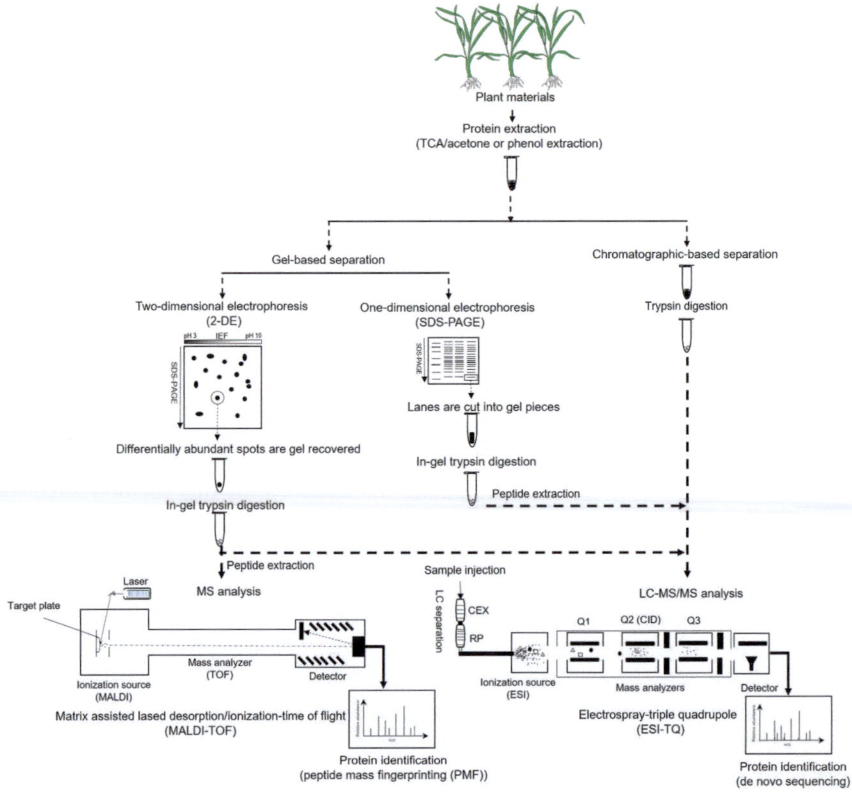

Fig. 5.1 A general workflow of comparative proteomics analyses in plants. In a typical comparative plant proteomics analysis depending on the tissue of the interest, proteins are extracted using TCA/acetone or phenol extraction methods. Extracted proteins are subjected to separation using gel-based (SDS-PAGE or 2-DE) or gel-free (chromatographic separation) approaches. In the gel-based separation, proteins are fractionated based on their MW (SDS-PAGE) or a combination of MW and pI (2-DE) and proteins in excised 2-DE spots or SDS-PAGE gel bands are trypsin digested and the resulting tryptic peptides are extracted and subjected to MS or MS/MS analysis for protein identification. In the gel-free chromatographic separation, extracted proteins are trypsin digested and tryptic peptides are separated using a combination of cation exchange (CEX) and reverse phase (RP) chromatography. In this approach, the separated peptides are directly injected into the ionization source of mass spectrometer for MS/MS based protein identification

analysis, the first mass analyzer scans for the mass to charge ratio of tryptic peptides coming from the ionization source and then selects specific precursor peptide ions (usually more intense ions) to be subjected to fragmentation using CID. The second mass analyzer then measures the accurate masses of the fragmented ions (Fig. 5.1). A computer software interprets the mass spectrum of the fragmented ions for determining the amino acid sequence of the precursor peptide which is then used for database search and protein identification [29]. LC-MS/MS based proteomic analysis can also be used for quantitative measurement of proteome through chemically or metabolically labeling proteins and/or label-free approaches which measure signal intensity of precursor ions or count the number of peptides assigned to a particular protein in a MS/MS scan (spectral counting) [30, 31].

MS-based proteomics largely depends on the availability of the genome sequences of the species of interest or their relatives for protein identification. In recent years, with the availability of genome sequences for most plant species even those with limited sequence annotations, our ability to study the proteome responses of plants to diverse environmental adversities has greatly been expanded.

5.3 Cold Stress Response Proteins in Plants

Most proteomic analyses conducted to decipher the response of crop plants to low temperatures are comparative studies that either compared the proteins between cold-stressed and control plants or compared the proteins among genotypes with different susceptibility toward cold stress. Most proteomics studies have used leaf as plant material for protein extraction and analysis [8, 32–46] and only limited numbers studied the responsive proteins of crown [33, 47–49], root [40, 50], and reproductive tissues [51] to cold or freezing stresses (Table 5.1). In addition, most proteomics studies have explored the proteins of wheat and rice under cold stress which specifies the importance of cold stress tolerance in these two economically-important crop species (Table 5.1).

Comparative proteomics analysis have shown that the abundances of proteins belonging to different metabolic, regulatory, and structural components of cell or membrane are changed in crop plants in response to low or sub-optimal temperatures [8, 37, 39, 40, 42, 43, 45, 46, 48, 49, 51, 52]. Many of the genes that are responsive to low temperature also respond to other environmental stresses such as salinity and drought. This might be attributed to the fact that all of these environmental stresses cause cellular dehydration and ROS accumulation indicating that common responses are required to develop tolerance to these environmental adversities [10]. Figure 5.2 schematically displays what happens during cold stress exposure in a plant cell from cold signal perception to changes that occur in gene expression and the proteome of the cell.

5.3.1 *Cold-Regulated Proteins*

Wheat is one of the major economically important cereal plants, which grows in temperate regions. Common wheat species can be divided into winter and spring wheat species based on their requirements to a long period of low temperatures to become competence for flowering during a process known as vernalization [12]. Winter wheat species delay transition from vegetative to reproductive stage until

Table 5.1 A summary of comparative proteomic studies that have been performed to dissect the proteome response of different crop species to short and long-term low temperature treatments

Crop species	Cultivar	Treatment conditions	Developmental stage at the onset of treatment	Plant tissue	Proteomics approach	Reference
Wheat	Norstar and Azar2	0, 14, 28, 42, and 56 days at 2 °C	14 day old seedlings	Leaf	2-DE	[46]
	Samanta and Sandra	0, 3, 21 days at 4 °C	21 day old seedlings	Crown	2D-DIGE	[49]
	Cheyenne	63 days at 4 °C	14 day old seedlings	Leaf	2-DE	[45]
	–	0, 3, 21, and 84 days at 6 °C	Three-leaf stage	Crown	2D-DIGE	[48]
	Shixin 828 and Shiluan 02-1	5 h at −8 °C, with or without pretreatment with soluble sugars and amino acids	14 day old seedlings	Leaf	2-DE	[44]
	Kohdasht	0 and 42 days at 4 °C	14 day old seedlings	Leaf	2-DE	[43]
	Yumai 34	0, 1, and 3 days at −5 °C	Seven leaf stage (anther connective tissue formation phase)	Leaf	2-DE	[119]
	Jackson	21 days at 3 °C followed by 6 h, 1 and 3 days at −3 °C	35 day old seedlings	Crown	2-DE	[120]
	Triticum urartu	28 days at 4–6 °C followed by 12 h at −2 °C	35 day old seedlings	Leaf	2-DE	[121]
	Mironovskaya 808 and Bezostaya 1	21 days at 2 °C	Three leaf stage (∼21 day old seedlings)	Leaf	2-DE	[41]
Rice	Nipponbare	48 h at 5 °C	14 day old seedlings	Leaf and root	2-DE	[40]
	Doongara	72 h at 20/12 or 12/5 °C (day/night)	24 day old seedlings	Leaf	SDS-PAGE-LC-MS/MS	[39]
	Nipponbare	24 h at 6 °C	21 day old seedlings	Leaf	2-DE	[122]

(continued)

Table 5.1 (continued)

Crop species	Cultivar	Treatment conditions	Developmental stage at the onset of treatment	Plant tissue	Proteomics approach	Reference
	Doongara and HSC55	1, 2, and 4 days at 12 °C	At the time of panicle initiation	Anther	2-DE	[51]
	Dongjin	24 and 72 h at 10 °C	14 day old seedlings	Root	2-DE	[123]
	–	24 h at 15, 10, and 5 °C	14 day old seedlings	Leaf	2-DE	[124]
	Dongjin	12, 24, and 36 h at 5 and 10 °C	21 day old seedlings	Leaf	2-DE	[125]
	Liangyoupeijiu	6 h at 4 °C	14 day old seedlings	Leaf	2-DE	[85]
	Nipponbare	0, 48, 72, and 96 h at 14/12 °C (day/night) temperatures	4–5 leaf stage	Leaf	LC-MS/MS	[126]
Barely	Luxor	21 days at 3 °C followed by 1 day at −3 °C	Two-leaf stage	Leaf and crown	2D-DIGE	[33]
Rye	Maskateer	28 days at 2 °C	14 day old seedlings	Leaf	SDS-PAGE-LC-MS/MS	[32]
Oat	New Almighty	28 days at 2 °C	14 day old seedlings	Leaf	SDS-PAGE-LC-MS/MS	[32]
Maize	Reciprocal hybrids	11 days at 12.5 °C	4 day old seedlings	Embryo axes	2-DE	[127]
Soybean	Z22	24 h at 4 °C	Imbibing seeds	Embryo axes	2-DE	[128]
Sunflower	Hopi, PI 543006 and BSD-2-691	7 days at 15/5 °C day/night temperatures	42 day old seedlings	Leaf	SDS-PAGE-LC-MS/MS	[129]
Pea	Champagne and Terese	11 days at 10/2 °C followed by 4 days at 6/−8 °C day/night temperatures	11 day old seedlings	Roots, stems, and leaves	2-DE	[130]

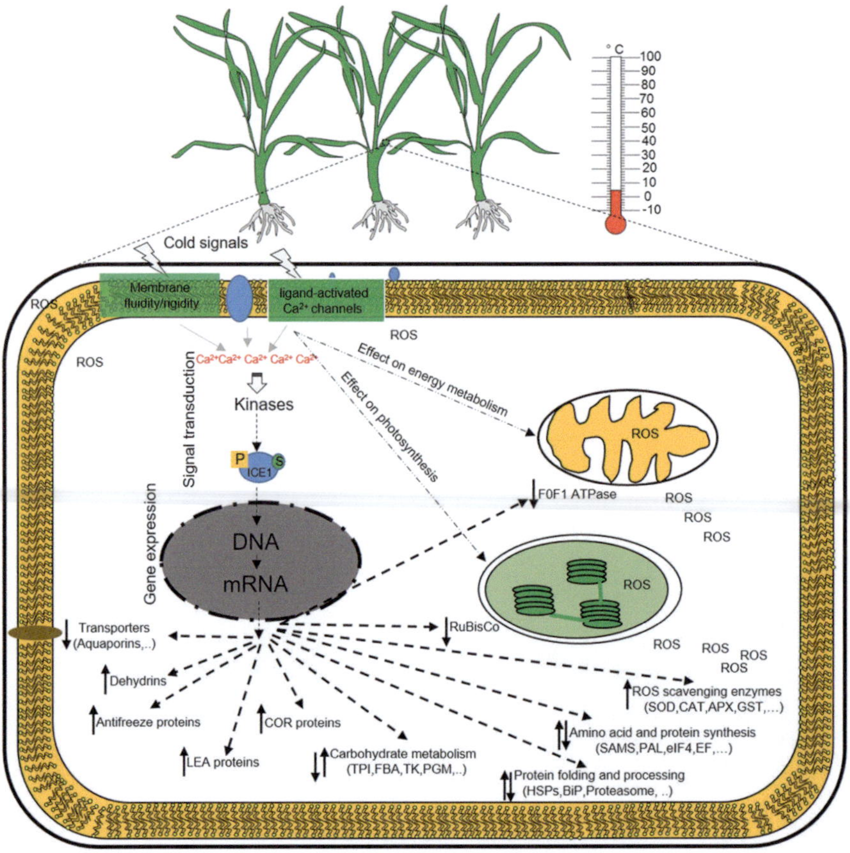

Fig. 5.2 Schematic illustration of proteome responses of plants to cold stress. The cold signal is probably sensed through changes in membrane rigidification/fluidity and/or other cellular changes. Changes in physical properties of cell membrane lead to an increased cytoplasmic Ca^{2+} concentration which is thought to be mediated by membrane rigidification-activated mechano-sensitive or ligand-activated Ca^{2+} channels [2, 3]. Ca^{2+} signature initiates a cascade of kinase signaling pathway leading to the activation of constitutively-expressed ICE1 transcription factor through phosphorylation (P) and sumoylation (S) [2]. Activated ICE1 induces the expression of c-repeat binding transcription factors (CBFs). CBFs then activate the expression of COR, LEA, antifreeze, and dehydrin genes leading to cold acclimation. Proteomics analyses have shown that cold acclimation is also associated with changes in abundance of proteins related to carbohydrate metabolism, amino acid biosynthesis, and protein synthesis and processing. Cold stress also affects photosynthesis and energy metabolism by largely declining the efficiency of plant in photosynthesis and energy production. Cold stress elicits profound changes in cellular homeostasis leading to the generation of excess amounts of ROS. Cold-acclimated plants express elevated levels of ROS scavenging enzymes to cope with detrimental effects of ROS. *Upward and downward solid arrows* indicate increased and decreased abundances during cold stress, respectively

they fulfill their vernalization requirements. These winter habit species go through winter under vegetative growth during which they have the capacity to tolerate subzero temperatures since only under this stage of growth that full expression of cold tolerant genes occurs [53]. In contrast, spring wheat species do not have vernalization requirements and therefore, have faster development [54]. The level of tolerance to cold stress in winter habit plants is usually correlated with increased abundance of cold and freezing tolerance associated proteins including dehydrins (DHN), wheat cold specific (WCS) proteins, COR proteins, responsive to abscisic acid, and LEA proteins. Since they mostly belong to the LEA family of proteins and are responsive to cold stress, they are collectively referred to as LEA/COR proteins [11]. The exact function of these proteins is yet to be realized but they might serve as cryoprotectant preventing the denaturation and aggregation of proteins and may also contribute to membrane stabilization during cold stress [5, 55]. The abundance of COR proteins such as LEA proteins, dehydrin proteins, WCS19, WCOR18, WCOR17, WCOR719, cold-responsive LEA/RAB-related COR proteins (Wrab17), WCOR615, WCOR14c, WCOR719, and WCOR14a has been shown to increase during cold acclimation [8, 41, 42, 45–48, 56–58]. Increased abundance of these proteins directly correlates with cold acclimation and freezing tolerance and they have been known as markers of cold stress tolerance. Studies have also shown that the level of expression of LEA/COR genes including WCS19, WCS120, WCOR14 and 15, and WDHN13 differs between winter wheat species (freeze tolerant) and spring wheat species (freeze sensitive) being expressed at higher levels in the former ones [59–61]. Interestingly, overexpression of WCS19 in Arabidopsis plants improves freezing tolerance only after cold acclimation [59].

Dehydrins are soluble, heat-stable, and hydrophilic proteins belonging to group 2 LEA proteins which are changed in abundance under cold, drought, and salinity conditions and seem to protect proteins and membranes against unfavourable structural changes caused by dehydration [62]. Dehydrin proteins are characterized by the presence of a specific lysine-rich amino acid sequence motif known as the K-segment [63]. Several dehydrins specifically accumulate during cold stress and their accumulation is thought to be an important part of cold acclimation. Interestingly, WCOR410 and WCS120 are dehydrin proteins which accumulate in the meristem of wheat crown tissues during cold acclimation and serve as cryoprotective agents for other proteins [64, 65]. Interestingly, the accumulation of WCOR410 has been positively correlated with the level of freezing tolerance in different wheat genotypes. Some dehydrin proteins such as COR19 from *Citrus unshiu* improves cold stress tolerance and displays ROS scavenging activity when ectopically expressed in tobacco plants [66].

The abundance of some LEA/COR proteins such as WCOR18 has been shown to be maintained at higher levels under prolonged cold stress treatment suggesting a possible involvement in the vernalization process [45]. WCOR18 is a member of highly-conserved phosphatidylethanolamine-binding family proteins whose members are known to be involved in the control of flowering time in both mono and dicots [67]. WCS19 and WCO14a are chloroplast-targeted proteins belonging to group 3 LEA proteins which their abundance are known to be under the control of light and cold stress [59]. In addition, ectopic overexpression of members of this family of proteins has been shown to increase freezing tolerance in model plants [59]. In vernalization-requiring crops such as winter wheat, it has been shown that the abundance of cold acclimation associated proteins (LEA/COR proteins) correlates with the vernalization fulfillment [46]. In another words, the expression of COR genes is maintained at high levels until the time of vernalization fulfillment and then declines with transition from vegetative to reproductive stage of growth. This is thought to be an adaptive mechanism utilized by cereal plants to increase the length of vegetative growth and to extend the time during which the expression of cold tolerance genes is maintained at high levels.

Proteomic studies have also reported an increased abundance of proteins involved in the vernalization process including VER2 in crown and leaf tissues of winter wheat plants exposed to cold stress [45, 49]. VER2 is a mannose binding lectin protein containing a jacalin-like domain at the C-terminus that is known to play an important role in saccharide signaling [68]. VER2 specifically localizes at the wheat shoot apex and the surrounding young leaves and is known to be under the control of both vernalization and jasmonic acid signaling [45, 69]. VER2 may be involved in vernalization through participating in intracellular glycoprotein trafficking and O-linked N-acetylglucosamine (O-GlcNAc) signaling [70]. A 200 kDa cold induced protein showed a marked increase in abundance in crown tissue of wheat [47].

An increased abundance of glycine rich RNA binding protein (GRP) was repeatedly reported in leaf and crown tissues of wheat in response to low temperature [43, 45, 47, 49]. GRPs contain an N-terminal RNA recognition motif and a C-terminal glycine-rich region and have been shown to accumulate during cold acclimation. The exact function of GRPs in stress tolerance has remained elusive, but recent evidences suggest that GRPs enhance cold and freezing tolerance in Arabidopsis plants [71]. In addition, ectopic expression of Arabidopsis GRPs in *E. coli* mutants lacking cold shock proteins enhanced growth and survival under cold stress suggesting RNA chaperone activity for GRP proteins during the cold acclimation process [72].

5.3.2 Antifreeze Proteins

In temperate regions during winter, plants are contentiously exposed to freezing temperatures and the formation of ice crystals in the intercellular and intracellular spaces is unavoidable if they have no strategies to withstand freezing. Ice formation in intercellular spaces can cause cellular dehydration since growing extracellular ice promotes the migration of water molecules to the ice crystals [6]. Winter habit plants secret antifreeze proteins into the apoplastic spaces where they inhibit intercellular ice nucleation through binding to the ice crystals [6]. In other words, antifreeze proteins decrease the freezing temperature in cold-acclimated plants and after freezing of plants under frost temperatures, they inhibit the recrystallization of intercellular ice [73]. Plant antifreeze proteins are structurally homologous to PR proteins especially those that are responsive to low temperatures including β-1,3-glucanases (PR-2), chitinases (PR-11), thaumatin-like proteins (PR-5), and polygalacturonase inhibitor protein [73]. PR proteins are normally secreted into apoplastic space in response to pathogen infection and either degrade fungal cell walls or inhibit fungal enzymes and therefore, impede fungal pathogenicity [73]. Even though the majority of antifreeze proteins are targeted to extracellular spaces, some intracellular LEA/COR proteins such as dehydrins may also show antifreeze activity and therefore, may prevent intracellular ice formation [56]. Proteomic analysis reported an increased abundance of proteins with antifreeze function including PR-4, chitinase, thaumatin-like protein, β-1,3-glucanase during cold acclimation in winter wheat [46, 47, 49]. Interestingly, these proteins either specifically detected in hardy species of wheat in response to cold stress or being expressed at higher abundance in response to cold stress [46]. PR proteins are primarily involved in pathogen defense and their accumulation in cold acclimated plants is thought to immunize them against pathogens when they are prone to infection due to freezing damages [73, 74]. For example, the induction of chitinase may not be an important adaptation to cold stress, however, it may confer resistance to fungal pathogens such as snow moulds that attack plants upon freezing damage [47].

Down regulation of aquaporins, proteins channels that regulate water efflux across plasma membrane, may also serve as an important adaptation to cold stress since the flow of water to apoplastic spaces increases intercellular ice formation and causes cellular dehydration [5]. Accordingly, a decreased abundance of aquaporin proteins was detected in oat and rye under cold acclimation [32]. In addition, a decreased abundance of V-type ATPase was observed in rice leaf [40], sunflower [58], oat and rye challenged with cold stress [32]. By transferring protons across membrane, V-type ATPases participate in the acidification of intracellular compartments and are thought to contribute to cold acclimation through intracellular pH regulation.

5.3.3 Oxidative Stress and ROS Scavenging Related Proteins

In plant cells, ROS are produced at a steady rate and low level as byproducts of various metabolic processes and are distributed in different cellular compartments including organelles of highly oxidizing metabolic activity or intense rate of electron flow, such as mitochondria, chloroplast, and peroxisome [75]. Under normal conditions, the rate of ROS generation is balanced by the rate of their removal. Cold, drought, salt, and other environmental adversities enhance the production of ROS in cells by disrupting cellular homeostasis and uncoupling of metabolic processes. It has now been well-documented that ROS serve as signals causing cells to adjust energy and metabolic fluxes and to reprogram their gene expression to quickly adapt to stress conditions [76]. For example, Arabidopsis mutants defective in the Fe–S subunit of complex I (NADH dehydrogenase) of mitochondria electron transfer chain which accumulates excess level of ROS, show impaired expression of COR genes and increased hypersensitivity to chilling and freezing stresses suggesting a critical role for ROS in regulating cold acclimation-associated genes [2]. In addition, the superoxide anion produced by NADPH oxidase during low temperature stress triggers stress response pathways that leads to the development of defense mechanisms [10]. However, excess accumulation of ROS has cytotoxic consequences for plant cells since they cause oxidative damages to proteins, DNA, and membrane lipids, and induce cells to undergo programmed cell death [75, 77]. Therefore, the ability of plants to alleviate the intracellular buildup of ROS under low temperatures is an important component of cold tolerance mechanisms that are activated during cold acclimation. During evolution, plants have evolved sophisticated and complex network of enzymatic and non-enzymatic mechanisms to maintain the intracellular concentration of ROS under tight control [77]. In plant cells, ascorbate, glutathione (GSH), and vitamin E as well as secondary metabolites such as tocopherol, flavonoids, alkaloids, and carotenoids are the primary non-enzymatic antioxidants. The role of ascorbate and GSH in detoxification of ROS has been well understood but little is known about the antioxidative potential of flavonoids and carotenoids in plant cells [77]. Superoxide dismutase (SOD), ascorbate peroxidase (APX), glutathione peroxidase (GPX), and catalase (CAT) are major ROS scavenging enzymes that are responsible for removal of ROS under normal and stressful conditions. The activity and abundance of these proteins have been reported to be significantly elevated in plants under various biotic and abiotic stresses.

Proteomics analyses showed an increased abundance of Cu/Zn-SOD in wheat leaf [8, 42, 46], maize embryonic axes [34], and decreased abundance in rice leaf [40] under cold stress treatment. Mn-SOD also showed an increased abundance in wheat leaf challenged with low temperature stress [46, 52]. However, in a study by Xu et al. [44], Mn-SOD showed a decreased abundance in leaves of wheat under low temperatures, suggesting that the expression of this enzyme was affected by the genotype and treatment conditions. SODs constitute the first line of defense against ROS by converting highly toxic superoxide anions to less toxic H_2O_2 [77, 78]. Based on the type of metal cofactor required for their activities, SOD enzymes are classified into three groups: Fe-SODs, Mn-SODs, and Cu/Zn-SODs. In addition, their subcellular localization also differs; Fe-SODs are chloroplastic proteins, Mn-SODs are mainly localized in the mitochondrion and the peroxisome, while Cu/Zn-SODs are generally detected in the chloroplast, the cytosol, and possibly the extracellular spaces [78]. An increased abundance of APX was detected in rice [36] and wheat leaves [42, 43] in response to low temperatures. APX is a plant-specific enzyme that plays an important role in the protection of cells against the H_2O_2 produced in the cytosol, mitochondria, chloroplast and peroxisomes under stressful conditions [79]. APX catalyzes the reduction of H_2O_2 to water using ascorbate as an electron donor. In plant cells, H_2O_2 is also detoxified by the enzymatic action of

CAT and GPX, which are mainly localized in peroxisomes and cytosol, respectively [77]. CAT showed an increased abundance in rice leaf exposed to low temperature [40] while GPX showed a decreased abundance in wheat leaf under low temperatures [48]. The GPX pathway for ROS scavenging is thought to be the major enzymatic defense system against ROS-mediated membrane damage.

Glutathione S-transferase (GST) showed an increased abundance in rice [36, 37], wheat [46, 48], sunflower [58], and soybean [57] challenged with cold stress. GST is predominantly localized in the cytosol where it catalyzes the conjugation and detoxification of xenobiotics and organic hydroperoxides generated as byproducts of oxidative stress using GSH as cofactor [80]. If organic hydroproxides are not detoxified by GST, they could be converted into aldehyde derivatives which are highly toxic for cells. Increased abundance of GST under low temperature conditions marked it as an important part of cold acclimation process. In addition, the abundance of glyoxalase I and II decreased in maize embryonic axes in response to low temperatures [34] while glyoxalase I showed an increased abundance in rice roots under cold stress [50]. Both enzymes are involved in the glyoxalase pathway, which mediates glutathione-based detoxification of methylglyoxal. The glyoxalase pathway is known to be involved in regulation of cell cycle as well as in protection against oxoaldehyde toxicity [81].

Proteomics analyses showed an increased abundance of thioredoxins (Trxs) including Trx h in rice leaves [36] and Trx m in wheat leaves [42] suggesting their critical role in modulating the plant response to cold stress. Trx F showed a decreased abundance in a chilling sensitive genotype of sunflower [58]. Trxs are evolutionary conserved proteins that supply reducing power to reductase enzymes which are responsible for the detoxification of lipid hydroperoxides or the repair of oxidized proteins [82]. Excess accumulation of ROS will always result in the oxidation of proteins which may lead to their inactivation and accumulation in cells. For example, Trxs play a critical role in reactivation of thiol-containing enzymes including those of the Calvin cycle such as glyceraldehyde 3-phosphate dehydrogenase (GAPDH), fructose 1,6-bisphosphatase (FBP), sedoheptulose 1,7-bisphosphatase (SBPase) and phosphoribulokinase (PRK) which are inactivated upon oxidation. In addition, an increased abundance of thioredoxin peroxidase [36, 49, 58] and 2-Cys peroxiredoxin [46], either newly-expressed in response to cold stress or increased during cold acclimation, was noted in response to low temperatures. Interestingly, 2-Cys peroxiredoxin plays a role in detoxification of excess H_2O_2, alkyl hydroperoxides, and peroxinitrites [83].

An increased abundance of germin E was also detected in the crown tissues of winter wheat during cold acclimation [49]. Since meristemic tissues in the crown of winter cereals play a critical role in regeneration and recovery from freezing stresses, increased abundance of this protein is thought to contribute to the cold acclimation process. Germin proteins are known to be involved in plant embryogenesis and response to environmental stresses. Some germin proteins show dual Mn-SOD and oxalate oxidase activities suggesting their possible implication in oxidative stress response especially in defense against extracellular superoxide radicals [84]. A significant change in abundance of oxalate oxidase was also reported in wheat in response to low temperatures [8, 46]. Oxalate oxidase catalyzes the conversion of oxalate to CO_2 and H_2O_2 in the presence of O_2 and therefore, may serve as a source of H_2O_2 especially in certain plant-pathogen interactions.

Nucleoside diphosphate kinase (NDPK) displayed species dependent changes to low temperatures showing a decreased abundance in wheat [46] and maize [34], and increased abundance in rice [40, 85] and pea [86]. NDPK is a housekeeping enzyme responsible for maintaining the intracellular balance between ATP and other nucleoside triphosphates. Evidences have suggested that NDPK proteins play regulatory roles in signaling pathways leading to the oxidative stress response especially by regulating the gene expression of antioxidants [87, 88]. In addition, oxidative stress-induced overexpression of NDPK2 in potato plants suggested that NDPK proteins protect plants from oxidative stresses associated with multiple environmental stressors including high temperatures and salt stress [88]. Accordingly, Hosseini et al. [89] found that NDPK1 was differentially regulated in rice genotypes with contrasting responses to salt stress; showing an increased abundance in a salt-tolerant genotype (FL478) and decreased abundance in a salt-sensitive genotype (IR29) under salinity suggesting a critical role for NDPK1 in conferring tolerance to salt stress [89].

5.3.4 *Photosynthesis Related Proteins*

In plants, photosynthesis plays a critical role in energy production by converting light energy into chemical energy in the form of ATP and reducing power in the form of NADPH. Maintaining photosynthesis during low temperature conditions is necessary since cold acclimation is an active process that requires energy provided by the photosynthesis process. During cold stress, the balance between the capacity for light harvesting and the capacity for light energy consumption is disrupted resulting in excess photosystem II (PSII) excitation which leads to irreversible inactivation of PSII and damages to the D1 reaction center protein [90, 91]. This process is known as photoinhibition which may lead to the transfer of energy-rich electrons to molecular oxygen resulting in over production of ROS and occurrence of oxidative burst. Under cold stress, the activity of enzymes required for the Calvin cycle is limited which may lead to excess photoinhibition even under relatively low irradiations [92]. In addition, cold-induced limitation in photosynthetic carbon fixation will result in decreased NADPH consumption and depletion of the major electron acceptor of PSI (e.g. $NADP^+$) which ultimately leads to the transfer of electron to molecular oxygen and generation of ROS and occurrence of oxidative stress [92]. It is thought that during cold acclimation, plants are also photosynthetically acclimated to excess light and can therefore tolerate photoinhibition [91]. In cold acclimated plants, excess light is eliminated by thermal dissipation during a process known as nonphotochemical quenching [10]. It should be noted that light and photosynthetic activities are required for cold acclimation process since full cold acclimation is not achieved under dark conditions even though that cold associated genes are up-regulated [93]. In addition, studies have shown that the number of cold responsive genes are doubled when plants are cold stressed in the light compared with that under the same conditions in the dark [94]. Interestingly, it has been shown that the relative redox status of PSII appears to control the expression of COR genes as documented in winter rye [95]. Proteomics analyses significantly improved our understanding about photosynthetic pathways and reactions affected by low temperatures.

During light reactions of photosynthesis, light is trapped by light-harvesting complex (LHC) and its energy is utilized by an oxygen evolving complex (OEC) to split water to oxygen, proton, and electron. Interestingly, proteomics analysis of wheat plants showed a significant down regulation of components of OEC including 33 kDa oxygen-evolving complex proteins [47], oxygen-evolving enhancer protein 1, 2, and 3 [42, 43, 45], and LHC I [45]. In contrast, an increased abundance of the same proteins was also noted in wheat (oxygen-evolving enhancer protein 1 and 2, [8, 46]), rice (oxygen-evolving complex proteins [37]), and barley (33 kDa oxygen evolving protein [33]). Similarly in barley, chlorophyll a-b binding protein showed a decreased abundance in response to cold acclimation [33], while in cold acclimated wheat and pea, it showed an increased abundance [42, 86]. Extensive increase and decrease in the abundance of proteins associated with PSII suggests that it is profoundly affected by cold stress. These contrasting abundances of proteins in response to the same treatment suggested that the response of these proteins is largely dependent on the species, genotype, photoperiod, as well as the duration and severity of cold stress treatment.

The electrons released from PSII is subsequently transferred to PSI via cytochrome b6f complex (Cyt b6f complex). Proteomics analyses showed that the abundance of iron–sulfur subunit of the Cyt b6f complex and plastocyanin significantly decreased during cold acclimation [45, 86]. Decreased abundance of proteins associated with Cyt b6f complex may affect the efficiency of electron transfer between PSII and PSI and, thereby affecting photosynthetic ATP and NADPH production. The abundance of proteins related to downstream reactions of photosynthesis including ferredoxin-NADP(H) oxidoreductase (FNR) also significantly changed in response to cold stress in different plant species. More specifically, it showed an increased abundance in wheat [43, 44] and rice [37] under low temperature treatments. However, in a study by Rinalducci et al. [46], the abundance of this protein significantly decreased in response to cold acclimation. FNR catalyzes the reversible transfer of electrons from reduced ferredoxin to $NADP^+$ as the final step of the photosynthetic electron transfer chain. This reaction is a rate-limiting step in the photosynthesis that generates NADPH as reducing power for several biosynthetic pathways including CO_2 fixation. FNR plays a critical role in photosynthesis since transgenic down regulation or knock down of this protein in tobacco resulted in a significant reduction in growth and net photosynthesis [96] as well as enhanced susceptibility to photo-oxidative damages and excess accumulation of singlet oxygen when FNR-deficient plants were exposed to moderate irradiation [97]. Increased abundance of this protein may lead to an enhanced photosynthesis and may be part of the cold acclimation process. Proteomics analysis also showed a decreased abundance of subunits of chloroplastic ATP synthase during cold stress [43] which may affect ATP production during photosynthesis.

Change in abundance of proteins associated with dark reactions of photosynthesis (i.e., the Calvin cycle) has also been reported in proteome analysis of cold stress responses. Interestingly, a marked change in abundance (mostly decreased) of RuBisCO small and large subunits was noted in wheat [8, 42–47], rice [38, 39], and pea [86]. In addition, proteome analyses showed that the subunits of RuBisCO were subjected to extensive degradation during cold stress treatment and were detected in multiple spots with varying MW and p*I* [42, 43, 45]. This degradation is speculated to be enhanced by ROS which are known to accumulate during cold stress. In fact, evidence suggests that ROS trigger non-enzymatic degradation of RuBisCO large subunit [98]. A significant increase/decrease in abundance of RuBisCO activase was also reported in wheat and rice plants exposed to low temperatures [43, 45, 46, 85]. When plants are subjected to dehydration, the intracellular level of CO_2 is diminished due to stomata closure. Under these conditions, the inhibitory sugar phosphates such as ribulose-1,5-bisphosphate tightly bind to the active site of RuBisCO causing its inactivation [99]. The function of RuBisCO activase is to mediate the release of these inhibitors through its chaperone-like activity. An increase in abundance of RuBisCO activase may be required to maintain RuBisCO in an active state under low temperatures when the concentration of CO_2 is declined due to stomata closure and dehydration stress.

5.3.5 Carbohydrate Metabolism Related Proteins

Adaptation to low and freezing temperatures always requires reprogramming gene expression and reorganizing metabolic and physiologic processes. Carbohydrates play an important role in freezing tolerance as it has been shown that the accumulation of simple sugars such as trehalose, raffinose and sucrose correlates with enhanced freezing tolerance. Compared with spring cereals, the ability of winter cereals to tolerate and survive subzero temperatures largely depends on their ability to accumulate sugars during cold acclimation [93]. Under cold stress conditions, a significant deregulation of proteins related to the Calvin cycle and other glycolytic pathways has been reported suggesting an immense role for photosynthetic carbon fixation and carbohydrate metabolism in cold acclimation. As noted above, the abundance and stability of RuBisCO as the first and important enzyme of the Calvin cycle are significantly affected by low temperatures. Even though they are mainly involved in the photosynthesis carbon fixation, the remaining enzymes of the Calvin cycle are discussed under the category of carbohydrate metabolism since they also participate in the other carbohydrate metabolism related reactions. Among the Calvin cycle related enzymes, the abundance of carbonic anhydrase (CA) decreased in cold acclimated wheat [48]. CA catalyzes the conversion of atmospheric CO_2 to bicarbonate and consequently concentrates and facilitates its diffusion in the mesophyll. Phosphoglycerate kinase (PGK), the second enzyme of the Calvin cycle and the seventh enzyme of the glycolysis pathway, showed an increased abundance in wheat [45, 48] and rice [40] in response to low temperatures. GAPDH, the third enzyme of the pathway and the sixth enzyme of the glycolysis, showed contrasting responses in wheat under low temperatures [43, 45, 48, 49]. Studies using transgenic down regulation of GAPDH have shown that it has little control over photosynthetic carbon fixation [100]. Similarly, triose phosphate isomerase (TPI), another enzyme of the pathway, showed an increased abundance in maize [34], soybean [57], and wheat [48] in response to low temperatures suggesting it as a candidate cold responsive protein. Fructose bisphosphate aldolase (FBA) and transketolase (TK) also showed profound changes in abundance in response to cold stress. FBA showed contrasting responses by being increased in wheat [46, 49] and rice [40] and decreased in abundance in wheat [8, 48] and maize [34] in response to low temperatures. TK is a critical enzyme in the Calvin cycle and pentose phosphate pathway that exerts significant control over carbon flux in the photosynthetic carbon assimilation [100]. TK showed a marked down regulation in wheat and rice plants subjected to cold stress [43, 85], suggesting that carbon assimilation has significantly been perturbed under low temperatures. Another enzyme that is also known to play a key regulatory role over carbon assimilation is sedoheptulose-1,7-bisphosphatase (SBPase) that showed a significant increase in abundance under low temperatures [44]. Overall, proteomic analyses showed that photosynthetic carbon assimilation is sensitive to cold stress as revealed by up/down regulation of proteins associated with the Calvin cycle.

Proteomics results also showed a significant up/down regulation of enzymes related to other catabolic and anabolic pathways of carbon metabolism. Interestingly, enzymes associated with the final steps of the glycolysis such as triose phosphate isomerase (TPI), PGK, and phosphoglycerate mutase (PGM) showed an increased abundance in wheat [48, 49] and rice [38] in response to low temperatures. Increased abundance of these enzymes may lead to an increased production of ATP as an immediate energy source for cold acclimation. Enolase (ENO), the enzyme catalyzing penultimate reaction of glycolysis, displayed contrasting responses showing an increased abundance in wheat [48], rice [38, 50], and pea [86], and decreased abundance in maize [34] and rice [40] in response to low temperatures. In addition to its key role in carbon metabolism, ENO has been shown to localize in nucleus and functions as transcription regulator controlling gene expression. In Arabidopsis plants, it has been shown that ENO can localize in nucleus and serves as negative regulator of the expression of zinc-finger transcription factor STZ/ZAT10 [101], a repressor of cold-inducible CBF regulatory pathway, and therefore, may be involved in the regulation of cold responsive genes. In addition, an increased abundance of alcohol dehydrogenase 1 (ADH1), which catalyzes the last step in the ethanolic fermentation process, was also reported in a spring wheat and embryonic axes of soybean exposed to low temperatures [49, 57].

Depending on the species, sucrose synthase (SUS) displayed contrasting responses to low temperatures. In wheat, it showed a decreased abundance [45, 48, 49], while in maize [34] and rice [37, 38, 50], it showed an increased abundance in response to low temperatures. Sucrose-binding protein also showed an increased abundance in embryonic axes of germinating seeds of soybean [57]. SUS catalyzes the reversible conversion of sucrose to fructose and UDP-glucose. Fructose is phosphorylated by fructokinases to fructose-6-phosphate. Fructose-6-phosphate is subsequently utilized as the main substrate for several metabolic processes including starch biosynthesis, glycolysis, and oxidative pentose phosphate. Therefore, the reaction catalyzed by fructokinases serves as a gateway for fructose metabolism [102]. Similar to SUS, fructokinase showed species-dependent changes in abundance under low temperatures. It showed a decreased abundance in wheat [45, 48, 49] and increased abundance in maize [34] and barley [33] in response to low temperatures. Interestingly, decreased abundance of SUS in wheat plant under low temperatures increases intracellular buildup of sucrose as a cryoprotectant and consequently improves freezing tolerance. In addition, decreased abundance of SUS may also lead to depletion of UDP-glucose which is mainly utilized as precursor for the biosynthesis of cellulose, the dominant polysaccharide of plant cell walls. In addition, the changes in abundance of enzymes involved in the biosynthesis of UDP-glucose and other precursors of cell wall polysaccharides including UDP-glucose pyrophosphorylase [37, 38, 40, 48, 57], UDP-glucosyltransferase [48], and UDP-glucose 6-dehydrogenase [33] were reported under low temperatures. The abundance of beta-glucosidase, an enzyme involved in plant cell wall degradation, increased in embryonic axes of maize [34] and decreased in crown tissue of wheat [49] in response to low temperatures. It is now well-documented that during cold stress, plant growth is largely restricted which is in accordance with the decreased abundance of enzymes involved in the biosynthesis of cell wall polysaccharides [48].

Under low temperatures, the protein abundance of enzymes involved in the pentose phosphate pathway significantly increased including phosphoribulokinase [37, 44], 6-phosphogluconolactonase [44], and phosphogluconate dehydrogenase [50]. However, in a study by Rinalducci et al. [45], the abundance of phosphoribulokinase decreased under low temperature suggesting that the expression of this protein fluctuates under different low temperature treatment conditions. Since the oxidative pentose phosphate pathway is the major source of reducing power (e.g. NADPH) and metabolic intermediates for other biosynthetic pathways, an increased abundance of enzymes involved in this pathway may correspond to an increased demand for reducing power or metabolite intermediates for cold acclimation. In addition, under low temperatures, enzymes involved in the tricarboxylic acid (TCA) cycle including malate dehydrogenase (MDH) [34, 44, 45, 57], aconitate hydratase [37, 50], and NADP-specific isocitrate dehydrogenase [50] showed an increased abundance. Aconitate hydratase also showed a decreased abundance in rice in response to low temperatures [38]. Aconitate hydratase is known to be sensitive to ROS and was shown to be down-regulated by oxidative stress in Arabidopsis. The TCA cycle plays a profound role in the production of ATP and carbon skeletons for a range of biosynthetic processes. The activation of enzymes associated with this pathway may be part of the cold acclimation which is an energy-demanding process. In addition, MDH also plays an important role in transfer and distribution of reducing powers among different cellular compartments through malate/oxaloacetate shuttle [103].

An interesting enzyme that showed an increased abundance in rice [35, 37], wheat [8, 46, 47], and pea [86] in response to low temperatures was thiamine biosynthetic enzyme. Thiamine (vitamin B1) plays a fundamental role as a cofactor for several enzymes involved in carbohydrate metabolism including glycolysis, the pentose phosphate pathway/Calvin cycle, and the TCA cycle. Recent evidence indicates that Arabidopsis plants subjected to environmental stresses such as cold, salinity, drought, and high light accumulate excess amount of thiamin and it is thought that thiamin confers enhanced tolerance to oxidative stresses associated with these environmental stressors [104]. Increased abundance of thiamine biosynthetic enzyme is associated with an increased intracellular buildup of thiamin, which might be part of the cold acclimation process to protect plants from oxidative stresses imposed by low temperatures.

5.3.6 Proteins Related to Protein Synthesis and Metabolism

Acclimation to cold stress is an active process that is associated with changes in gene expression and usually depends on the synthesis of new proteins and removal of old and damaged proteins. Low temperatures and their associated oxidative stresses increase risk of damages such as unfolding, misfolding, degradation, and oxidation to proteins and promote their accumulation in cells. Therefore, increased abundance of proteins with protective functions against protein damages is thought to be an important and indispensable part of cold acclimation process. In support of this, proteomics analyses have shown increased abundance of proteins such as chaperones, chaperonins, and heat shock proteins (Hsps) under low temperatures. Interestingly, the abundance of general stress responsive protein heat shock protein 70 kDa (Hsp70) significantly increased in wheat [48, 49], rice [37, 50], barley [33], maize [34], pea [86], and sunflower [58] under low temperatures. In addition, the abundance of heat shock cognate protein 70 (Hsc70), which is a member of Hsp70 family proteins, was also increased under low temperatures [34, 48]. In cells, Hsp70 along with its cognate protein, Hsc70, assist folding of newly synthesized or refolding of non-native and misfolded proteins and prevent their aggregation under both normal and stressful conditions [105]. In addition, they are also involved in protein translocation across membranes, and in targeting misfolded or unstable proteins to lysosomes or proteasomes for degradation. Members of Hsp60 chaperonin family proteins are also increased in rice [37], barley [33], and wheat [48, 49] in response to low temperatures. Hsp60s are largely localized in the chloroplast where they assist newly synthesized or newly translocated proteins to fold or assemble correctly [105]. Interestingly, the assembly of RuBisCO holoenzyme is mediated by chaperonin functions of chloroplast Hsp60 (cpn60), Hsp21 (cpn21), and an ATP-dependent chloroplast protein with chaperonin function named RuBisCO binding protein (RBP) [106, 107]. The abundance of both cpn21 and RBP was increased under low temperatures in wheat [45, 48, 49] but in case of RBP, it showed a decreased abundance in sunflower [58], suggesting a key role for these proteins in maintaining chloroplastic proteins in an active state under low temperatures.

Hsp90 showed a decreased abundance in wheat under low temperatures [48]. Although, Hsp90 family proteins are constitutively expressed and are responsive to stress conditions but their abundance in Arabidopsis might be developmentally regulated [108]. Hsp82, a member of Hsp90 family of molecular chaperones, showed an increased abundance in maize and sunflower under low temperatures [34, 58]. Small heat shock proteins including Hsp20 also showed increased abundance under low temperatures [37, 47]. Members of this family of proteins have the capacity to bind non-native proteins through hydrophobic interactions and prevent their aggregation but do not have the capacity to refold them by themselves [105]. Thereby, they facilitate refolding of non-native and/or misfolded proteins and maintain them in a competent conformation for folding by other chaperone proteins such as Hsp70/Hsp100 complexes [109]. An increased abundance of luminal binding protein (BiP) was also reported in wheat plants subjected to cold stress [49]. BiP is an ER-resident molecular chaperone that facilitates folding of newly synthesized proteins and also involves in ER quality control which is responsible for recognizing misfolded or abnormally-folded proteins and targeting them out of the ER for proteasome-dependent degradation [110]. Increased abundance of BiP under low temperature nominates it as candidate cold responsive protein that might be directly involved in alleviating cells from stress conditions by preventing accumulation of misfolded and cold stress damaged proteins. Protein disulfide isomerase (PDI) also showed an increased [48] and decreased [42, 57] abundance in wheat and soybean plants exposed to low temperatures, respectively. PDIs are ER resident molecular chaperones containing thioredoxin domains which aid protein folding through catalyzing the formation of proper disulfide bonds [111]. A change in abundance of cyclophilin was also reported in wheat and pea plants exposed to prolonged cold stress treatment [45, 46, 86]. Cyclophilins are conserved proteins belonging to immunophilin super family of proteins, which have peptidyl-prolyl cis-trans isomerase activity and are involved in protein folding.

Misfolded or damaged proteins that could not be refolded or restored and reactivated by the action of chaperone proteins need to be eliminated, since the accumulation and aggregation of unfolded proteins have detrimental consequences for cells. This is in accordance with changes in abundance of proteins that are involved in protein degradation during cold stress. Interestingly, an increased abundance of Clp protease, aspartic protease, and subunits of proteasome was noted in cold stressed rice [37] and wheat [48], suggesting that misfolded or unfolded proteins are actively degraded during cold stress and the removal of damaged or unnecessary proteins is thought to be part of cold acclimation process.

Proteins synthesis or translation is a fundamental step in gene expression and is mainly fine-tuned at the initiation step. The initiation of translation involves the recruitment of the 5′-untranslated region (UTR) of the mRNA to the small subunit (40S) of the ribosome, a reaction that is catalyzed by the eukaryotic translation initiation factors 4 (eIF4) [112]. The main function of eIF4 families of translation initiation factors is to promote binding of ribosome to the target mRNA and to facilitate scanning by 40S ribosome for the start codon. In cold-exposed wheat [48] and rice [37, 39], an increased abundance of eIF4A was reported suggesting an important cold-regulated role for it. eIF4A is an ATP-dependent RNA helicase that unwinds secondary structures in the 5′-UTR of the target mRNA to facilitate scan for the initiation codon by 40S ribosomal subunit [112]. An increased abundance of eIF5A1 and A2 [42, 49] as well as eIF3 [49] was also reported in wheat plants under low temperatures. In addition, the abundance of translational elongation factor P (EF-P) decreased in wheat under cold stress [46]. The eIF5A and its bacterial homologue, EF-P, are small conserved proteins and are the only cellular proteins that are subjected to hypusination and lysinylation, respectively [113]. In contrast to its name, eIF5A is thought to be involved in translation elongation by enhancing peptide bond formation between two poorly-reactive prolines in A and P sites of ribosome and might therefore be involved in translation of proteins containing consecutive prolines in their amino acid sequences [113]. Ectopic overexpression of an eIF5A gene from *Rosa chinensis* in Arabidopsis resulted in enhanced tolerance to heat, oxidative and osmotic stresses, suggesting a critical role for this translation factor in acquiring tolerance to environmental stresses [114]. Organelle and nuclear specific EFs such as EF-Tu [37, 58], EF-G [44], EF2 [34], EF-1 alpha [47] also showed an increased abundance under low temperatures suggesting an active protein synthesis during cold stress. Ploy (A)-binding protein which is an essential protein required for the protein synthesis process showed an increased abundance in wheat plants subjected to low temperatures [47].

The metabolism of amino acids and certain polyamines is also responsive to low temperature. Accordingly, changes in abundance of enzymes associated with amino acid metabolism were reported in proteome analyses of cold stress. Within this category, enzymes such as phenylalanine ammonia-lyase (PAL), methionine synthase (MTR), *S*-adenosylmethionine synthetase (SAMS), cysteine synthase (CYSL), and glutamate semialdehyde aminotransferase (GSA) showed an increased abundance in rice, wheat, and sunflower in response to low temperatures [37, 38, 42, 43, 45, 85, 111]. Methionine synthase catalyzes the formation of methionine while SAMS catalyzes the conversion of methionine to S-adenosylmethionine (AdoMet). AdoMet serves as universal methyl donor for most methylation reactions including DNA methylation and as precursor for the production of stress related metabolites such as polyamines and glycine betaine and also for the production of plant growth regulator ethylene. Polyamines including glycine betaine accumulates during cold acclimation and thought to increase plasma membrane stability during cold stress and therefore, to increase tolerance to low temperatures. Cysteine synthase catalyzes the final step of cysteine biosynthesis which is a rate-limiting step in the production of GSH, a thiol containing tripeptide involved in resistance to biotic and abiotic stresses. Therefore, increased cysteine synthase may lead to an increased abundance of GSH which is required for detoxifying byproducts of oxidative stress associated with cold stress. An increased abundance of glutamine synthetase (GS) was also reported in cold-acclimated plants [36, 38, 42, 43] suggesting it as a candidate cold responsive protein. GS catalyzes the ATP-dependent condensation reaction of inorganic ammonium with glutamate to yield glutamine which is subsequently utilized as nitrogen source for other metabolic processes. GS is also implicated in the biosynthesis of precursors of proline, an osmolyte involved in adaptation to cellular dehydration associated with environmental stressors such as drought, salt, and cold.

5.3.7 Proteins Related to Energy Metabolisms

Cold acclimation is an active adaptive process that is associated with broad morphological, physiological, and molecular changes that demand sustained energy production as indicated by profound changes in energy metabolism related proteins. In plants, energy metabolism is usually adjusted with the rate of photosynthesis and growth. Interestingly, proteomics analysis showed that the abundance of subunits of mitochondria electron transport chain complexes and ATP synthesizing complex F0F1 was changed with most of them being significantly decreased during cold stress treatment [37, 38, 40, 45, 47, 49, 57, 58, 86]. This suggests that energy production was significantly perturbed which was in agreement with reduced photosynthesis and growth under such conditions.

5.3.8 Signaling and Gene Regulatory Proteins

In plants, cold stress signal is mainly sensed at the cell membrane through low temperature induced changes in membrane fluidity and/or rigidity (Fig. 5.2) [2]. However, changes in abundance and conformation of proteins and nucleic acid and/or metabolites may also serve as signals for sensing cold stress. Changes in physical properties of cell membrane lead to an increase in cytoplasmic Ca^{2+} concentration which is thought to be mediated by membrane rigidification-activated mechano-sensitive or ligand-activated Ca^{2+} channels [2, 3]. Transient increase in cytosolic Ca^{2+} triggers cascades of signaling pathways that lead to changes in gene expression and is usually followed by synthesis and accumulation of LEA/COR proteins. Therefore, the abundance of proteins involved in signaling pathways is expected to change in plants in response to low temperature. In support of this, proteomics analyses showed changes in abundance of proteins with possible implication in cold stress signaling. Interestingly, the abundance of 14-3-3 protein was decreased in wheat and rice plants exposed to low temperature [40, 41, 48]. 14-3-3 proteins comprise a large group of highly conserved proteins whose members are responsive to various environmental stresses including cold, salinity, and drought, but their exact role in plant response to such stresses are largely unknown. Evidence suggest that 14-3-3 proteins are involved in the regulation of gene expression and modulation of signaling pathways through physical protein-protein interactions with transcription factors and signaling proteins [115]. Recent evidences suggest that Arabidopsis *RCI1A* gene which encodes a 14-3-3 psi isoform is a negative regulator of constitutive freezing tolerance and cold acclimation by controlling the gene expression of COR genes [116]. C2H2-type zinc finger protein showed an increased abundance in wheat plants subjected to short-term freezing stress (1 and 3 days at −5 °C) [42]. C2H2-type zinc finger proteins constitute a large family of transcription factors whose members are known to be involved in defense and acclimation to different environmental stressors including cold stress [117]. For example, constitutive overexpression of soybean C2H2-type zinc finger protein SCOF1, which is induced by low temperature and ABA but not by dehydration and salinity, elicited the expression of COR genes and increased tolerance to low temperature in non-acclimated Arabidopsis and tobacco plants [118].

5.4 Conclusion and Future Perspectives

Exposure to cold stress elicits profound changes in the proteome of plants. Proteomics studies have shown that the abundance of proteins associated with critical metabolic and biosynthetic pathways including photosynthesis, carbohydrate metabolism, energy metabolism, protein synthesis and processing, and ROS scavenging change to adjust to cold stress conditions. In particular, proteomic analyses have clearly demonstrated the accumulation of proteins with protective role against cold stress including dehydrins, WCS proteins, COR proteins, and LEA proteins in various crop species. Studying plant proteome in response to cold stress has greatly improved our understanding about the molecular mechanisms of cold acclimation. Despite its current limitations in the detection of low abundant proteins and PTMs, proteomics will continue to expand our knowledge regarding the molecular basis of cold acclimation. In-depth characterization of plant proteome in response to cold stress could lead to the identification of proteins that play critical roles in cold stress tolerance. Proteomics findings can therefore be used for the improvement of crop tolerance to cold stress through molecular breeding approaches. However, there is a large gap between the proteomics findings and their application for improving crop tolerance to low temperature that needs to be filled in the future.

References

1. Thomashow MF (1998) Role of cold-responsive genes in plant freezing tolerance. Plant Physiol 118:1–8
2. Chinnusamy V, Zhu J, Zhu JK (2007) Cold stress regulation of gene expression in plants. Trends Plant Sci 12:444–451
3. Lissarre M, Ohta M, Sato A, Miura K (2010) Cold-responsive gene regulation during cold acclimation in plants. Plant Signal Behav 5:948–952
4. Chinnusamy V, Zhu JK, Sunkar R (2010) Gene regulation during cold stress acclimation in plants. Methods Mol Biol 639:39–55
5. Janska A, Marsik P, Zelenkova S, Ovesna J (2010) Cold stress and acclimation—what is important for metabolic adjustment? Plant Biol (Stuttg) 12:395–405
6. Pearce RS (2001) Plant freezing and damage. Ann Bot 87:417–424
7. Atici O, Nalbantoglu B (2003) Antifreeze proteins in higher plants. Phytochemistry 64:1187–1196
8. Gharechahi J, Alizadeh H, Naghavi MR, Sharifi G (2014) A proteomic analysis to identify cold acclimation associated proteins in wild wheat (*Triticum urartu* L.). Mol Biol Rep 41:3897–3905
9. Sharma P, Sharma N, Deswal R (2005) The molecular biology of the low-temperature response in plants. BioEssays 27:1048–1059
10. Ouellet F, Charron J-B (2001) Cold acclimation and freezing tolerance in plants. In: eLS. Wiley, Chichester
11. Thomashow MF (1999) Plant cold acclimation: freezing tolerance genes and regulatory mechanisms. Annu Rev Plant Physiol Plant Mol Biol 50:571–599
12. Amasino R (2010) Seasonal and developmental timing of flowering. Plant J 61:1001–1013
13. Ream TS, Woods DP, Amasino RM (2012) The molecular basis of vernalization in different plant groups. Cold Spring Harb Symp Quant Biol 77:105–115
14. Winfield MO, Lu C, Wilson ID, Coghill JA, Edwards KJ (2010) Plant responses to cold: transcriptome analysis of wheat. Plant Biotechnol J 8:749–771
15. Yates JR, Ruse CI, Nakorchevsky A (2009) Proteomics by mass spectrometry: approaches, advances, and applications. Annu Rev Biomed Eng 11:49–79
16. Chen S, Harmon AC (2006) Advances in plant proteomics. Proteomics 6:5504–5516
17. Isaacson T, Damasceno CM, Saravanan RS, He Y, Catala C, Saladie M et al (2006) Sample extraction techniques for enhanced proteomic analysis of plant tissues. Nat Protoc 1:769–774
18. Damerval C, Devienne D, Zivy M, Thiellement H (1986) Technical improvements in two-dimensional electrophoresis increase the level of genetic-variation detected in wheat-seedling proteins. Electrophoresis 7:52–54
19. Hurkman WJ, Tanaka CK (1986) Solubilization of plant membrane proteins for analysis by two-dimensional gel electrophoresis. Plant Physiol 81:802–806
20. Vander Mijnsbrugge K, Meyermans H, Van Montagu M, Bauw G, Boerjan W (2000) Wood formation in poplar: identification, characterization, and seasonal variation of xylem proteins. Planta 210:589–598
21. Wang W, Scali M, Vignani R, Spadafora A, Sensi E, Mazzuca S et al (2003) Protein extraction for two-dimensional electrophoresis from olive leaf, a plant tissue containing high levels of interfering compounds. Electrophoresis 24:2369–2375
22. Sharifi G, Ebrahimzadeh H, Ghareyazie B, Gharechahi J, Vatankhah E (2012) Identification of differentially accumulated proteins associated with embryogenic and non-embryogenic calli in saffron (*Crocus sativus* L.). Proteome Sci 10:3
23. Saravanan RS, Rose JK (2004) A critical evaluation of sample extraction techniques for enhanced proteomic analysis of recalcitrant plant tissues. Proteomics 4:2522–2532
24. Ghosh D, Xu J (2014) Abiotic stress responses in plant roots: a proteomics perspective. Front Plant Sci 5:6

25. Washburn MP, Wolters D, Yates JR 3rd (2001) Large-scale analysis of the yeast proteome by multidimensional protein identification technology. Nat Biotechnol 19:242–247
26. Zhang Z, Wu S, Stenoien DL, Pasa-Tolic L (2014) High-throughput proteomics. Annu Rev Anal Chem (Palo Alto Calif) 7:427–454
27. Domon B, Aebersold R (2006) Mass spectrometry and protein analysis. Science 312:212–217
28. Pappin DJ, Hojrup P, Bleasby AJ (1993) Rapid identification of proteins by peptide-mass fingerprinting. Curr Biol 3:327–332
29. Aebersold R, Goodlett DR (2001) Mass spectrometry in proteomics. Chem Rev 101:269–295
30. Ong SE, Mann M (2005) Mass spectrometry-based proteomics turns quantitative. Nat Chem Biol 1:252–262
31. Neilson KA, Ali NA, Muralidharan S, Mirzaei M, Mariani M, Assadourian G et al (2011) Less label, more free: approaches in label-free quantitative mass spectrometry. Proteomics 11:535–553
32. Takahashi D, Kawamura Y, Uemura M (2013) Changes of detergent-resistant plasma membrane proteins in oat and rye during cold acclimation: association with differential freezing tolerance. J Proteome Res 12:4998–5011
33. Hlavackova I, Vitamvas P, Santrucek J, Kosova K, Zelenkova S, Prasil IT et al (2013) Proteins involved in distinct phases of cold hardening process in frost resistant winter barley (*Hordeum vulgare* L.) cv Luxor. Int J Mol Sci 14:8000–8024
34. Kollipara KP, Saab IN, Wych RD, Lauer MJ, Singletary GW (2002) Expression profiling of reciprocal maize hybrids divergent for cold germination and desiccation tolerance. Plant Physiol 129:974–992
35. Neilson KA, Mariani M, Haynes PA (2011) Quantitative proteomic analysis of cold-responsive proteins in rice. Proteomics 11:1696–1706
36. Lee DG, Ahsan N, Lee SH, Kang KY, Lee JJ, Lee BH (2007) An approach to identify cold-induced low-abundant proteins in rice leaf. C R Biol 330:215–225
37. Cui S, Huang F, Wang J, Ma X, Cheng Y, Liu J (2005) A proteomic analysis of cold stress responses in rice seedlings. Proteomics 5:3162–3172
38. Yan SP, Zhang QY, Tang ZC, Su WA, Sun WN (2006) Comparative proteomic analysis provides new insights into chilling stress responses in rice. Mol Cell Proteomics 5:484–496
39. Gammulla CG, Pascovici D, Atwell BJ, Haynes PA (2011) Differential proteomic response of rice (*Oryza sativa*) leaves exposed to high- and low-temperature stress. Proteomics 11:2839–2850
40. Hashimoto M, Komatsu S (2007) Proteomic analysis of rice seedlings during cold stress. Proteomics 7:1293–1302
41. Vitamvas P, Saalbach G, Prasil IT, Capkova V, Opatrna J, Ahmed J (2007) WCS120 protein family and proteins soluble upon boiling in cold-acclimated winter wheat. J Plant Physiol 164:1197–1207
42. Han Q, Kang G, Guo T (2013) Proteomic analysis of spring freeze-stress responsive proteins in leaves of bread wheat (*Triticum aestivum* L.). Plant Physiol Biochem 63:236–244
43. Rinalducci S, Egidi MG, Karimzadeh G, Jazii FR, Zolla L (2011) Proteomic analysis of a spring wheat cultivar in response to prolonged cold stress. Electrophoresis 32:1807–1818
44. Xu J, Li Y, Sun J, Du L, Zhang Y, Yu Q et al (2013) Comparative physiological and proteomic response to abrupt low temperature stress between two winter wheat cultivars differing in low temperature tolerance. Plant Biol (Stuttg) 15:292–303
45. Rinalducci S, Egidi MG, Mahfoozi S, Godehkahriz SJ, Zolla L (2011) The influence of temperature on plant development in a vernalization-requiring winter wheat: a 2-DE based proteomic investigation. J Proteomics 74:643–659
46. Sarhadi E, Mahfoozi S, Hosseini SA, Salekdeh GH (2010) Cold acclimation proteome analysis reveals close link between the up-regulation of low-temperature associated proteins and vernalization fulfillment. J Proteome Res 9:5658–5667

47. Herman EM, Rotter K, Premakumar R, Elwinger G, Bae H, Ehler-King L et al (2006) Additional freeze hardiness in wheat acquired by exposure to −3 °C is associated with extensive physiological, morphological, and molecular changes. J Exp Bot 57:3601–3618
48. Vitamvas P, Prasil IT, Kosova K, Planchon S, Renaut J (2012) Analysis of proteome and frost tolerance in chromosome 5A and 5B reciprocal substitution lines between two winter wheats during long-term cold acclimation. Proteomics 12:68–85
49. Kosova K, Vitamvas P, Planchon S, Renaut J, Vankova R, Prasil IT (2013) Proteome analysis of cold response in spring and winter wheat (*Triticum aestivum*) crowns reveals similarities in stress adaptation and differences in regulatory processes between the growth habits. J Proteome Res 12:4830–4845
50. Lee DG, Ahsan N, Lee SH, Lee JJ, Bahk JD, Kang KY et al (2009) Chilling stress-induced proteomic changes in rice roots. J Plant Physiol 166:1–11
51. Imin N, Kerim T, Weinman JJ, Rolfe BG (2006) Low temperature treatment at the young microspore stage induces protein changes in rice anthers. Mol Cell Proteomics 5:274–292
52. Janmohammadi M, Mock HP, Matros A (2014) Proteomic analysis of cold acclimation in winter wheat under field conditions. Icel Agric Sci 27:3–15
53. Janmohammadi M, Zolla L, Rinalducci S (2015) Low temperature tolerance in plants: Changes at the protein level. Phytochemistry 117:76–89
54. Danyluk J, Kane NA, Breton G, Limin AE, Fowler DB, Sarhan F (2003) TaVRT-1, a putative transcription factor associated with vegetative to reproductive transition in cereals. Plant Physiol 132:1849–1860
55. Renaut J, Hausman JF, Wisniewski ME (2006) Proteomics and low-temperature studies: bridging the gap between gene expression and metabolism. Physiol Plant 126:97–109
56. Wisniewski M, Webb R, Balsamo R, Close TJ, Yu XM, Griffith M (1999) Purification, immunolocalization, cryoprotective, and antifreeze activity of PCA60: a dehydrin from peach (*Prunus persica*). Physiol Plant 105:600–608
57. Cheng LB, Gao X, Li SY, Shi MJ, Javeed H, Jing XM et al (2010) Proteomic analysis of soybean [*Glycine max* (L.) Meer.] seeds during imbibition at chilling temperature. Mol Breeding 26:1–17
58. Balbuena TS, Salas JJ, Martinez-Force E, Garces R, Thelen JJ (2011) Proteome analysis of cold acclimation in sunflower. J Proteome Res 10:2330–2346
59. Bestel-Corre G, Dumas-Gaudot E, Poinsot V, Dieu M, Dierick JF, Van TD et al (2002) Proteome analysis and identification of symbiosis-related proteins from Medicago truncatula Gaertn. by two-dimensional electrophoresis and mass spectrometry. Electrophoresis 23:122–137
60. Kobayashi F, Takumi S, Nakata M, Ohno R, Nakamura T, Nakamura C (2004) Comparative study of the expression profiles of the Cor/Lea gene family in two wheat cultivars with contrasting levels of freezing tolerance. Physiol Plant 120:585–594
61. Fowler DB, Chauvin LP, Limin AE, Sarhan F (1996) The regulatory role of vernalization in the expression of low-temperature-induced genes in wheat and rye. Theor Appl Genet 93:554–559
62. Tunnacliffe A, Wise MJ (2007) The continuing conundrum of the LEA proteins. Naturwissenschaften 94:791–812
63. Kosova K, Vitamvas P, Prasil IT (2007) The role of dehydrins in plant response to cold. Biol Plant 51:601–617
64. Houde M, Daniel C, Lachapelle M, Allard F, Laliberte S, Sarhan F (1995) Immunolocalization of freezing-tolerance-associated proteins in the cytoplasm and nucleoplasm of wheat crown tissues. Plant J 8:583–593
65. Danyluk J, Perron A, Houde M, Limin A, Fowler B, Benhamou N et al (1998) Accumulation of an acidic dehydrin in the vicinity of the plasma membrane during cold acclimation of wheat. Plant Cell 10:623–638
66. Hara M, Terashima S, Fukaya T, Kuboi T (2003) Enhancement of cold tolerance and inhibition of lipid peroxidation by citrus dehydrin in transgenic tobacco. Planta 217:290–298

67. Chardon F, Damerval C (2005) Phylogenomic analysis of the PEBP gene family in cereals. J Mol Evol 61:579–590
68. Bertini L, Proietti S, Caporale C, Caruso C (2009) Molecular characterization of a wheat protein induced by vernalisation. Protein J 28:253–262
69. Yong WD, Xu YY, Xu WZ, Wang X, Li N, Wu JS et al (2003) Vernalization-induced flowering in wheat is mediated by a lectin-like gene VER2. Planta 217:261–270
70. Xing LJ, Li J, Xu YY, Xu ZH, Chong K (2009) Phosphorylation modification of wheat lectin VER2 is associated with vernalization-induced O-GlcNAc signaling and intracellular motility. PLoS ONE 4:e4854
71. Kim JY, Park SJ, Jang B, Jung CH, Ahn SJ, Goh CH et al (2007) Functional characterization of a glycine-rich RNA-binding protein 2 in *Arabidopsis thaliana* under abiotic stress conditions. Plant J 50:439–451
72. Kim JS, Park SJ, Kwak KJ, Kim YO, Kim JY, Song J et al (2007) Cold shock domain proteins and glycine-rich RNA-binding proteins from *Arabidopsis thaliana* can promote the cold adaptation process in *Escherichia coli*. Nucleic Acids Res 35:506–516
73. Griffith M, Yaish MW (2004) Antifreeze proteins in overwintering plants: a tale of two activities. Trends Plant Sci 9:399–405
74. Gaudet DA, Laroche A, Frick M, Davoren J, Puchalski B, Ergon A (2000) Expression of plant defence-related (PR-protein) transcripts during hardening and dehardening of winter wheat. Physiol Mol Plant Pathol 57:15–24
75. Miller G, Shulaev V, Mittler R (2008) Reactive oxygen signaling and abiotic stress. Physiol Plant 133:481–489
76. Suzuki N, Koussevitzky S, Mittler R, Miller G (2012) ROS and redox signalling in the response of plants to abiotic stress. Plant, Cell Environ 35:259–270
77. Apel K, Hirt H (2004) Reactive oxygen species: metabolism, oxidative stress, and signal transduction. Annu Rev Plant Biol 55:373–399
78. Alscher RG, Erturk N, Heath LS (2002) Role of superoxide dismutases (SODs) in controlling oxidative stress in plants. J Exp Bot 53:1331–1341
79. Asada K (1992) Ascorbate peroxidase—a hydrogen peroxide-scavenging enzyme in plants. Physiol Plant 85:235–241
80. Dixon DP, Lapthorn A, Edwards R (2002) Plant glutathione transferases. Genome Biol 3: REVIEWS3004
81. Singla-Pareek SL, Reddy MK, Sopory SK (2003) Genetic engineering of the glyoxalase pathway in tobacco leads to enhanced salinity tolerance. Proc Natl Acad Sci USA 100:14672–14677
82. Vieira Dos Santos C, Rey P (2006) Plant thioredoxins are key actors in the oxidative stress response. Trends Plant Sci 11:329–334
83. Dietz KJ, Jacob S, Oelze ML, Laxa M, Tognetti V, De Miranda SM et al (2006) The function of peroxiredoxins in plant organelle redox metabolism. J Exp Bot 57:1697–1709
84. Woo EJ, Dunwell JM, Goodenough PW, Marvier AC, Pickersgill RW (2000) Germin is a manganese containing homohexamer with oxalate oxidase and superoxide dismutase activities. Nat Struct Biol 7:1036–1040
85. Yang PF, Li XJ, Liang Y, Jing YX, Shen SH, Kuang TY (2006) Proteomic analysis of the response of Liangyoupeijiu (super high-yield hybrid rice) seedlings to cold stress. J Integr Plant Biol 48:945–951
86. Dumont E, Bahrman N, Goulas E, Valot B, Sellier H, Hilbert JL et al (2011) A proteomic approach to decipher chilling response from cold acclimation in pea (*Pisum sativum* L.). Plant Sci 180:86–98
87. Otero AS (2000) NM23/nucleoside diphosphate kinase and signal transduction. J Bioenerg Biomembr 32:269–275
88. Tang L, Kim MD, Yang KS, Kwon SY, Kim SH, Kim JS et al (2008) Enhanced tolerance of transgenic potato plants overexpressing nucleoside diphosphate kinase 2 against multiple environmental stresses. Transgenic Res 17:705–715

89. Hosseini SA, Gharechahi J, Heidari M, Koobaz P, Abdollahi S, Mirzaei M et al (2015) Comparative proteomic and physiological characterisation of two closely related rice genotypes with contrasting responses to salt stress. Funct Plant Biol 42:527
90. Tyystjärvi E (2012) Photoinhibition of photosystem II. Int Rev Cell Mol Biol 300:243–303
91. Miura K, Furumoto T (2013) Cold signaling and cold response in plants. Int J Mol Sci 14:5312–5337
92. Takahashi S, Murata N (2008) How do environmental stresses accelerate photoinhibition? Trends Plant Sci 13:178–182
93. Gusta LV, Wisniewski M (2013) Understanding plant cold hardiness: an opinion. Physiol Plant 147:4–14
94. Soitamo AJ, Piippo M, Allahverdiyeva Y, Battchikova N, Aro EM (2008) Light has a specific role in modulating Arabidopsis gene expression at low temperature. BMC Plant Biol 8:13
95. Ndong C, Danyluk J, Huner NP, Sarhan F (2001) Survey of gene expression in winter rye during changes in growth temperature, irradiance or excitation pressure. Plant Mol Biol 45:691–703
96. Hajirezaei MR, Peisker M, Tschiersch H, Palatnik JF, Valle EM, Carrillo N et al (2002) Small changes in the activity of chloroplastic NADP(+)-dependent ferredoxin oxidoreductase lead to impaired plant growth and restrict photosynthetic activity of transgenic tobacco plants. Plant J 29:281–293
97. Palatnik JF, Tognetti VB, Poli HO, Rodriguez RE, Blanco N, Gattuso M et al (2003) Transgenic tobacco plants expressing antisense ferredoxin-NADP(H) reductase transcripts display increased susceptibility to photo-oxidative damage. Plant J 35:332–341
98. Feller U, Anders I, Demirevska K (2008) Degradation of rubisco and other chloroplast proteins under abiotic stress. Gen Appl Plant Physiol 1:5–18
99. Reddy AR, Chaitanya KV, Vivekanandan M (2004) Drought-induced responses of photosynthesis and antioxidant metabolism in higher plants. J Plant Physiol 161:1189–1202
100. Raines CA (2003) The Calvin cycle revisited. Photosynth Res 75:1–10
101. Lee H, Guo Y, Ohta M, Xiong L, Stevenson B, Zhu JK (2002) LOS2, a genetic locus required for cold-responsive gene transcription encodes a bi-functional enolase. EMBO J 21:2692–2702
102. Pego JV, Smeekens SC (2000) Plant fructokinases: a sweet family get-together. Trends Plant Sci 5:531–536
103. Minarik P, Tomaskova N, Kollarova M, Antalik M (2002) Malate dehydrogenases—structure and function. Gen Physiol Biophys 21:257–265
104. Tunc-Ozdemir M, Miller G, Song L, Kim J, Sodek A, Koussevitzky S et al (2009) Thiamin confers enhanced tolerance to oxidative stress in Arabidopsis. Plant Physiol 151:421–432
105. Wang W, Vinocur B, Shoseyov O, Altman A (2004) Role of plant heat-shock proteins and molecular chaperones in the abiotic stress response. Trends Plant Sci 9:244–252
106. Demirevska-Kepova K, Holzer R, Simova-Stoilova L, Feller U (2005) Heat stress effects on ribulose-1,5-bisphosphate carboxylase/oxygenase, Rubisco binding protein and Rubisco activase in wheat leaves. Biol Plant 49:521–525
107. Nishimura K, Ogawa T, Ashida H, Yokota A (2008) Molecular mechanisms of RuBisCO biosynthesis in higher plants. Plant Biotechnol 25:285–290
108. Krishna P, Gloor G (2001) The Hsp90 family of proteins in *Arabidopsis thaliana*. Cell Stress Chaperones 6:238–246
109. Lee GJ, Vierling E (2000) A small heat shock protein cooperates with heat shock protein 70 systems to reactivate a heat-denatured protein. Plant Physiol 122:189–198
110. Ma Y, Hendershot LM (2004) ER chaperone functions during normal and stress conditions. J Chem Neuroanat 28:51–65
111. Houston NL, Fan C, Xiang JQ, Schulze JM, Jung R, Boston RS (2005) Phylogenetic analyses identify 10 classes of the protein disulfide isomerase family in plants, including single-domain protein disulfide isomerase-related proteins. Plant Physiol 137:762–778

112. Hernandez G, Vazquez-Pianzola P (2005) Functional diversity of the eukaryotic translation initiation factors belonging to eIF4 families. Mech Dev 122:865–876
113. Rossi D, Kuroshu R, Zanelli CF, Valentini SR (2014) eIF5A and EF-P: two unique translation factors are now traveling the same road. Wiley Interdiscip Rev RNA 5:209–222
114. Xu JY, Zhang BL, Jiang CH, Ming F (2011) RceIF5A, encoding an eukaryotic translation initiation factor 5A in Rosa chinensis, can enhance thermotolerance, oxidative and osmotic stress resistance of *Arabidopsis thaliana*. Plant Mol Biol 75:167–178
115. Roberts MR (2003) 14-3-3 proteins find new partners in plant cell signalling. Trends Plant Sci 8:218–223
116. Catala R, Lopez-Cobollo R, Mar Castellano M, Angosto T, Alonso JM, Ecker JR et al (2014) The Arabidopsis 14-3-3 protein RARE COLD INDUCIBLE 1A links low-temperature response and ethylene biosynthesis to regulate freezing tolerance and cold acclimation. Plant Cell 26:3326–3342
117. Ciftci-Yilmaz S, Mittler R (2008) The zinc finger network of plants. Cell Mol Life Sci 65:1150–1160
118. Kim JC, Lee SH, Cheong YH, Yoo CM, Lee SI, Chun HJ et al (2001) A novel cold-inducible zinc finger protein from soybean, SCOF-1, enhances cold tolerance in transgenic plants. Plant J 25:247–259
119. Han Q, Kang G, Guo T (2013) Proteomic analysis of spring freeze-stress responsive proteins in leaves of bread wheat (*Triticum aestivum* L.). Plant Physiol Biochem 63:236–244
120. Herman EM, Rotter K, Premakumar R, Elwinger G, Bae R, Ehler-King L et al (2006) Additional freeze hardiness in wheat acquired by exposure to −3 °C is associated with extensive physiological, morphological, and molecular changes. J Exp Bot 57:3601–3618
121. Gharechahi J, Alizadeh H, Naghavi MR, Sharifi G (2014) A proteomic analysis to identify cold acclimation associated proteins in wild wheat (*Triticum urartu* L.). Mol Biol Rep 41:3897–3905
122. Yan S-P, Zhang Q-Y, Tang Z-C, Su W-A, Sun W-N (2006) Comparative proteomic analysis provides new insights into chilling stress responses in rice. Mol Cell Proteomics 5:484–496
123. Lee D-G, Ahsan N, Lee S-H, Lee JJ, Bahk JD, Kang KY et al (2009) Chilling stress-induced proteomic changes in rice roots. J Plant Physiol 166:1–11
124. Cui S, Huang F (2005) A proteomic analysis of cold stress responses in rice seedlings. Proteomics 5:3162–3172
125. Lee D-G, Ahsan N, Lee S-H, Kang KY, Lee JJ, Lee B-H (2007) An approach to identify cold-induced low-abundant proteins in rice leaf. CR Biol 330:215–225
126. Neilson KA, Mariani M, Haynes PA (2011) Quantitative proteomic analysis of cold-responsive proteins in rice. Proteomics 11:1696–1706
127. Kollipara KP, Saab IN, Wych RD, Lauer MJ, Singletary GW (2002) Expression profiling of reciprocal maize hybrids divergent for cold germination and desiccation tolerance. Plant Physiol 129:974–992
128. Cheng L, Gao X, Li S, Shi M, Javeed H, Jing X et al (2010) Proteomic analysis of soybean [*Glycine max* (L.) Meer.] seeds during imbibition at chilling temperature. Mol Breeding 26:1–17
129. Balbuena TS, Salas JJ, Martínez-Force E, Garcés R, Thelen JJ (2011) Proteome analysis of cold acclimation in sunflower. J Proteome Res 10:2330–2346
130. Dumont E, Bahrman N, Goulas E, Valot B, Sellier H, Hilbert J-L et al (2011) A proteomic approach to decipher chilling response from cold acclimation in pea (*Pisum sativum* L.). Plant Sci 180:86–98

Chapter 6
How Proteomics Contributes to Our Understanding of Drought Tolerance

Mehdi Mirzaei, Yunqi Wu, Paul Worden, Ante Jerkovic and Brian J. Atwell

Abstract Drought is a ubiquitous threat to plant performance, whether in crops or natural ecosystems. Acclimation responses to drought have been loosely categorised as escape, avoidance or tolerance, with a gradual reversal of these events once water is re-supplied. Many analytical disciplines have been brought to bear on plant organs and tissues experiencing water deficit, including biophysics, physiology and the various -omics technologies. One such technique, proteomics, is supported by an ever-expanding array of technologies that have evolved from two-dimensional gel electrophoresis and sequencing to various label and label-free techniques that yield progressively deeper insights into gene expression. These experiments reveal major changes in abundance of ABA-responsive proteins, heat shock proteins and proteins involved in defence against oxidative damage. More surprising is the commonly reported changes in proteins participating in energy metabolism and down-regulation of photosynthetic proteins. Such experiments have been greatly enabled by the exploitation of known genetic variation in 'drought tolerance' in cereals, with the acknowledgement that yet more drought resistance mechanisms are certain to exist in unrelated arid-zone species. The next step will be to interrogate large-scale data sets and construct gene networks (interactomes) for a more meaningful understanding of the drought phenotype.

Keywords Abiotic stress · Cereals · Drought tolerance · Quantitative proteomics · Rice

---

Mehdi Mirzaei, and Yunqi Wu have equally contributed in this work.

M. Mirzaei · Y. Wu · P. Worden · A. Jerkovic
Department of Chemistry and Biomolecular Sciences, Macquarie University, Sydney, NSW, Australia

B.J. Atwell (✉)
Department of Biological Sciences, Macquarie University, Sydney, NSW, Australia
e-mail: brian.atwell@mq.edu.au

G.H. Salekdeh (ed.), *Agricultural Proteomics Volume 2*,
DOI 10.1007/978-3-319-43278-6_6

6.1 Introduction

An adequate water supply is essential for cell expansion and all stages of plant development. Increasing oscillations in climatic conditions and rising demand for food call for a deeper understanding of the mechanisms of drought tolerance in the dominant world food crops [1]. While crop plants have a finite capacity to acclimate to water deficits without manifesting substantial yield reductions, there is abundant evidence for ecotypes adapted to arid regions surviving during long-term drought. This raises the question of whether the genes expressed in drought-tolerant genotypes are targets for yield improvement during transient exposure to drought. Plants have evolved multiple mechanisms at the morphological, physiological, cell and molecular levels that all contribute to reduced drought stress. The drought-response mechanism in crop species can notionally be divided into four basic strategies. These are neither mutually exclusive nor easily defined conceptually. However, even though they have fallen from common usage in ecophysiology, these terms might form a useful framework for design and analysis of proteomics experiments. They are: drought escape (DE), drought avoidance (DA), drought tolerance (DT), and drought recovery (DR) [2–6] and will be more carefully defined below (see Fig. 6.1).

How plants survive and resist drought is governed by a vast array of traits. The specific suite of traits in any plant community will depend upon the drought regime (e.g. time courses, intensity, soil type) and thus the precise nature of the selective pressure on the plant population. For instance, plants from extremely arid places might have 'sophisticated' mechanisms to survive sustained and extreme drought

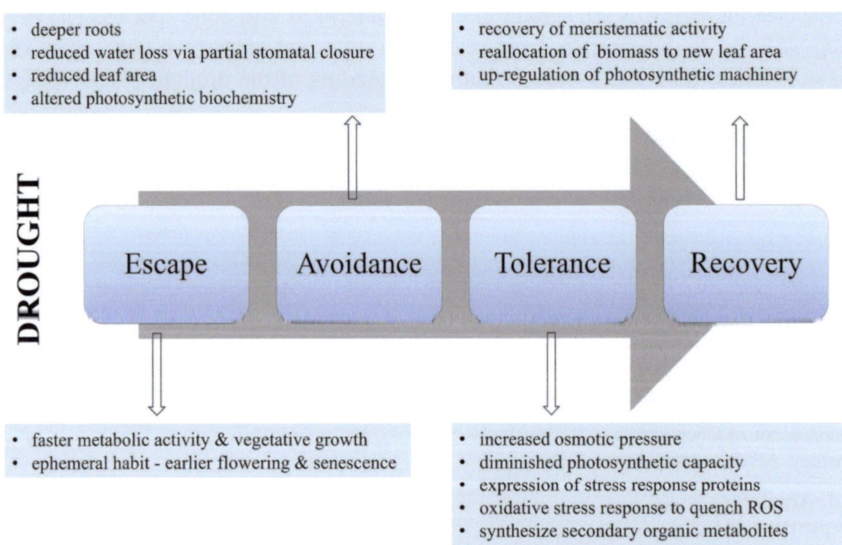

Fig. 6.1 Some common plant responses to drought and recovery after re-watering

but these traits might not be operative in a crop species subjected to transient drying regimes. Further complexity comes from key drought-responsive traits being expressed in either shoots or roots, thus necessitating tissue-specific sampling for proteomics experiments. Common traits that play a vital role in survival and growth during droughts are: osmotic adjustment and generation of organic osmotic agents, hydraulics, rate of ABA accumulation and ABA responsiveness, stomatal behaviour, leaf orientation and reflectance, root response to drought and root-to-shoot signalling, and biomass partitioning. The wealth of physiological data on drought responses in many ecosystems and taxa should allow us to target traits of interest and identify the gene expression changes that characterise trait changes under drought. This requires a systematic alignment of physiological observations with gene expression analysis. This chapter is aimed at highlighting some key proteomic observations in three major cereal species (rice, wheat and barley), with a view to linking them to physiological drought responses.

Drought escape: *the ecological strategy by which plants modify their life cycle to evade times of most severe water shortage.* Plants employ drought escape to cope with restricted and unpredictable water availability. For such species, high metabolic activity and rapid growth and earlier onset of flowering are hypothesized to confer a fitness advantage, enabling plants to complete their life cycle before the most intense period of drought [7]. Raorane et al. [8] studied the effects of drought stress on two contrasting genotypes of rice-$qDTY_{12.1}$ 481-B (drought tolerant) and Vandana. They found that in general, the total number of proteins was greater in tissues from 481-B than in Vandana under drought stress. This may be related to 481-B having more vigorous growth, more lateral root branching and more meristematic tissue than Vandana. Accordingly, cytoskeleton proteins including actins, tubulins and expansins, which are highly expressed in actively growing roots and root hairs, were more abundant in 481-B. Key proteins identified in carbohydrate metabolism, such as adenylyltransferase and sucrose synthase, were found to be more severely down-regulated in Vandana than in 481-B. This suggests that 481-B has stronger source and sink activity, which may result in earlier completion of its life cycle before severe drought takes effect.

Drought avoidance: *the capacity of plants to evade the direct impact of physiological drought by morphological or metabolic acclimation.* A prime example is increased depth of rooting, morphological and physiological changes in leaves, and adjustments to photosynthetic pathways, such as a shift to C4 or CAM metabolism. Rooting depth, volume and density are mainly influenced by the distribution of soil moisture and fertility through the profile. In the case of soil water deficit, plants dynamically adapt and modify their root system by changing their root growth in different manners [9–11]. Mirzaei et al. [12] used proteomics to infer that drought signal(s) can be transported remotely from part of a root system under drought stress to the other half of the root system that is well-watered. Such long-distance drought signals resulted in down regulation of tubulins, while chitinases, which have been reported to play a role in altering root system architecture in response to various environmental conditions, were up-regulated. Another drought avoidance strategy is morphological and physiological changes to leaves, reducing water loss and enhancing water use efficiency. Shorter-term responses to leaf water deficit include the partial closure of stomata to reduce transpiration. Events leading to stomatal closure might also be registered through the proteome but this remains less clear than the changes that lead to longer-term acclimation. The proteomic profile of isolated guard cells and analysis of xylem sap might provide clues to the gene expression responses under sustained drought. For example, a protein similar to coronatine-insensitive 1 (COI1), which was thought to be involved in stomatal closure, was found to be up-regulated in a drought-adapted rice cultivar NSG19 after osmotic stress treatment. In addition, expression of a potassium transporter also increased in response to osmotic stress [13]. Further interacting protein partners that play a role is regulation of stomatal behaviour are likely to be discovered when highly reproducible proteomic techniques (protein–protein interaction) can be applied to ever smaller sample sizes.

Drought tolerance: *the acclimation of plants to drought through changes in gene or protein expression that enable metabolism to proceed in tissues with low water potentials*. Osmotic adjustment and antioxidant defence systems are the most common mechanisms associated with drought tolerance in plants. Plants accumulate a variety of organic and inorganic substances (such as sugars, polyols, amino acids, alkaloids, and inorganic ions) to increase the concentration of solutes and thereby reduce the osmotic potential to maintain the gradient of cell water potential during water stress. This phenomenon is defined as osmotic adjustment [14] but is generally at the cost of carbon skeletons, therefore imposing a growth penalty in higher plants. Proteomic analysis of rice leaves indicates indirectly that rice also utilizes osmotic adjustment to adapt to drought environments. Mirzaei et al. [15] identified nine aquaporins by imposing different drought stress regimes to rice. Aquaporins are selective water transport channels in plant cells and regulate the rapid transmembrane transport of water during the processes of seed germination, cell elongation, stomatal movements and abiotic stress responses [16, 17]. In addition to aquaporins, late embryogenesis abundant (LEA) proteins were identified by proteomics during drought. This family of proteins is usually associated with developing seeds but was found in droughted leaves under dry conditions. Ke et al. [18] also reported that LEA proteins were up-regulated in leaves when rice seedlings were exposed to drought. LEAs are hyperhydrophilic proteins with extremely high thermal stability and they can remain in the aqueous state even under boiling conditions. LEA proteins can protect biological macromolecules, redirect intracellular water distribution and bind to inorganic ions to avoid the damage attributed to the accumulation of high concentrations of ions under drought stress conditions. This ultimately prevents excessive dehydration of plant tissues, and controls the expression of other genes by binding to nucleic acids [19]. Thus their expansive roles in plants is likely to be elucidated by tissue-specific proteomic experiments, allowing tissue function to be assigned more accurately to the 'dehydration response'.

Oxidative stress commonly occurs along with drought stress. Antioxidant defence systems are one of the tolerance mechanisms in plants to counter dry conditions. Not surprisingly, rice has evolved an antioxidant defence system to protect itself from polymer damage during drought stress. By using proteomics, Liu et al. [20] discovered that rice plants express different peroxidases during drought stress. Peroxidases are the protective enzymes that can protect cells against damage from excessive reactive oxygen species (ROS) which if left unchecked is a large part of plant abiotic stress impact. Thus, through proteomics, we have seen that rice tissues can mount an antioxidant defence to enhance drought tolerance.

Drought recovery: *the capacity of plants to resume normal metabolism, growth and development after an episode of drought stress.* Reversing the effects of drought, even when plants have acclimated to it by avoidance or tolerance, is critically important to the drought-affected plant and therefore an important field of proteomic study. The resulting data hopefully provide *bona fide* evidence that the genes postulated to acclimate the plant to drought are a direct response to the stress event if they are equally down-regulated during re-watering. Many proteomic studies have shown that rice can resume normal metabolism when re-watered after moderate but non-lethal drought. Mirzaei et al. [15] noticed that proteins induced by drought stress, like stress-responsive proteins, signalling and transport proteins, were repressed by re-watering. In a different study, Salekdeh et al. [21] found that all proteins, differentially expressed in rice during water deficit could completely or largely recover to the abundance of the well-watered control by 10 days after re-watering, whether the rice was a lowland or upland variety.

Drought resistance in plants is highly polygenic. By using proteomics, we can identify novel elements of the physiological response to drought when careful sampling is combined with tightly controlled drought treatments. However, our understanding of the regulatory networks and the crosstalk between the signalling pathways under drought-prone conditions is still fragmentary. We have a long way to go to comprehensively and intensively understand the biological functions related to drought response, and draw a clear picture of the complete regulatory network.

6.2 Rice and Water Deficits

Rice is one of the most widely consumed staple foods with around half of the world's population dependent on it for survival. As a model plant, it is also a surprisingly powerful experimental tool because of its vast and well-documented genetic diversity, including a degree of drought tolerance in some upland landraces and hybrids. In recent years, scientists have identified numerous proteins involved in drought response of rice, and have made substantial progress in the genetic improvement of drought resistance. Drought and rising temperatures are a major concern for rice crops in particular, as they are sensitive to drought [22]. These abiotic events are among the most significant environmental factors affecting rice yield worldwide. Despite improved technology in forecasting drought, it is an unavoidable feature of our global climate and thus mitigation through breeding and biotechnology is a high priority. The small genome, well documented biodiversity and gene synteny between rice and other important cereals makes it an ideal model for crop research. There is now substantial literature on quantitative proteomic analysis of drought stress in rice, pointing to effects on metabolic and regulatory networks that link defence against stress to development, growth, and reproduction. The following paragraphs describe recent experimental methods, proteomic approaches and findings related to drought stress in rice.

6.2.1 Panicle Exsertion and Spikelet Sterility

Intermittent drought causes the failure of panicle exsertion and increases spikelet sterility in rice [23]. The rice cultivar IR64 was grown to maturity in liberally watered 3-kg pots of fine-textured soil until three days before heading, when drought stress was imposed by removing standing water and withholding water for another three days, followed by re-watering. Protein was extracted from the peduncle, followed by separation using 2-DE (2-dimensional gel electrophoresis). This approach reproducibly resolved over 500 protein spots with 31 spots showing significant differential expression under the simulated drought conditions. These drought-induced proteins were identified by MALDI-TOF mass spectrometry. Among the drought-responsive proteins, six of them were ABA-responsive, such as group 6 LEA, LEA type 1 protein and ABA/WDS-induced protein. Induction of LEA proteins is associated with a rise in the level of ABA and growth retardation. This may be connected with the observation that peduncle growth arrests during drought. Another protein putatively connected with cell elongation and cell wall biosynthesis is xyloglucan endotransglycosylase (XET), and it was down-regulated by drought cycles. In addition to ABA-responsive proteins and xyloglucan endotransglycosylase, S-adenosylmethionine (SAM) synthetases were found to be an essential group affecting peduncle growth. Vascular tissues are highly lignified and the biosynthesis of lignin consumes SAM. As a result of the down-regulation of SAM synthetases under water deficit, the availability of SAM for lignin biosynthesis is constrained and consequently impairment of peduncle growth is seen.

To further investigate the proteomic response to abiotic stress in spikelets, Jagadish et al. [24] imposed a combination of water-deficit with high temperature stress on an indica rice (cv. Nagina N22). Total soluble protein extracted from spikelets were analysed by 2-DE and then identified by MALDI-TOF mass spectrometry. Approximately 500 protein spots were revealed by 2-DE, of which 29 spots showed differential expression to water deficit, heat, and/or combined stress. Mass spectrometry identified pollen allergens, low-molecular-weight heat shock proteins (HSPs), beta expansin, soluble inorganic pyrophosphatase, putative fructokinase and two unknown proteins. The expression levels of these proteins was characteristic of water deficit, heat stress and/or combined stress. However, heat shock proteins had much higher expression levels under combined stress than with independent water deficit and heat stress. Further proteomic studies are essential to consolidate the conclusions drawn from the 2-DE expression data, which are complex and require cautious interpretation.

Crop yield is most sensitive to water deficits during the reproductive stage and the most sensitive yield component is spikelet fertility. To discover the mechanism by which rice responds to drought during the reproductive stage, Liu and Bennet [20] examined the effect of drought on the anther proteome of two contrasting rice genotypes to water stress. Fresh anthers from the top of four rachis branches at various stages in IR64 (drought sensitive) and Moroberekan (tolerant), or mature pollen grains at anthesis after drought/re-watering in IR64 were collected. Protein from these tissues were extracted and separated by 2-DE. Ninety-three proteins were observed to change in abundance in the drought stressed plants across two rice genotypes. Upon re-watering, expression levels of 24 of these protein spots were unchanged in both genotypes. Illustrating the cultivar contrast, sixty protein spots were unchanged in IR64 but returned to control levels in Moroberekan after re-watering, while only nine protein spots maintained their abundance in Moroberekan when the homologous proteins in IR64 returned to control levels. The most interesting candidates are the 14 drought-induced proteins that were novel in IR64 and did not change on re-watering. By contrast, of the 13 drought-induced proteins in Moroberekan, ten returned to control levels on re-watering, including six drought-induced protein spots that were not reversed by re-watering in IR64. Such a proteomic analysis provides some insights into drought responses, specifically showing that the drought-tolerant genotype, Moroberekan, recovered better following re-watering at the anther proteome level than the drought-sensitive genotype IR64.

6.2.2 Panicle Proteome

The panicle proteomes of two upland rice varieties exposed to hyperosmotic stress were investigated by Huang et al. [25]. Seedlings of the two varieties, Zhonghan 3 and IR29 were grown to the sixth stage of panicle primordial differentiation. They were then treated with 15 % PEG 6000 for 10 days followed by cultivation in Kimura B solution until harvest. Proteins were extracted from panicle tissue followed by 2-DE separation. Over 800 protein spots were reproducibly detected. Of these spots, 38 showed significant changes in abundance under drought stress. The proteins were analysed by MALDI-TOF mass spectrometry, which identified 22 out of the 38 differentially expressed proteins. The identified proteins were categorised into four functional groups. They include (1) energy production, (2) metabolism, (3) stress defence and (4) protein transport. Five of the identified proteins were related to energy production. Among them, three were involved in glycolysis and the other two were involved in photosynthesis. Proteins involved in metabolism include mitochondrial glycine decarboxylase, arginase, UDP-glucose-6-dehydrogenase (UGDH) and UDP-D-glucuronic acid decarboxylase (UXS). Stress defence proteins were the major group of differentially expressed proteins under osmotic stress. These identified proteins were HSP70, a glyoxalase 1 and five ascorbate peroxidases (APX's), three APX1s and two APX2s. Importin was the only protein identified in the protein transport category. While the proteins identified are functionally disparate, this study highlights the importance in maintaining the energy supply and removing ROS under hyperosmotic stress conditions.

6.2.3 Grain Filling and Inferior Spikelets

The mechanism of rice grain filling and the response of inferior spikelets to drought stress was investigated by Dong et al. [26]. Rice seedlings were exposed to severe drought. Panicles were harvested 7 and 14 days after anthesis then iTRAQ labelling was performed followed by LC-MS/MS analysis. Proteins involved in starch biosynthesis showed impaired activity under drought stress. The low activities of ADP-glucose pyrophosphorylase, granule-bound starch synthase, sucrose synthase (SuSase) and other starch biosynthesis-related proteins in inferior spikelets were the likely cause of the slow grain filling rate under drought stress. Impaired grain-filling is ascribed to inefficient starch biosynthesis, possibly enhancing the concentration of soluble carbohydrates. In contrast to starch biosynthesis, various stress responses were enhanced under excessive water stresses.

6.2.4 Yield

Raorane et al. [8] studied the effects of drought stress on rice yield of two different genotypes, $qDTY_{12.1}$ 481-B (drought tolerant) and Vandana. Seedlings were exposed to simulated drought by an automated field rainout shelter, where irrigation was stopped at 35 days after sowing. Developing spikelets, flag leaves and root crowns were sampled after 71 days of growth. Samples were labelled with six different Tandem Mass Tags (TMT's) and analysed by nano-LC/MS/MS. In general, the total number of proteins was greater in tissues from 481-B than in Vandana under drought stress. This may be related to 481-B having more vigorous growth, more lateral root branching and more meristematic tissue than Vandana. Accordingly, proteins including actins, tubulins and expansins, which are highly expressed in actively growing roots and root hairs, were more abundant in 481-B. On the other hand, a lower photosynthetic rate in the flag leaves of 481-B was found under drought conditions, consistent with a lower abundance of proteins involved in photosynthesis. Proteins like the membrane extrinsic subunit of Photosystem II (PsbP) and oxygen-evolving enhancer protein, which take part in the light-dependent photosystem reactions, were less abundant in 481-B than in Vandana. Enzymes involved in the photorespiratory pathway such as glycolate oxidase, glycine dehydrogenase and serine hydroxymethyltransferase were higher in abundance in 481-B. As mentioned, antioxidant enzymes play an important role in neutralizing ROS under stress conditions. Also, increased amount of ascorbate peroxidase and glutathione peroxidase were reported in 481-B under drought. These proteins help support ROS detoxification and signalling process to maintain cellular homeostasis. Maintenance of carbohydrate metabolism under drought conditions is important for grain yield. Key proteins identified in carbohydrate metabolism, such as adenylyltransferase and sucrose synthase, were found to be more severely down-regulated in Vandana than in 481-B. This suggests that 481-B has stronger source and sink activity, which results in the better carbohydrate delivery to grain and improved yield under drought conditions when compared with Vandana.

6.2.5 *Flag Leaf Proteome*

To explore the biochemical response to water deficits in the field during the reproductive stage in rice, Ji et al. [27] compared the proteomes of a drought susceptible (Zhenshan97B) and a drought tolerant (IRAT109) cultivar. Seedlings of both cultivars were grown for 30 days then drought stress was applied by draining water slowly at the booting stage. Flag leaves were collected for 2-DE separation followed by MALDI-TOF-MS analysis. Orthophosphate dikinase, glycine dehydrogenase, ribulose bisphosphate carboxylase (Rubisco), glycine hydroxylmethyltransferase and ATP synthase were all down-regulated for Zhenshan97B in response to drought stress. This implies that there was a reduction in the capacity of carbon assimilation in this rice cultivar, either as a direct or downstream effect of drought. In the drought-stressed IRAT109, transketolase and rubisco were down-regulated; however, rubisco activase and peptidyl-prolyl *cis-trans* isomerase—which might enhance the activity of rubisco by drought stress—were both up-regulated. The increase in abundance of chloroplastic superoxide dismutase [Cu–Zn] and dehydroascorbate reductase indicate antioxidant protection for IRAT109 during dehydration.

6.2.6 *Leaf Sheath Proteome*

The mechanisms of how rice sheaths respond to different osmotic stresses were investigated by Zang et al. [28]. They monitored protein expression in the basal part of rice leaf sheaths under various osmotic stresses. Rice cultivars Nipponbare and Zhonghua were grown for two weeks and were then transferred to containers with mannitol solution (0.5–2.5 MPa osmotic pressure) for 24–120 h. Similarly, seedling pots were transferred to plastic containers containing the solutions of 150 mM NaCl or 50 μM ABA. In the drought experiment, water was withheld at two weeks after sowing and the samples were collected when plants had lost 50 % fresh weight. Protein extracted from leaf sheath was separated by 2-DE, followed by

MALDI-TOF MS analysis. The expression pattern of 15 proteins had significantly changed under osmotic stress induced by mannitol, where at least six proteins were identified as important components in the glyoxalase, lipid accumulation and proteasome regulatory pathways through endoplasmic reticulum stress. Similar changes in protein expression patterns were seen in rice that was exposed to NaCl, ABA or drought stress, in spite of the dramatically different time frames over which stresses were applied. The similarity in response for the different stress treatments, different tissues and different cultivars, underscores the likelihood that the proposed pathways are involved in plant response to osmotic stress.

6.2.7 Leaf Proteome

Salekdeh et al. [21] examined two contrasting genotypes-IR62266-42-6-2 (lowland indica) and CT9993-5-10-1-M (upland japonica) to determine the identities of differentially expressed proteins and whether expression is reversed on rewatering. The contrast of drought-sensitive and drought-tolerant accessions was intended to add insights to drought tolerance. Drought was initiated by withholding water at 20 days after sowing (DAS). At 43 DAS, stressed plants were rewatered. Leaf samples were collected from both stressed and rewatered plants. Proteins were extracted from leaf samples and separated by 2-DE followed by MALDI-MS analysis. Forty-two spots out of more than 1000 protein spots showed significant differential expression under stress, with 27 of them exhibiting a different response pattern between the two cultivars. Reassuringly, only one protein (chloroplast Cu–355 Zn superoxide dismutase) changed in both the cultivars when grown under drought stress, increasing in one cultivar while decreasing in the other. The up-regulated proteins seen in the droughted CT9993 were not affected in IR62266. Similarly, proteins were down-regulated in IR62266 remained unchanged in CT9993. At 10 days after rewatering, all protein levels had returned to normal, as seen in the well-watered control. Mass spectrometry enabled the identification of 16 drought-responsive proteins, including an actin depolymerizing factor, which was one of three proteins detectable under stress in both cultivars but undetected in well-watered plants or in plants 10 days after rewatering.

Protein phosphorylation has been recognized as an important mechanism for stress signalling. To identify the differential expression of proteins and phosphoproteins induced by drought, Ke et al. [18] imposed drought stress on two-week-old rice seedlings. Total protein extract were separated by 2-DE and probed by immunoblotting using a phosphor-amino acid-specific antibody followed by mass spectrometry analysis. Three drought-responsive proteins were identified: late embryogenesis abundant (LEA)-like protein and chloroplast Cu–Zn superoxide dismutase (SOD) were up-regulated, whereas Rieske Fe–S precursor protein was down-regulated. In addition, ten drought-responsive phosphoproteins were identified: NAD-malate dehydrogenase, OSJNBa0084K20.14 protein, abscisic acid (and stress-inducible protein), ribosomal protein, drought-induced S-like ribonuclease, ethylene-inducible protein, guanine nucleotide-binding protein beta subunit-like protein, r40c1 protein, OSJNBb0039L24.13 protein and germin-like protein 1. Despite the identification of these drought-responsive proteins, their function in the drought stress response remains to be determined. Post-translational modifications are likely to play a major role in short-term stress responses and deserve close attention in future studies of the basis of drought tolerance.

The patterns of gene expression in *Oryza sativa* cv. Nipponbare leaf under water deficit were investigated by Mirzaei et al. [15]. Thirty-five-day-old plants were subjected to gradual drought by withholding water for 14 days. Proteins extracted from the youngest fully expanded leaf samples were separated on 10 % Bis-Tris polyacrylamide gels before being analysed by nanoflow LC-MS/MS using an LTQ-XL ion-trap mass spectrometer. A total of 1548 proteins were reproducibly identified across all treatments. Three major groups of protein including aquaporins, small G-proteins, small heat-shock proteins (HSPs) and V-ATPases, appeared to be responding to drought and re-watering. Aquaporins, which include six isoforms of plasma membrane intrinsic proteins (PIP1-1, PIP1-2, PIP2-1, PIP2-2, PIP2-6 and PIP2-7) and three isoforms of tonoplast intrinsic protein (TIP1-1, TIP1-2 and TIP2-2), were all differentially expressed. These proteins were lower in abundance in response to moderate drought stress: however, they had considerably increased as the drought became more severe. Nine small GTP-binding proteins were identified, of which six were abundant and unique to the extreme drought conditions. Also, nine signal transduction proteins (small G-proteins) showed increased expression levels in drought conditions. Only six of these proteins were identified in drought, while the other three proteins were identified in all conditions. The abundance of these proteins was higher in the extreme drought conditions. Furthermore, this study identified four HSPs, three of which were expressed consistently across all treatments, while heat-shock cognate 70 kDa protein was down-regulated in extreme drought. V-ATPases, which are known as enzymes that play a role in stress response and undergo stress-related modifications, were also identified in this study. Among these, vacuolar proton pump subunit d protein was detected only in extreme drought, while the other four proteins were found in all treatments, predominantly in extreme drought conditions. In short, changes observed in multiple isoforms within individual protein families are interesting and important as a clue to the functional roles that proteins such as aquaporins play in drought response.

The shoot proteome was also studied by Mirzaei et al. [29] when part of the root system experienced water deficit while the other parts had sufficient access to soil water. The roots of rice seedling were separated into two compartments of a pot for split-root culture. After 14 days of growth, watering was stopped in one-half of each pot, producing a root system that is partially wet and partially dry. Shoot proteins were extracted and then separated on SDS-PAGE. Extracted gel tryptic digests were analysed by nanoLC-MS/MS and resulted in identification of 1383 reproducible proteins across all conditions. The patterns observed were interesting in that in categories such as protein metabolism and oxidation-reduction, substantial numbers of proteins were greatest when leaves were receiving signals from "wet" and "dry" roots. In other categories such as transport, several key transporters were surprisingly abundant in leaves supported by partially or completely droughted root systems, especially plasma membrane and vascular transporters. Stress-related proteins behaved very consistently by increasing in stressed plants but notably some proteins were most abundant when roots of the same plants were grown in both wet and dry soils. Changes in carbohydrate-processing proteins were consistent with the passive accumulation of soluble sugars in shoots under drought, with hydrolysis of sucrose and starch synthesis both enhanced. This study suggests a complex of drought signals interacting in the leaf and shows the power of classical physiological techniques such as split-root systems to elucidate novel proteins and the signals that induce them.

To illustrate a comprehensive network of rice in response to drought, Shu et al. [30] applied genetic, proteomic and metabolic approaches to study rice seedlings exposed to drought stress. IRAT109, an upland rice cultivar, was used in the study. Thirty days after sowing, drought was initiated by withholding water for period of up to nine days. Leaf samples were collected and extracted before SDS-PAGE separation. Protein spots were then analysed by MALDI-TOF/TOF. Genetic and metabolic analyses were performed using the same sample tissue. Results indicated that 71 protein spots were significantly altered, with 60 spots successfully identified. The greatest down-regulation protein functional category was translation, presumably compromising the acclimation to stress. The major up-regulated proteins were related to protein folding and assembly. In addition, many proteins involved in metabolism showed change in level of expression. cDNA microarray and GC-MS analysis showed 4756 differentially expressed mRNAs and 37 differentially expressed metabolites. This study suggests that increased energy consumption from storage substances occurred during drought. In addition, increased expression of the enzymes involved in anabolic pathways accorded with an increase in the content of six amino acids. Analysis of these basic metabolic networks provides an opportunity to understand how rice plants acclimate to drought conditions.

6.2.8 Root Proteome

In a separate study to the leaf proteome study, analysis of long-distance drought signalling in rice roots was also performed by Mirzaei et al. [12]. Rice (*Oryza sativa* L. cv. IR64) was grown in split-root systems to analyse long-distance drought signalling within root systems. Thirteen-day-old seedlings were transplanted into pots for split-root culture. An internal wall divided each pot into two equal parts allowing for equal distribution of soil and water between the parts. Equal proportions of the root system of each seedling were distributed between the compartments. Water was withheld in one compartment and the other compartment was well-watered for 14 days. Protein extracted from roots were separated by 1-DE and the protein bands were analysed by nano LC-MS/MS. There were 1487 non-redundant proteins identified in the root tissues. By comparing proteins in well-watered root tissue and adjacent droughted roots, 126 proteins were down-regulated and 90 proteins were up-regulated. Surprisingly, there was a large change in the proportion of total gene expression between the well-watered roots with and without a vascular connection to droughted roots. This leads to the conclusion that water supply can alter gene expression remotely, either by inhibiting exported signals or eliciting drought response by long-distance signalling.

Rabello et al. [31] analysed the roots of rice using both genomic and proteomic approaches. Two contrasting genotypes, Prata Ligeiro (drought tolerant) and IRAT20 (drought sensitive) were selected for the study. Drought stress was applied for 21 days from anthesis. Total protein was extracted from roots of both genotypes before separation by 2-DE. Protein spots were then analysed by MALDI-TOF. The proteomic approach allowed the identification of 307 overlapping spots from both genotypes, with 156 proteins exclusive to the tolerant genotype and 215 proteins exclusive to the susceptible genotype. In general, the tolerant genotype produced higher abundance of proteins associated with cell protection against oxidative damage. On the other hand, the susceptible genotype showed a higher diversity in the protein profile, surprisingly revealing more unique proteins than the tolerant one. This reinforces the importance of detailed functional analysis, including protein interactivity in understanding tolerance mechanisms, rather than simple counts of proteins or peptides.

Recently, Paul et al. [32] looked at the protein abundance changes in roots from transgenic rice cultivars under drought stress. Rice cultivar BRs 29 and an *Arabidopsis* DREB1A-overexpressing homozygous transgenic BR-29 was used as a wild-type drought-sensitive control and transgenic drought-tolerant line respectively. Drought stress was imposed by withholding water for seven days. Proteins extracted from root tissue were separated by 2-DE and then analysed by MALDI TOF MS/MS. The majority of identified proteins that changed in abundance belonged to carbohydrate and energy metabolism. Stress and defence-related proteins including peroxidase 2, L-ascorbate peroxidase 1 and chitinase were especially up-regulated under drought stress in both wild-type and the transgenic plants. In addition, a novel protein, R40C1 was observed to be up-regulated in roots of the transgenic plant, which may play an essential role in acquisition of drought tolerance.

6.2.9 Nuclear Proteome

The nucleus is the regulatory hub of the eukaryotic cell, with the resident DNA under continual modulation by signal molecules which control transcription, nucleotide modifications and thus, the transcriptome. To reveal molecular mechanisms of drought-responsive adaptation in rice, Choudhary et al. [33] published the first comprehensive nuclear proteome of rice. Seeds of eight rice varieties (Azucena, Anjali, Buddha, IR-64, IR-20, Moroberekan, Rasi and Vandana) were grown in well-watered soil then gradually dehydrated for four weeks by withdrawing water: tissues were harvested up to 120 h later. Nuclear protein extracts were separated by 2-DE and then analysed by electrospray ionization LC/MS/TOF. The differential display of nuclear proteins in the tolerant variety under drought stress revealed 150 spots that showed alteration in their intensities by more than 2.5-fold. Proteomics analysis led to the identification of 109 differentially expressed proteins because of drought, putatively involved in a variety of functions such as transcriptional regulation and chromatin remodelling, signalling and gene regulation, cell defence and rescue, and protein degradation. The dehydration-responsive nuclear proteome displayed a coordinated response, including both regulatory and functional proteins, revealing some of the molecular mechanisms of acclimation to dehydration. However, shock-induced responses are always a risk in such short-term stress treatments.

Jaiswal et al. [34] found a functional link between dehydration-responsive pathways in the nucleus by investigating the rice nucleus proteome under drought stress. Rice seedlings of both the rice cultivars (*Oryza sativa* L.; cv. IR-64 and Rasi) were grown in mixture of soil and soilrite. Dehydration stress was imposed on four-week-old seedlings by withdrawing water, and the aerial parts were harvested sequentially up to 120 h. Nuclear proteins were then prepared for 2-DE separation followed by mass spectrometric analysis. The comparative analysis of dehydration-responsive nuclear proteome revealed relatively lower abundance of dehydration-responsive proteins (DRPs) in cv. IR-64 compared with Rasi. Among the identified proteins, 78 were predicted to be targeted to the nucleus. The total number of protein spots detected in cv. IR64 was higher when compared with that of Rasi, while the number of DRPs was found to be smaller in the intolerant rice cultivar. Forty-three percent of the DRPs were found to be different between the two cultivars, indicating distinctive nuclear proteomes. In addition, a functional association network of the DRPs of cv. IR-64 was constructed to show that a significant number of the proteins are capable of interacting with each other.

6.2.10 Summary

In this section, we have shown a wide range of proteomic studies focusing on rice under drought stress, mainly using classical 2-DE. The opportunities afforded by accessions and cultivars of rice with contrasting tolerance to drought have been heavily exploited, yet more remain. Improved proteomic platforms will enable multiple genotype comparisons and more confidence in the identification of key markers and/or pathways. Similarly, the march of technologies and ever-improved databases are reducing the uncertainties about protein identity and expanding the range of tissues that can plausibly be sampled. For example, a primary target has to be pollen, a vulnerable tissue in plant development. Patterns of gene expression response to drought from these disparate studies are emerging, with the most common stress responses being oxidation-reduction, metabolism, and less so, change to carbohydrate turnover, cytoskleletal changes, transport, and energy metabolism. One hopes that this will lead to markers for drought tolerance. While the number of genes that appears to be altered in activity during drought might be small when compared with the entire proteome, indications are that the use of multiple markers will still be required to make gains in drought resistance in rice.

6.3 Wheat and Water Deficits

More wheat is harvested worldwide than any other grain. As a dryland crop that extends into low rainfall zones, there has been much focus in recent years into the effects of drought stress [35–37]. Drought stress is especially detrimental to the reproductive and grain-filling stages, resulting in depressed yields [38]. It is likely that further oscillations in climatic conditions could lead to catastrophic food supply shortages through periodic failure of dryland cereal crops.

6.3.1 Wheat Grain

Proteomic analysis of wheat and barley was first performed as early as 2001 and 2002 [39, 40] but the first proteomic study specifically into the effect of drought on wheat was performed in 2007 by Hajheidari et al. [41]. In this study, two drought-susceptible genotypes (Arvand and Kelk Afghani) and one drought-tolerant genotype (Khazar-1) were investigated. Mature seeds were harvested after drought treatment and protein extract from the seeds were separated using 2-DE. There were 121 distinct protein spots detected, with 57 of them being identified by MALDI-TOF/TOF. The largest functional class of the differentially expressed proteins (27 out of 57 total) were involved in abiotic defence. While most of these proteins were up-regulated in all cultivars, the expression levels were highest in the drought-tolerant cultivar. The majority of these proteins are involved in ROS scavenging. There was also a number of isoforms of α-amylase inhibitors that were up-regulated in the tolerant and down-regulated in the susceptible genotypes. Protein synthesis and assembly (thirteen proteins) and metabolism (ten proteins) were the next largest functional groups detected.

Jiang et al. [42] investigated the proteome of developing wheat grain exposed to drought. One drought-tolerant (Kauz) and one susceptible (Janz) wheat cultivar were subjected to drought. To simulate drought after heading, soil water content was maintained at one-third of the water content in the control pots. Grain samples from the middle spike were collected at 10, 15, 20 and 25 days post-anthesis. Protein extract from grain samples were separated by linear and non-linear 2-DE. There were 153 differentially expressed protein spots observed, which were identified by MALDI-TOF and MALDI-TOF/TOF. The results showed that detoxification and defence proteins (ROS scavengers, peptidase inhibitors and salt-stress-responsive proteins), proteins involved in carbohydrate metabolism (ADP-glucose pyrophosphorylase, sucrose synthase, and aldose reductase), and proteins involved in signal transduction (WD40 and G-beta like protein), were all more abundant in the tolerant cultivar under drought simulation. The photosynthetic oxygen evolving enhancer proteins were up-regulated in both cultivars at the first two sampling stages, while the expression of oxygen evolving enhancer protein subunit 1 was lower in the susceptible cultivar in the last two stages. These differentially expressed proteins could be related to the biochemical pathways for stronger drought resistance of Kauz.

In another time point study, Ge et al. [43] investigated the effects of drought on the wheat grain proteome over time in order to understand proteomic changes in developing wheat kernels. Ningchun 4 (tolerant) and Chinese Spring (susceptible) were subjected to a watering regime of one-third the level of the control, starting 12 days prior to heading. Samples were collected 10, 14, 18, and 26 days after flowering (DAF). One hundred and fifty-two significant protein spots were observed by 2-DE, with 96 identified with MALDI-TOF. From these, the three largest functional protein classes seen were carbohydrate metabolism, stress/defence, and photosynthesis (39, 18, and 13 %, respectively). Common stress defence proteins were detected in both cultivars (SOD, CAT, APX), while the growth regulator translationally controlled tumor protein was generally up-regulated only over time and only in the tolerant cultivar. In the early stages of grain development, HSP70 was increased, while LEA increased at later stages—more so for stressed samples than controls. Enzymes involved in glycolysis and tricarboxylic acid cycle (TCA) were also upregulated under drought, such as GAPDH, cytosylic 3-phosphoglycerate kinase, and malate dehydrogenase. β-amylase was up-regulated in both cultivars, while sucrose synthases were up-regulated more in the tolerant cultivar, the latter potentially giving the tolerant cultivar more energy for stress responses. Moreover, higher expression of triosephosphate isomerase in drought-tolerant wheat, which may act in Ningchuan 4 to enhance energy generation and promote water potential via the sucrose biosynthesis pathway, indicates that this cultivar also has a greater tolerance to water stress. Homeostasis of photosynthesis also appears more effective in the tolerant cultivar, with general up-regulation of the RuBisCO large subunit and an early up-regulation of the oxygen evolving complex (OEC), followed by a gradual decrease over time.

6.3.2 *Wheat Leaf*

In addition to the drought-affected wheat grain proteomic studies, Caruso et al. [44] performed an experiment looking at the proteome of leaves in drought-affected wheat. In this study, durum wheat seedlings (*T. durum* cv. Ofanto) were subjected to seven days severe drought in order to identify proteins involved in drought-stress response. Leaf protein extract was separated by 2-DE. Thirty-six proteins were identified as being differentially expressed, all of which were identified by MALDI-TOF. The 36 proteins identified belonged to six different functional classes; Calvin cycle (9 %), glycolysis and gluconeogenesis (18 %), amino acid biosynthesis (12 %), ROS scavenging (15 %), defence (6 %) and post-transcriptional regulation (3 %). In the drought-affected wheat seedlings, proteins involved in photosynthetic mechanisms, such as photosystem II, were up-regulated, possibly as part of the shift in the light/dark reaction activity. RuBisCO subunits and other associated proteins were generally down-regulated, possibly to reduce ROS production. In contrast, ROS scavenging proteins were all up-regulated, as were most proteins involved in amino acid and amine biosynthesis (S-adenosylmethionine synthetase, glutamine synthetase). Amino acid and amine biosynthesis is important to provide substrates for protein production, osmoregulation and ROS scavenging.

Ford et al. [44] investigated the response to drought over time in three Australian wheat cultivars, Kukri (drought susceptible) and Excalibur and RAC875 (both drought tolerant). All three were placed under increasing drought stress after emergence of the first flag leaf. Samples of mature leaves were taken on day 5, 14 and 24 and after re-watering on day 25. One hundred and fifty-nine proteins were identified by 8-plex iTRAQ, and LC-MS/MS. Proteins involved in photosynthesis were slightly down-regulated in the tolerant cultivars. Upon re-watering, the most drought-tolerant cultivars displayed the least number of differentially expressed photosynthesis-related proteins. This indicated that it had the fastest response in returning to normal state. Proteins involved in ROS scavenging (CAT, SODs, chloroplastic and cytosolic Cu/Zn-SOD) were detected in all of the treated cultivars. In the tolerant cultivars, ALDH and geranylgeranyl reductase increased only under drought. Glycolysis and gluconeogenesis, protein folding, transport, and defence, were up-regulated in all cultivars during drought, where the proteomes in the tolerant varieties recovered the fastest. RAC875 (tolerant) showed the highest number of Calvin cycle related proteins at the first time-point; Kukri (susceptible) had the highest at the second time-point; and Excalibur (tolerant) had the highest at the third time-point. This result suggests that RAC875 (tolerant) has strategies to respond to drought quickly through increased Dark Reaction activity. Kukri (susceptible) is slightly slower to respond, whereas Excalibur (tolerant) might have other physiological factors that allow it to resist early drought effects for longer.

To investigate drought response in leaves, as well as to identify candidate genes for breeding programs, Budak et al. [45] compared the proteomes of two wild emmer wheats (TR39477 and TTD22) and durum wheat (Kiziltan) under drought conditions. In their experiment, plants were grown for four weeks under normal conditions, followed by nine days without watering after which the leaves were harvested. Seventy-five differentially expressed protein spots were observed in the 2-DE gels, of which, 66 were identified using nanoLC-ESI-MS/MS. The major functional class of proteins that were up-regulated during drought were those involved in carbohydrate transport and metabolism. Other proteins that were up-regulated were proteins involved in photosynthetic processes. Proteins such RuBisCO isoforms were at higher levels in all cultivars, especially in the wild emmer wheats. The next largest class was those involved in energy production and conversion, with ion transporter proteins up-regulated in wild emmer, which conversely appeared to down-regulate parts of its photosynthetic machinery. Amino acid transport and metabolism made up another large functional class. The more susceptible variety had higher methionine synthase levels, possibly because selection favoured processes involving growth over stress tolerance in durum wheat. Polyamine oxidase levels were higher in the tolerant variety, despite being a source of H_2O_2. These results suggest that the wild tolerant variety may have an ancient alternative line of defence against abiotic stress.

In understanding the importance of phosphorylation in response to drought, Zhang et al. [46] investigated the phosphoproteome in the leaves of two drought tolerant wheat cultivars (Hanxuan 10 and Ningchun 47). Wheat was grown normally until the three-leaf stage and then exposed to drought conditions (20 % PEG) for 48 h. At the end of the drought period, the second leaf was harvested and proteins were extracted. Phosphorylated peptides were endoproteinase-digested and the subsequent peptides were enriched using TiO_2 beads. Quantification and identification of the phosphopeptides was by LC-MS/MS. It was found that both tolerant cultivars had similar numbers of significantly expressed phosphorylated proteins: 173 for Hanxuan-10 and 227 for Ningchun-47. Approximately 96–97 % of the proteins detected were class I phosphorylated, so class II and III were ignored in this study. Of class I, the most numerous were related to signal perception and transduction, which include kinases and phosphatases that respond to ABA as well as the Ca^{2+} second messenger (phosphorylated SnRK homolog, PP2C, kinases, calcium responsive proteins, transcription factors, basic helix-loop-helix and CCCH domain-containing). Phosphorylated transcription factors were another class of proteins often detected and included a number of proteins with zinc finger CCCH motifs, which are speculated to be involved in RNA processing (TaABI5-1, MYB1R1, bHLH, and zinc finger containing CCCH domain). Another common set of phosphoproteins observed were those involved in the transport of water, sugar and hydrogen ions (aquaporins, NOD26-like membrane integral proteins, monosaccharide sensing protein 2, H^+-ATPase and CCCH containing proteins). Lastly, phosphoproteins involved in stress defence, osmotic protection and ROS scavenging, were also detected (stress-associated protein, *Aspergillus giganteus* antifungal protein, three E3 ubiquitin ligases, hydrophilic osmoprotectants, WCOR and LEA proteins, glutamate decarboxylase 1 and glutathione peroxidase 4). From these results, it appears that ABA and Ca^{2+} cause a phosphorylation cascade to signal perception and transduction proteins, which in turn phosphorylate a number of transcription factors. The latter induce or suppress the production of another set of proteins involved in the drought stress response, with a subset of these proteins being themselves phosphorylated.

6.3.3 Wheat Roots

The proteome of wheat roots exposed to simulated drought stress was investigated by Alvarez et al. [47]. In their experiment, the plant hormone abscisic acid (ABA) was used to initiate a simulated drought response in the drought-tolerant (Nesser) and drought-sensitive (Opata) wheat varieties. Wheat was grown for ten days under normal conditions, followed by exposure to ABA (100 μM) for 6 h. The roots were harvested after the treatment period. Root proteins were extracted and trypsin digested, followed by 4-plex iTRAQ labelling and LC-MS/MS detection. Eight-hundred-and-five differentially expressed proteins were detected after ABA treatment. While 151 of these were common to both cultivars, 421 were cultivar-independent and ABA responsive. The tolerant cultivar had 131 differentially expressed proteins, which represented the following functional classes: defence, heat shock proteins and signal transduction pathways (kinases, phosphatases, GTP-binding proteins, and 14-3-3 protein homologs). Proteins in both cultivars that showed differential expression (cultivar-specific and non-specific) had much more significant expression levels in the tolerant cultivar (166) than in the susceptible (67). The tolerant cultivar had more types of heat shock protein, proteins involved in secondary metabolism, and cell wall biogenesis proteins. These observations may suggest that Nesser has a greater number of pathways providing a stronger ability to respond to drought stress.

6.3.4 Wheat Root and Leaf

The effects of drought on the root and leaf proteomes of a drought-tolerant hybrid wheat cultivar (Shanrong No. 3) and its drought-susceptible wheat parent (Jinan 177) were investigated by Peng et al. [22]. Because the second parent of the hybrid was the wheatgrass *Thinopyrum ponticum,* the authors believed that the genetic diversity between the hybrid and its wheat parent would help them understand the effect of drought on the root and leaf proteome of both cultivars. Shanrong No. 3 and Jinan 177 were both exposed to 24 h of osmotic drought (18 % polyethylene glycol) followed by harvesting of the roots and leaves. This period is too short to allow for most acclimation events to be registered in the proteome. Ninety-three root and 65 leaf differentially expressed protein spots were revealed by 2-DE. The protein spots were identified using MALDI-TOF-TOF. The results showed that the majority of differentially expressed proteins were shared by the two cultivars, with the major protein classes being signal transduction, transport, detoxification, and carbon and nitrogen metabolism. However, the tolerant cultivar had a generally higher induction of differentially expressed proteins. Specifically, proteins involved in ROS scavenging (including antioxidant production), were higher in the tolerant cultivar, as were enzymes such as V-ATPases, which are important in maintaining ion and water balance. In contrast, the susceptible cultivar displayed a more fragmented set of RuBisCO subunit isoforms, while the tolerant had an increase in a number of chlorophyll protector proteins. This indicates that the susceptible cultivar was less likely to maintain normal photosynthetic function in drought, while the tolerant cultivar was better able to protect these mechanisms. Lastly, higher levels of proteins involved in the gibberellin pathway and lower levels for proteins involved in the ethylene synthesis pathway in the tolerant cultivar (compared with susceptible), implied the promotion of growth in the tolerant cultivar and senescence in the susceptible.

6.3.5 Wheat Stem

With the aim of identifying molecular mechanisms of reserve mobilisation in wheat stems, Bazargani et al. [48] subjected two cultivars of wheat with differing capacity to remobilise reserves form stems under drought conditions. After anthesis, drought susceptible (N14) and tolerant (N49) wheat were exposed to drought by maintaining their soil at 50 % field capacity. The tillers were harvested at 10, 20, and 30 days after drought exposure. One hundred and thirty-six differentially expressed protein spots were observed by 2-DE, 82 of which were identified by MALDI-TOF-TOF MS/MS. Overall, the tolerant cultivar (N49) showed a higher level of differently expressed proteins, with the peak for both cultivars being at twenty days after anthesis (DAA). The tolerant cultivar had 18, 74 and 23 differentially expressed proteins at 10, 20 and 30 DAA, while for the same time series the susceptible cultivar had 25, 38, and 21 differentially expressed proteins. Interestingly, although the tolerant cultivar had a greater number of differentially expressed proteins overall, most of these were down-regulated, with up-regulation in the tolerant cultivar only more prevalent for proteins involved in energy metabolism and ROS removal. RuBisCO (large and small subunit), RuBisCO activase and oxygen evolving proteins, were all less abundant in the tolerant variety, while SAM synthase (ethylene production and senescence) was up-regulated. Also, the tolerant variety showed an up-regulation of nine proteins involved in ROS removal as opposed to two for the susceptible cultivar, with both cultivars showing differential expression of signalling proteins, including 14-3-3, MFP-1, and MAF1. From these results the authors suggest that the tolerant cultivar senesces faster and is more efficient at protecting cells within the stem, while the stem reserves for the tolerant cultivar are more effectively transported from the stem to the developing grain.

6.4 Barley and Water Deficits

A study by Wendelboe-Nelson et al. [49] marked the first investigation into the effects of drought on barley. Their aims were to compare how the proteomes of a drought-tolerant (Basrah) and susceptible cultivar (Golden Promise) adapted to drought. Seeds were germinated and grown for seven days. They were then subjected to drought (watering withheld) for another seven days, followed by harvesting of the roots and shoots. Sixty-six differentially expressed protein spots were detected in the leaves and 77 in the roots using 2-DE-DIGE. The protein spots were identified by MALDI-TOF. The results showed a number of proteins constitutively expressed at higher levels by the tolerant variety. The proteins identified were 70 kDa heat shock protein, RuBisCO large subunit binding protein and cyclophilin A. Proteins associated with ROS scavenging, cellular defence or osmoregulation were generally up-regulated in the tolerant cultivar, while photosynthesis and carbohydrate metabolism were down-regulated in the susceptible cultivar. Also, the susceptible cultivar proteome displayed a down-regulation in enzymes involved in the Yang cycle (ethylene synthesis and senescence).

Similar to the investigation by Wendelboe-Nelson et al., another study of the barley proteome under drought was performed by Ashoub et al. [50]. In this study, the authors used the drought-tolerant barley landrace 15141 and the drought-sensitive barley landrace 15163 to examine the potential for genetic adaption to drought stress within Egyptian barley landraces. The plants were maintained at 70 % field capacity until four extended leaves developed. Watering was then stopped for two days until 10 % field capacity had been reached, which was maintained for another five days. The samples were harvested and physiological tests confirmed that landrace 15141 was the more drought-tolerant cultivar. Thirty-six protein spots were observed by 2D-DIGE, with the protein spots of interest identified with MALDI-TOF-MS. The results showed proteins involved in protective functions, such as methionine synthase, lypoxygenase, NADP-ME, sucrose synthase and betaine aldehyde, were more abundant in the tolerant genotype. Also more abundant were proteins with chaperone and disaggregation functions, such as heat shock proteins, ATP dependent Clp protease, zinc metalloprotease, HsP90 and HsP100. The importance of mobilising nutrients from senescing leaves to fresh growth was seen in the up-regulation of energy metabolism for both tolerant and sensitive landraces. The tolerant variety also displayed greater differential expression patterns for proteins involved in the photosynthetic mechanism (RuBisCO large subunit, transketolase, PPDK). Lastly, the more tolerant landrace had a lower induction of proteins involved in osmoregulation (betaine aldehyde dehydrogenase), thus implying it regulated water relations more tightly.

Rollins et al. [51] applied proteomic analysis to two drought-tolerant barley cultivars from different geographical origins to examine the biomolecular (genetic) diversity of abiotic stress responses. The plants were exposed to drought at heading (15 % field capacity), which was maintained until maturity, with a subset exposed to 36 °C/32 °C light/dark heat stress for one week. The harvested leaf proteins were analysed by 2D-DIGE and protein spots of interest were identified by MALDI-TOF/TOF. The results for this study showed no significant difference in the proteomes of either of the two tolerant cultivars under drought stress. However, there were substantial physical differences in biomass and spike number between the control and drought-affected plants. In contrast the application of heat to plants that had already been stressed by drought did elicit protein expression differences. The functional classes of proteins identified were photosynthesis, detoxification, energy metabolism and protein biosynthesis. RuBisCO B was up-regulated under heat stress (possibly to replace damaged protein), while RuBisCO activase A—known to be heat labile—was down-regulated.

Ashoub et al. [50] looked at the proteome of wild barley (*Hordeum spontaneum*) to investigate protein expression level changes when exposed to drought. At the two-leaf stage, drought conditions were imposed for three weeks, after which the third leaves were then harvested. Differentially expressed protein spots were observed by 2D-DIGE. Sixty-three protein spots had an expression level difference equal to or greater than 1.5-fold between control and treated, with forty-five of these identified by MS/MS. This showed proteins that were up-regulated were involved in cell detoxification, water homeostasis, amino acid synthesis, lipid metabolism, heat shock and chaperone functions. Proteins that were down-regulated were related to nitrogen metabolism.

6.5 Summary

Although it is widely recognized that drought stress is probably the greatest constraint for crop growth, knowledge about the mechanisms by which plants avoid or tolerate drought is still evolving. The congruence of approaches to drought—physiology, genetics, transcriptomics, proteomics, genomic and metabolomics—reveal some basic phenomena engaged by plants to evade or tolerate water deficits (Fig. 6.1). Proteomics, as one of the high-throughput technologies, has played an important role in identifying the underlying mechanisms for stress response. In particular, in the past two decades proteomics has made a major contribution to identification of the main families of drought-responsive proteins, including ABA-responsive proteins, heat shock proteins, detoxification and defence proteins. This chapter shows that proteomic studies also implicate energy metabolism and down-regulation of photosynthesis as drought-tolerance mechanisms. Such experiments have been enhanced by the exploitation of known genetic variation in 'drought tolerance' in cereals, with the acknowledgement that yet more drought resistance mechanisms are certain to exist in unrelated arid-zone species. As we see the emergence of whole-genome sequence data, large collections of publicly available transcriptomic data sets (including microarray and RNA-seq) and substantial proteomic data sets, the possibilities of integrating the findings of disparate studies grows ever greater. The next step will be to interrogate large-scale data sets and construct gene networks (interactomes) for a deeper analysis of drought response. Gene networks, along with functional assays, should greatly facilitate our understanding of gene function and provide a path for development of biomarkers for drought stress. This in turn is hoped to lead to yield enhancement in commercial crops.

References

1. Somerville C, Briscoe L (2001) Genetic engineering and water. Science 292:2217
2. Lawlor DW (2013) Genetic engineering to improve plant performance under drought: physiological evaluation of achievements, limitations, and possibilities. J Exp Bot 64:83–108
3. Snaydon RW (1980) Responses of plants to environmental stresses. Academic Press, New York
4. Yao F, Huang J, Cui K, Nie L, Liu X, Wu W et al (2012) Agronomic performance of high-yielding rice variety grown under alternate wetting and drying irrigation. Field Crops Res 126:16–22
5. Yue B, Xue W, Xiong L, Yu X, Luo L, Cui K et al (2006) Genetic basis of drought resistance at reproductive stage in rice: separation of drought tolerance from drought avoidance. Genetics 172:1213
6. Turner NC, Siddique NC, Turner GC, Wright GC, Siddique GC (2001) Adaptation of grain legumes (pulses) to water-limited environments. Adv Agron 71:193–231
7. Mckay JK, Richards JH, Mitchell-Olds T (2003) Genetics of drought adaptation in Arabidopsis thaliana: I. Pleiotropy contributes to genetic correlations among ecological traits. Mol Ecol 12:1137–1151
8. Raorane M, Pabuayon I, Varadarajan A, Mutte S, Kumar A, Treumann A et al (2015) Proteomic insights into the role of the large-effect QTL qDTY 12.1 for rice yield under drought. New Strateg Plant Improv 35:1–14
9. Herder GD, Van Isterdael G, Beeckman T, De Smet I (2010) The roots of a new green revolution. Trends Plant Sci 15:600–607
10. Smith S, De Smet I (2012) Root system architecture: insights from Arabidopsis and cereal crops. Philos Trans R Soc B 367:1441–1452
11. Malamy JE (2002) A draft sequence of the rice genome (*Oryza sativa* L. ssp. *indica*). Science 296:79–92
12. Mirzaei M, Soltani N, Sarhadi E, Pascovici D, Keighley T, Salekdeh GH et al (2012) Shotgun proteomic analysis of long-distance drought signaling in rice roots. J Proteome Res 11: 348–358
13. Maksup S, Roytrakul S, Supaibulwatana K (2014) Physiological and comparative proteomic analyses of Thai jasmine rice and two check cultivars in response to drought stress. J Plant Interact 9:43–55
14. Morgan JM (1984) Osmoregulation and water stress in higher plants. Annu Rev Plant Physiol 35:299–319
15. Mirzaei M, Pascovici D, Atwell B, Haynes PA (2012) Differential regulation of aquaporins, small GTPases and V-ATPases proteins in rice leaves subjected to drought stress and recovery. Proteomics 12:864–877
16. Maurel C (1997) Aquaporins and water permeability of plant membranes. Annu Rev Plant Biol 48:399–429
17. Maurel C, Verdoucq L, Luu D-T, Santoni V (2008) Plant aquaporins: membrane channels with multiple integrated functions. Annu Rev Plant Biol 59:595–624
18. Ke Y, Han G, He H, Li J (2009) Differential regulation of proteins and phosphoproteins in rice under drought stress. Biochem Biophys Res Commun 379:133–138
19. Close TJ (1996) Dehydrins: emergence of a biochemical role of a family of plant dehydration proteins. Physiol Plant 97:795–803
20. Liu J, Bennett J (2011) Reversible and irreversible drought-induced changes in the anther proteome of rice (*Oryza sativa* L.) genotypes IR64 and Moroberekan. Mol Plant 4:59–69
21. Salekdeh GH, Siopongco J, Wade LJ, Ghareyazie B, Bennett J (2002) Proteomic analysis of rice leaves during drought stress and recovery. Proteomics 2:1131–1145
22. Peng Z, Wang M, Li F, Lv H, Li C, Xia G (2009) A proteomic study of the response to salinity and drought stress in an lntrogression strain of bread wheat. Mol Cell Proteomics 8:2676–2686

23. Muthurajan R, Shobbar Z-S, Jagadish S, Bruskiewich R, Ismail A, Leung H et al (2011) Physiological and proteomic responses of rice peduncles to drought stress. Part B Appl Biochem Biotechnol 48:173–182
24. Jagadish S, Muthurajan R, Rang Z, Malo R, Heuer S, Bennett J et al (2011) Spikelet proteomic response to combined water deficit and heat stress in rice (*Oryza sativa* cv. N22). Rice 4:1–11
25. Huang W, Bi T, Sun W (2010) Comparative analysis of panicle proteomes of two upland rice varieties upon hyper-osmotic stress. Select Publ Chin Univ 5:546–555
26. Dong M, Gu Jr, Zhang L, Chen P, Liu T, Deng J et al (2014) Comparative proteomics analysis of superior and inferior spikelets in hybrid rice during grain filling and response of inferior spikelets to drought stress using isobaric tags for relative and absolute quantification. J Proteomics 109:382–399
27. Ji K, Wang Y, Sun W, Lou Q, Mei H, Shen S et al (2012) Drought-responsive mechanisms in rice genotypes with contrasting drought tolerance during reproductive stage (report). J Plant Physiol 169:336
28. Zang X, Komatsu S (2007) A proteomics approach for identifying osmotic-stress-related proteins in rice. Phytochemistry 68:426–437
29. Mirzaei M, Soltani N, Sarhadi E, George IS, Neilson KA, Pascovici D et al (2014) Manipulating root water supply elicits major shifts in the shoot proteome. J Proteome Res 13:517–526
30. Shu L, Lou Q, Ma C, Ding W, Zhou J, Wu J et al (2011) Genetic, proteomic and metabolic analysis of the regulation of energy storage in rice seedlings in response to drought. Proteomics 11:4122–4138
31. Rabello AR, Guimarães CM, Rangel PHN, Da Silva FR, Seixas D, De Souza E et al (2008) Identification of drought-responsive genes in roots of upland rice (*Oryza sativa* L). BMC Genome 9:485
32. Paul S, Gayen D, Datta SK, Datta K (2015) Dissecting root proteome of transgenic rice cultivars unravels metabolic alterations and accumulation of novel stress responsive proteins under drought stress. Plant Sci 234:133–143
33. Choudhary M, Basu D, Datta A, Chakraborty N, Chakraborty S (2009) Dehydration-responsive nuclear proteome of rice (*Oryza sativa* L.) illustrates protein network, novel regulators of cellular adaptation, and evolutionary perspective. Mol Cell Proteomics 8:1579–1598
34. Jaiswal DK, Ray D, Choudhary MK, Subba P, Kumar A, Verma J et al (2013) Comparative proteomics of dehydration response in the rice nucleus: new insights into the molecular basis of genotype-specific adaptation. Proteomics 13:3478–3497
35. Kosová K, Vítámvás P, Prášil IT, Renaut J (2011) Plant proteome changes under abiotic stress —contribution of proteomics studies to understanding plant stress response. J Proteomics 74:1301–1322
36. Farooq M, Hussain M, Siddique KHM (2014) Drought Stress in wheat during flowering and grain-filling periods. Crit Rev Plant Sci 33:331–349
37. Faghani E, Gharechahi J, Komatsu S, Mirzaei M, Khavarinejad RA, Najafi F et al (2015) Comparative physiology and proteomic analysis of two wheat genotypes contrasting in drought tolerance. J Proteomics 114:1–15
38. Pradhan GP, Prasad PVV, Fritz AK, Kirkham MB, Gill BS (2012) Effects of drought and high temperature stress on synthetic hexaploid wheat. Funct Plant Biol 39:190–198
39. Skylas DJ, Copeland L, Rathmell WG, Wrigley CW (2001) The wheat-grain proteome as a basis for more efficient cultivar identification. Proteomics 1:1542–1546
40. Østergaard O, Melchior S, Roepstorff P, Svensson B (2002) Initial proteome analysis of mature barley seeds and malt. Proteomics 2:733–739
41. Hajheidari M, Eivazi A, Majidi I, Salekdeh GH, Buchanan BB, Wong JH (2007) Proteomics uncovers a role for redox in drought tolerance in wheat. J Proteome Res 6:1451–1460
42. Jiang S-S, Liang X-N, Li X, Wang S-L, Lv D-W, Ma C-Y et al (2012) Wheat drought-responsive grain proteome analysis by linear and nonlinear 2-DE and MALDI-TOF mass spectrometry. Int J Mol Sci 13:16065–16083

43. Ge P, Ma C, Wang S, Gao L, Li X, Guo G et al (2012) Comparative proteomic analysis of grain development in two spring wheat varieties under drought stress. Anal Bioanal Chem 402:1297–1313
44. Caruso G, Cavaliere C, Foglia P, Gubbiotti R, Samperi R, Laganà A (2009) Analysis of drought responsive proteins in wheat (*Triticum durum*) by 2D-PAGE and MALDI-TOF mass spectrometry. Plant Sci 177:570–576
45. Budak H, Akpinar B, Unver T, Turktas M (2013) Proteome changes in wild and modern wheat leaves upon drought stress by two-dimensional electrophoresis and nano LC-ESI-MS/MS. Int J Mol Biol Mol Genet Biochem 83:89–103
46. Zhang M, Lv D, Ge P, Bian Y, Chen G, Zhu G et al (2014) Phosphoproteome analysis reveals new drought response and defense mechanisms of seedling leaves in bread wheat (*Triticum aestivum* L.). J Proteomics 109:290–308
47. Alvarez S, Roy Choudhury S, Pandey S (2014) Comparative quantitative proteomics analysis of the ABA response of roots of drought-sensitive and drought-tolerant wheat varieties identifies proteomic signatures of drought adaptability. J Proteome Res 13:1688–1701
48. Bazargani MM, Sarhadi E, A-aS Bushehri, Matros A, Mock H-P, Naghavi M-R et al (2011) A proteomics view on the role of drought-induced senescence and oxidative stress defense in enhanced stem reserves remobilization in wheat. J Proteomics 74:1959–1973
49. Wendelboe-Nelson C, Morris PC (2012) Proteins linked to drought tolerance revealed by DIGE analysis of drought resistant and susceptible barley varieties. Proteomics 12:3374–3385
50. Ashoub A, Beckhaus T, Berberich T, Karas M, Brüggemann W (2013) Comparative analysis of barley leaf proteome as affected by drought stress. Int J Plant Biol 237:771–781
51. Rollins JA, Habte E, Templer SE, Colby T, Schmidt J, Von Korff M (2013) Leaf proteome alterations in the context of physiological and morphological responses to drought and heat stress in barley (*Hordeum vulgare* L.). J Exp Bot 64:3201

Chapter 7
The Impact of Heat Stress on the Proteome of Crop Species

Andrew P. Scafaro and Owen K. Atkin

Abstract Heat stress impairs plant growth and reproduction. A future climate where more extreme heating events are accompanied by drought and higher atmospheric CO_2 concentrations will exacerbate heat stress considering such factors will reduce stomatal conductance, reducing transpiration-dependent evaporative cooling, leading to even higher leaf temperatures. Understanding the impact of heat on crops is therefore of paramount importance when factoring in future climate scenarios. The use of proteomic techniques will enable the mechanisms by which crops can withstand the detrimental impacts of heat to be further elucidated. The proteomic literature tells us that a wide range of physiological processes will be affected by heat, including photosynthesis, respiration and energy metabolism. Certain proteins are particularly responsive to heat including Rubisco and its chaperone partner Rubisco activase, ATP synthase, oxygen-evolving enhancer proteins and many heat shock proteins (HSPs). In particular, small heat shock proteins (sHSPs), HSP70 and Cpn60 increase in abundance with heat. Many of these most responsive proteins to heat interact with one another and understanding the nature of this interaction remains a priority of future research. Including more staggered application of heat over longer periods to account for acclimation processes and expanding studies to include more reproductive tissues will improve our ability to *heat-proof* crops in a warmer world.

Keywords Heat shock proteins (HSP) · Heat stress · ATP synthase · Oxygen-evolving enhancer protein (OEE) · Rubisco · Rubisco activase

A.P. Scafaro (✉) · O.K. Atkin (✉)
ARC Centre of Excellence in Plant Energy Biology,
Division of Plant Sciences, Research School of Biology,
Australian National University, Canberra, ACT 2601, Australia
e-mail: andrew.scafaro@anu.edu.au

O.K. Atkin
e-mail: owen.atkin@anu.edu.au

G.H. Salekdeh (ed.), *Agricultural Proteomics Volume 2*,
DOI 10.1007/978-3-319-43278-6_7

7.1 The Importance of Understanding Heat Stress

The implications of a warming atmosphere on crop production are already beginning to be realised. An increase in global atmospheric temperatures have reduced global crop yields, more so than changes in precipitation, by an estimated 5.5 and 3.8 % from 1980 to 2008 for wheat (*Triticum aestivum*) and maize (*Zea mays*), respectively [1]. Two forms of heat-stress can be considered: (1) a small but persistent increase in the average daily minimum and maximum temperature; and, (2) more frequent and extreme heat waves. The different forms of heat exposure will have different implications on crop yield. Further complexity arises as an increase in minimum and maximum daily temperatures may impact on yield in separate, even apposing ways. For example, rice (*Oryza sativa*) production is falling with increased minimum daily temperatures but rising with increased maximum daily temperature, presumably due to heat-induced respiratory loss of carbon at night and increased carbon assimilation during the day [2, 3]. More frequent and extreme heat waves are likely to impact negatively on yield. With exposure to extreme heat, many physiological processes, including photosynthesis, respiration and cellular components (e.g. membrane and protein structures) will be irreversibly damaged, leading to cell, tissue and in extreme cases, plant death. This susceptibility to extreme heat occurs despite high-temperature adaptation, with species such as the desert shrub *Larrea* having reduced physiological function above 45 °C [4, 5]. Similarly, another desert species, 'Living Rock Cactus' (*Ariocarpus fissuratus*) has 50 % cell death by 57 °C [6]. The inhibition of yield as a result of heat will be dependent on the developmental stage of the crop and its adaptation and acclimation potential to heat, with heat-dependent inhibition of growth and reproduction varying as widely as 25–45 °C between crop species [7]. Developmental inhibition at the vegetative stage will include heat-stress inhibition of photosynthesis, in conjunction with increased rates of respiratory CO_2 release [8]. Impacts at the reproductive stage will include impairment of anther and pollen development [9]. Importantly, not all crops will be affected by heat equally, reflecting the fact that many of the physiological processes susceptible to heat show adaptation to the environment in which the plant species has evolved. Variation in heat tolerance can arise from plasticity to the prevailing growth temperature [10, 11]. Considering the above factors, if we are to successfully improve the heat-stress tolerance of crops it will be necessary to: (1) understand both long-term incremental increases in the minimum and maximum temperature and periodical extreme heating events on crops; (2) determine susceptibility throughout all life-stages of the crop; (3) understand the basis of acclimated heat-stress tolerance; and, (4) in what genotypes, both intra and inter-specifically, heat-tolerance is found. This will enable modification of crops to maintain yield under supra-optimal temperatures. Given the multifaceted dimensions of heat-stress implications on crop yield, proteomics is, and will continue to be, instrumental in identifying the biochemical basis of heat-tolerance.

7.2 Interactions Between Heat and Other Abiotic Factors

Depending on the severity, most abiotic stresses will directly inhibit protein functionality and membrane stability and cause secondary damage to cellular processes through osmotic and oxidative stress [12]. Despite similar effects, combined stress such as heat and drought will reduce crop production more so than each stress individually and heat is frequently coupled with drought due to the prevalence of water limitations during times of heat [13]. Increasing temperature also leads to reduced plant water relations by drying the air and leading to a higher vapour pressure deficit between the leaf surfaces and surrounding air. This will drive higher rates of transpiration, increasing the loss of water from leaves and speed up the uptake of water by roots. There are predictions of severely reduced yield in major crops like corn due to the affect of increased air temperature on vapour pressure deficit and subsequent water limitations [14]. Reduced access to water by the roots will promote abscisic acid production, leading to stomatal closure and limiting transpiration. This will amplify thermal stress, as leaf temperatures will rise with reduced latent heat transfer as transpiration is reduced. This has been demonstrated in Arabidopsis (*Arabidopsis thaliana*), where a modification of an ABA receptor (PYR1), made it responsive to an agrochemical that, when applied, allowed for drought tolerance through closer of stomata, but at the same time significantly raised leaf temperature due to reduced transpiration [15].

As with water limitations, increased atmospheric CO_2 at a given air temperature reduces stomatal conductance and transpiration, subsequently leading to increases in leaf temperature [16]. Increasing atmospheric CO_2 may therefore raise the heat load experienced by leaves during heat waves. Paradoxically, increased leaf temperature due to increased atmospheric CO_2 and reduced transpiration leads to greater freezing damage as plants become less acclimated to lower temperatures, reducing protective measures during frosts [17]. Cold stress initiates many of the same protein response pathways as heat due to the similar affects of cold on cellular processes such as inhibition of photosynthesis and subsequent oxidative damage and membrane instability. Similarities between heat and cold stress extend to dehydration as ice formation resulting from chilling leads to cell dehydration and subsequent damage to cell membranes, cell solute concentration and protein stability [18], mirroring cellular dehydration damage caused by heat and drought. It is perhaps not surprising that many of the same protective proteins are induced in response to drought, chilling and heat stress. Protective proteins including the Heat Shock Proteins (HSPs) HSP70, HSP90 and Cpn60 respond to heat, drought and cold stress, as does the membrane bound protein complex ATP synthase [19, 20].

7.3 Quantitative Proteomic Profiling of Heat-Stress

Quantitative proteomic techniques for analysis of temperature stress are now routinely employed across both model and crop species, using a myriad of proteomic techniques [19]. What does an analysis of the proteomic literature on heat stress in plants demonstrate? Considered as a whole, short-term (less than 48 h) heat exposure of vegetative leaf tissue is highly represented in the literature (Fig. 7.1). Longer-term heat exposure over days to weeks is mostly limited to analysis of grains during filling. Almost all heat stress reports fall within a heat stress range of 10–20 °C above control temperatures. The vast majority of reports profile leaf and grain tissue rather than reproductive organs and root tissue.

Limited longer-term studies (e.g. >48 h) at a wider range of temperatures (above or below the 10–20 °C heating above control temperatures commonly used in proteomic studies) for vegetative tissue is a major gap in the heat-stress proteome literature. A longer exposure time is relevant considering the knowledge that many physiological processes, such as photosynthesis [21] and respiration [11] acclimate in response to the temperature at which a plant is grown. Such acclimation are likely to lead to different gene expression and protein changes than short-term heat-shock treatments, with the later more representative of a general stress response and the former providing insight into specific physiological adjustments to sustained heat. Acclimation over short time frames (such as pre-conditioning to an initially milder temperature) can *prime* plants to respond in a timely manner to more extreme subsequent temperature. For example, in Arabidopsis, one and a half hours at 38 °C prior to 2 h at 45 °C enables growth to continue when returned to the 22 °C control temperature, while growth is totally suppressed without the 38 °C pre-heating [22].

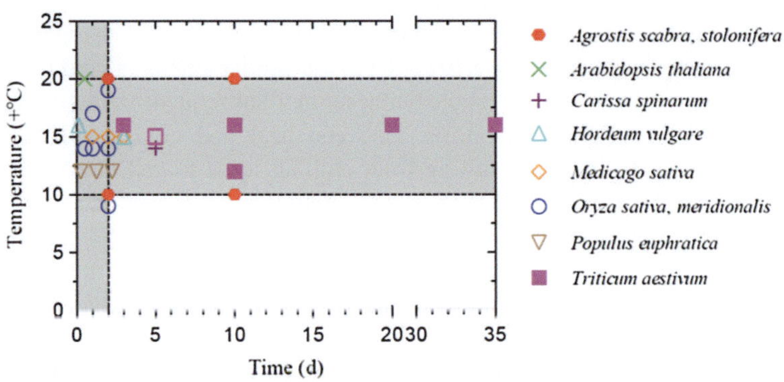

Fig. 7.1 The severity and duration of temperature treatments across heat stress proteomic studies. The magnitude of temperature increase (+°C) from the standard (control) growth temperature is given on the y-axis. The duration of time in days (d) that the plant was exposed to the heat treatment is given on the x-axis. *Open* and *filled symbols* represent studies of vegetative and reproductive tissue, respectively. Each *symbol shape* represents the species type provided in the key. The *shaded areas* demonstrate the temperature and duration of most reported treatments. The graph was generated analysing reports listed in Table 7.1

The acclimation is attributed to induction of HSP101 during 38 °C, which subsequently provides protection at 45 °C [23]. A progressive rather than rapid increase in temperature may enable better understanding of the acclimation process. This method of heat-application may be more reflective of certain heating events in nature, where temperatures increase incrementally over many days.

Functional characterisation of proteins responding to heat stress, pooled across the literature, is provided in Fig. 7.2. Proteins directly related to the stress response pathway, the majority of which are heat shock proteins, are a highly respondent

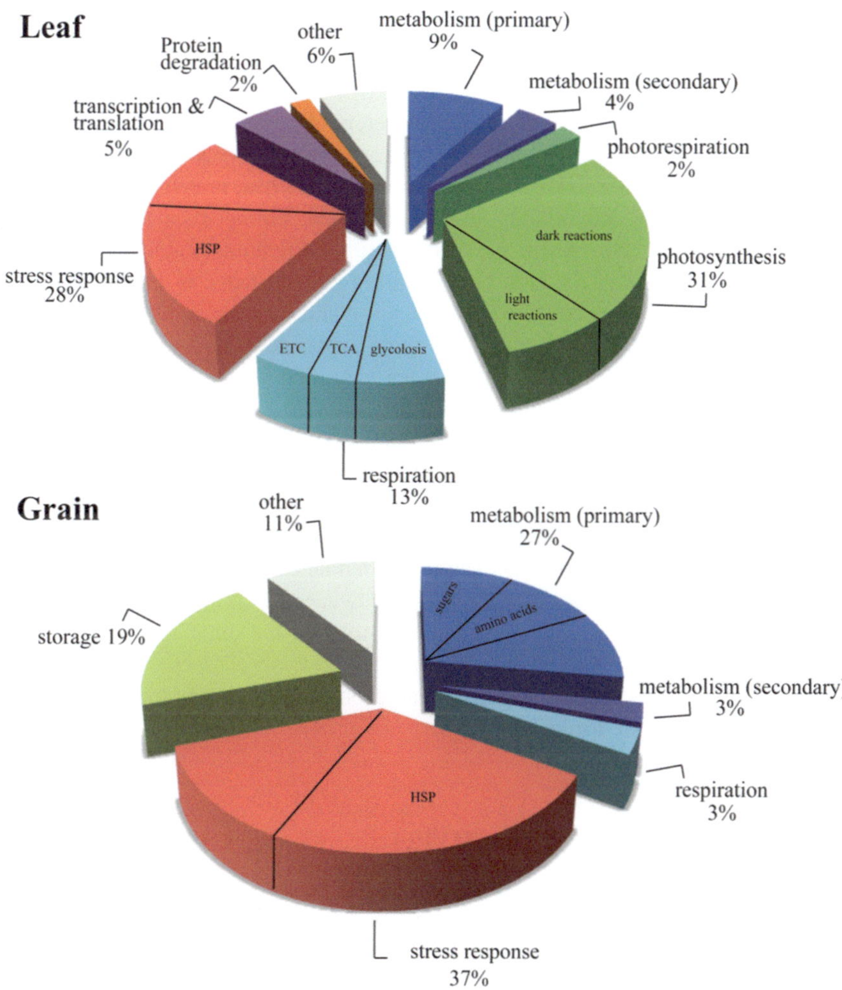

Fig. 7.2 The functional proportion, given as a percentage, of proteins differentially regulated by heat stress in leaf and grain tissue reported in the proteome literature. Metabolism, photosynthesis, respiration and stress response proteins are further segmented to show major categories of proteins within the functional group

functional group to heat in grain and leaves. In leaves, proteins related to photosynthesis respond to heat to the greatest extent, and combined with respiration make up 44 % of the proteome profile. Considering the well-established susceptibility of photosynthesis to heat and enhancement of respiration rates in response to increased temperature, it is unsurprising that many proteins relating to these two important physiological leaf functions change. With well-designed treatments, future proteomic studies may be an important tool in determining the biochemical basis of differences in heat shock and acclimation to photosynthesis and respiration.

7.4 Heat-Responsive Proteins

The information gained from profiling of differentially regulated proteins provides an initial indication of which proteins are responsive to heat and what marker proteins breeders can use to identify genotypes that are heat tolerant. It is therefore interesting to look at individually identified protein homologues that change in abundance with heat most frequently throughout the literature (Table 7.1). Most of the commonly identified proteins both increase and decrease in abundance with exposure to heat. In many cases proteins increase and decrease in abundance under the same treatment conditions within the same experiment. For example, in wheat grain endosperm 11 isoforms of HSP16.9 are identified [24], five isoforms each of Rubisco and Rubisco activase in wheat leaves [25] and seven isoforms of HSP70 in rice leaves [26]. Much of this may be due to post-translational modifications. Despite proteomic profiling not providing any information on the individual functionality of differentially regulated proteins, such *top-down* approaches can provide the basis for subsequent, more focused studies on the particular role of identified targets. Importantly, any change in abundance of the individual protein in response to heat does imply a relationship between that protein and heat stress. Thus, for the purposes of identifying target proteins, the directional change of a particular protein is a secondary consideration. Below is a characterisation of some of the most commonly reported proteins that respond to heat across multiple species and studies as listed in Table 7.1.

7.4.1 Heat Shock Proteins (HSPs)

Not surprisingly, four of the 21 most responsive proteins to heat stress are heat shock proteins (HSPs), including the most commonly reported protein that changes in response to heat, small heat shock proteins (sHSPs). Rather than being a single protein, sHSPs are actually a family of HSPs with molecular mass ranging from 16.9 to 42 kDa; however, for expediency HSPs are grouped as a single protein in Table 7.1. The common features of sHSPs are a conserved carboxyl-terminal domain (α-crystallin domain) and formation of 200–350 kDa holo-oligomers [27].

Table 7.1 Proteins most commonly identified as differing in abundance with heat stress application using two-dimensional protein separation, mass spectrometry and image analysis

Protein	Identified (up:down)[a]	Function	Species	Tissue	Study
sHSPs (Small Heat Shock Proteins ranging from 16.9 to 23 kDa)	49 (49:0)	Small heat shock proteins can prevent thermal denaturation of proteins; They also interact with denatured proteins prior to assisted refolding by larger HSPs	*Carissa spinarum*, *Hordeum vulgare*, *Medicago sativa*, *Oryza sativa*, *Triticum aestivum*	Leaf, grain	[26, 31, 84, 85, 87, 88]
Ribulose-1,5-bisphosphate carboxylase/oxygenase (Rubisco) large subunit (RbcL)	26 (11:15)	Rubisco is the only enzyme responsible for fixation of atmospheric CO_2 into sugars in all plants. Rubisco is made of eight large and eight small subunits	*Arabidopsis thaliana*, *Carissa spinarum*, *Medicago sativa*, *Oryza meridionalis*, *Oryza sativa*, *Triticum aestivum*	Leaf	[25, 51, 85–88]
ATP synthase α, β and ε chains (mitochondria, chloroplasts and vacuoles)	22 (13:9)	ATP synthase is the protein complex that is imbedded in cellular membranes and uses a electrochemical gradient driven flow of protons through its core to drive a motor that mechanically phosphorylates ADP to ATP, the chemical storage unit of all living cells	*Arabidopsis thaliana*, *Carissa spinarum*, *Medicago sativa*, *Oryza meridionalis*, *Oryza sativa*, *Populus euphratica*, *Triticum aestivum*	Leaf, grain	[25, 31, 51, 84–87, 89]
Rubisco activase (RCA)	21 (15:6)	Rubisco activase is a chaperone protein for Rubisco, using ATP to remove inhibitory sugar substrates blocking the active site of Rubisco. It belongs to the AAA^+ family of proteins. It is known to be heat labile	*Arabidopsis thaliana*, *Carissa spinarum*, *Medicago sativa*, *Oryza meridionalis*, *Oryza sativa*, *Populus euphratica*, *Triticum aestivum*	Leaf, grain	[25, 31, 51, 85–89]
HSP70 (Heat shock protein 70 kDa)	20 (20:0)	HSP70 binds to polypeptides during translation and enables the polypeptide to be translocated without incorrect folding	*Arabidopsis thaliana*, *Hordeum vulgare*, *Medicago sativa*, *Oryza meridionalis*, *Oryza sativa*, *Triticum aestivum*	Leaf, grain	[25, 26, 51, 84–86, 88, 90, 91]

(continued)

Table 7.1 (continued)

Protein	Identified (up:down)[a]	Function	Species	Tissue	Study
Cpn60 (chaperonin 60 α and β subunits)	17 (12:5)	Cpn60 is a member of the HSP family that uses ATP to create a hydrophilic environment within its interior that encapsulates substrate proteins and enabling these proteins to correctly fold. It is required to correct folding of RbcL	*Carissa spinarum*, *Oryza meridionalis*, *Oryza sativa*, *Populus euphratica*, *Triticum aestivum*	Leaf	[25, 51, 87–89]
Glyceraldehyde-3-phosphate dehydrogenase (GADPH)	13 (11:2)	An enzyme in the glycolysis pathway and Calvin cycle, interchangeably converting glyceraldehyde-3-phosphate to glycerate-1,3-bisphosphate	*Agrostis scabra*, *Agrostis stolonifera*, *Arabidopsis thaliana*, *Hordeum vulgare*, *Medicago sativa*, *Oryza meridionalis*, *Oryza sativa*, *Populus* euphratica, *Triticum aestivum*	Leaf, root	[25, 26, 51, 86, 88–90, 92]
Oxygen evolving enhancer protein (OEE)	12 (6:6)	Peptides associated with the oxygen-evolving complex (OEC) of Photosystem II, where water is split into O_2, H^+ and e^-	*Arabidopsis thaliana*, *Carissa spinarum*, *Medicago sativa*, *Hordeum vulgare*, *Triticum aestivum*	Leaf	[25, 26, 85–87, 90]
Ribulose-1,5-bisphosphate carboxylase/oxygenase (Rubisco) small subunit (RbcL)	9 (2:7)	The smaller subunit of the Rubisco complex with no active site but needed for Rubisco function in higher plants	*Arabidopsis thaliana*, *Oryza sativa*, *Medicago sativa*, *Triticum aestivum*	Leaf	[25, 26, 85, 86]
Fructose-bisphosphate aldolase (FBP)	6 (2:4)	Involved in gluconeogenesis/glycolysis and the Calvin cycle splitting 6 or 7-carbon sugars such as FBP and sedoheptulose-1,7-bisphosphate, or combing 3 or 4-carbon sugars, including G3P and erythrose-4-phosphate	*Agrostis scabra*, *Agrostis stolonifera*, *Arabidopsis thaliana*, *Hordeum vulgare*, *Medicago sativa*	Leaf, root	[26, 86, 90, 92]

(continued)

Table 7.1 (continued)

Protein	Identified (up:down)[a]	Function	Species	Tissue	Study
Malate dehydrogenase (MDH)	6 (2:4)	An enzyme that interconverts malate to oxaloacetate, a key component of the TCA cycle	*Agrostis scabra*, *Medicago sativa*, *Oryza sativa*	Leaf, root	[85, 88, 92]
Glycine dehydrogenase	5 (4:1)	An enzyme involved in glycine metabolism, converting glycine to glyoxylate a step in photorespiration	*Oryza meridionalis*, *Oryza sativa*	Leaf	[26, 51, 88]
GTP-binding protein	5 (2:3)	Involved in signal transduction, regulating kinases and facilitates protein import into nucleus and RNA export from nucleus	*Agrostis scabra*, *Agrostis stolonifera*, *Populus* euphratica, *Triticum aestivum*	Leaf, grain, root	[31, 89, 92]
Methionine synthase	5 (0:5)	Catalyses the conversion of homocysteine to methionine, the final step in synthesis of methionine, an essential amino acid	*Agrostis scabra*, *Agrostis stolonifera*, *Populus* euphratica, *Triticum aestivum*	Leaf, grain, root	[89, 91, 92]
Thiazole synthase (thi1)	5 (4:1)	An enzyme involved in the thiazole moiety in thiamine (vitamin B1) biosynthesis	*Oryza meridionalis*, *Oryza sativa*, *Populus* euphratica	Leaf	[26, 51, 89]
Triosephosphate isomerase	5 (1:4)	An enzyme that interconverts dihydroxyacetone phosphate to G3P, involved in glycolysis and the Calvin Cycle	*Arabidopsis thaliana*, *Carissa spinarum*, *Hordeum vulgare*, *Oryza meridionalis*, *Triticum aestivum*	Leaf, grain	[32, 51, 86, 87, 91]
S-adenosylmethionine synthase (SAMS)	5 (0:5)	Catalyses the production of *S*-adenosyl-L-methionine from L-methionine and ATP	*Agrostis scabra*, *Oryza meridionalis*, *Oryza sativa*, *Triticum aestivum*	Leaf, root	[25, 51, 88, 92]
Elongation factor (EF-TU)	4 (3:1)	Facilitates the elongation process during translation of RNA to polypeptides	*Arabidopsis thaliana*, *Hordeum vulgare*, *Triticum aestivum*	Leaf, grain	[31, 86, 90]

(continued)

Table 7.1 (continued)

Protein	Identified (up:down)[a]	Function	Species	Tissue	Study
HSP90 (heat shock protein 90 kDa)	4 (4:0)	Facilitates in the folding of newly synthesised polypeptides, stabilising proteins or refolding of denatured proteins, particularly signal transduction proteins	*Medicago sativa*, *Oryza meridionalis*, *Populus* euphratica	Leaf	[51, 85, 89]
Glutamine synthetase	3 (2:1)	Catalyses the conversion of glutamate and ammonia to glutamine	*Medicago sativa*, *Oryza sativa*, *Triticum aestivum*	Leaf, grain	[85, 88, 91]
Thioredoxin	3 (2:1)	Involved in redox signalling through cysteine thiol-disulphide exchange	*Arabidopsis thaliana*, *Populus* euphratica *Oryza sativa*	Leaf	[26, 86, 89]

[a]The number of protein spots changing in abundance with heat application and identified as the corresponding protein across the reported studies. The proportion of times the protein increased or decreased in abundance is given in brackets, respectively

sHSPs can interact directly with denatured proteins and improve thermal stability and functionality by preventing aggregation [28, 29]. The diverse functionality of sHSP during heat stress is evident in that sHSPs are responsive to heat in leaf, root, anther, carpel and seed tissue [30–33]. Apart from heat stabilisation of proteins, sHSPs work in tandem with larger HSPs, particularly HSP70 to facilitate protein folding and prevent protein aggregation. An example is seen in pea (*Pisum sativum*), where heat-denatured proteins that initially interact with sHSP18.1 have dramatically improved subsequent refolding by HSP70 [34].

Cpn60 in plants, known as HSP60 in animals and groEL in bacteria, is another highly represented heat-responsive protein and member of the chaperonin/heat shock protein (HSP) family. Cpn60 provides a hydrophilic cavity conducive to correct folding of polypeptides, limiting interference from components of the cellular environment and usually interacts with polypeptides subsequent to HSP70 interaction [35, 36]. Increased abundance of both HSP70 and Cpn60 indicates an overall upregulation of the protein folding and stabilisation pathway with heat stress.

Direct evidence of HSP requirement for heat stress tolerance and survival is surprisingly limited. One example is HSP101 in Arabidopsis, required for acclimated and basal thermal tolerance and when substantially suppressed leads to death of all plants within 6 days post heat exposure [23]. Expression of Arabidopsis HSP101 in rice provides thermal protection, with recovery of growth in seedling after 47 and 50 °C heat exposure only in plants expressing the transgenic HSP101 [37]. Homologues of the HSP101 family increase in abundance endogenously in rice exposed to 42 °C [26, 38]. In terms of sHSPs, rice seedlings exposed to 50 °C for 2 h and overexpressing sHSP17.7 survive, as determined by resumption of growth or maintained ion gradients of leaf cells, while seedlings not overexpressing sHSP17.7 do not [39]. Arabidopsis overexpressing wheat HSP26 have greater germination rates and superior growth rates than wild type plants when exposed to 35 °C [33]. The improved thermal tolerance was correlated with maintained efficiency of photosystem II. This is intriguing considering a recent study has shown heat induced sHSP26 interacts with oxygen-evolving enhancer proteins of PSII [40].

The regulation of HSPs is complex. There are interactions between variants of heat stress transcription factors, transcription factors and HSPs and interactions among HSPs, all influenced by genotype and environmental conditions [41]. The complexity in HSP expression is understandable considering the functional diversity of HSPs in maintaining active proteins through providing correct folding, inter-membrane transport and oligomer assembly of proteins [42]. This occurs irrespective of heat stress with heat stabilisation of proteins somewhat of a secondary role, at least for the larger HSPs. This complexity makes for a difficult holistic interpretation of heat stress response and makes it difficult to determine what genetic modifications to crops will be beneficial, both during exposure to heat stress and during homeostatic growth conditions (i.e. an improvement in heat tolerance through altered HSP expression may impede physiological function under non-stressed conditions). Proteomic analysis will be required to tease apart this complex yet subtle interplay of HSP adjustment to heat.

7.4.2 Rubisco and Rubisco Activase

Rubulose-1,5-bisphosphate carboxylase oxygnease (Rubisco) and its regulatory chaperone Rubisco activase (RCA) are two photosynthesis proteins identified in multiple proteomic studies. While Rubisco is thermally stable, there is a large body of work since its discovery documenting the heat-labile nature of Rubisco activase, even at moderately high temperature, below 40 °C [43–45]. RCA is the regulatory partner of Rubisco keeping it in an active state through ATP-driven remodelling of the conformation of Rubisco, clearing the active site of sugar-phosphate inhibitors [46]. In many species including the major crop species rice, wheat and maize, there are two isoforms of Rubisco activase [47]. The larger mass isoform (often denoted as the α isoform) is redox regulated by thioredoxin-f [48] and is preferentially expressed during heat stress [49–51]. Heat dependent changes in the redox state of the chloroplast may therefore explain the increased up-regulation of the α isoform of RCA. The susceptibility and heat-dependent regulation of RCA would account for the reoccurring identification of RCA in proteomic heat studies. It is becoming increasingly apparent that improving photosynthesis and subsequent crop yield will undoubtable involve improvements to RCA thermal stability [47, 52]. More in-depth proteomic analysis specifically targeting the response of RCA and its isoforms to heat is one method by which this may be achieved, especially considering the exact mechanism by which RCA reactivates the active site of Rubisco is not known.

Although temperature limitations in photosynthesis are usually attributed to heat-labile RCA and not the thermally stable Rubisco, abundance changes of both the small and large subunits of Rubisco indicate a dynamic regulation of Rubisco in response to heat. Indeed, Yamori et al. [53] found that spinach (*Spinacia oleracea*) grown at 30 and 15 °C had changes in Rubisco kinetics that enabled in vitro Rubisco activity from 30 °C plants to be more thermostable, although Rubisco activity from either growth temperature did not decline until above 40 °C, well above the temperature where net photosynthesis inhibition occurred. Moving the plants from heat stressed conditions to lower growth temperature for 2 weeks reversed the Rubisco kinetic changes. This demonstrates that although Rubisco, independent of its activation, is intrinsically thermostable and does not limit photosynthesis, the kinetic properties of the protein can change in response to temperature. This may explain the regulation of Rubisco isoforms in response to heat stress reported in the proteomic literature. A possible mechanism for this regulation may be phosphorylation, as the Rubisco of rice is phosphorylated in response to heat [54]. In many reports Rubisco falls in abundance, particularly the small subunit. In rice, temperatures of 45–50 °C reduces Rubisco content of leaves within hours, especially the small subunit [55]. Reduced Rubisco content in response to heat may be a general response to heat stress in many crop species and should be further explored.

7.4.3 ATP Synthase

The ATP synthase complex is driven by proton translocation and responsible for inorganic phosphate (P_i) phosphorylation of ADP to ATP, the essential electro-chemical energy currency of all cells [56]. Proteomic analysis shows the α, β, γ, and δ subunits of ATP synthase, from chloroplasts, mitochondria and vacuoles, change in abundance across many species and heat applications. It is likely proteomic adjustments in ATP synthase indicate changes to turnover rate of the protein due to susceptibility to heat degradation. The reactive oxygen species singlet oxygen and hydrogen peroxide, commonly induced under stress, inhibit ATP synthase activity and promote degradation of the complex [57, 58]. A loss of ATP synthase activity will have major impacts on the ability of a plant to photosynthesise and respire. For example, a drought-induced reduction in photosynthesis in sunflower (*Helianthus annuus*) was directly attributed to a reduction in ATP synthase, reducing ATP leaf content and inhibiting RuBP regeneration, more so than the commonly attributed limitations in CO_2 uptake or Rubisco activity [59]. In terms of heat, yeast (*Saccharomyces cerevisiae*) require a specific protein to facilitate the assembly and stability of ATP synthase during heat exposure [60]. Furthermore, a chloroplast 26-kDa sHSP interacts with ATP synthase subunits under heat stress in maize, presumably to protect against thermal damage [40]. When the sHSP26 was reduced in expression through RNAi, there was also a greater reduction in ATP synthase subunits with heat. This implies that ATP synthase is susceptible to high temperature and a reduced spot abundance correlates to thermal damage. Collectively, these observations suggest that ATP synthase is susceptible to heat and abundance changes in ATP synthase recorded in the proteome literature indicate a general susceptibility across species. Identifying heat-induced ATP synthase isoforms that are thermally stable will provide a mechanism for improving heat tolerance and should be a priority in future proteomic studies.

7.4.4 Oxygen-Evolving Enhancer Protein

Oxygen evolving enhancer proteins (OEE) found in the oxygen-evolving complex of PSII are another highly represented protein group in the heat proteome. Susceptibility of photosystem II to heat stress is well known [61]. Damage by stress leads to loss of Mn ions from the oxygen-evolving complex, followed by an overexcited chlorophyll reaction centre electron state and subsequent free radical production, with particular damage to the PSII reaction centre [62]. The susceptibility of PSII to stress is evident in the continual requirement to replace impaired DI protein of the PSII reaction centre under such conditions [63]. OEE inhibits loss of Mn ions from the oxygen-evolving complex, and when OEE is absent O_2 evolution is severely impaired [64, 65]. Loss of OEE from the PSII in direct consequence of heating to 50 °C results in a reduction in oxygen evolution [66]. Therefore, the representation of OEE in the proteome literature implies a particular susceptibility of the Mn cluster of the oxygen-evolving complex to heat and the potential to protect against this susceptibility through OEE regulation. OEE proteins have been shown in Arabidopsis to interact with a kinase and phosphorylation of OEE increased with pathogen stress [67]. Regulation of OEE through phosphorylation in response to stress may further explain the multiple identifications of this protein family in response to heat.

7.4.5 Thiamine

Intriguingly, a protein (thiamine biosynthesis protein or Thi1) involved in the synthesis of thiamine (an important metabolic co-factor), increases in abundance with heat application. The importance and function of Thi1 and thiamine in the heat-stress response of plants is currently tenuous. Independent of thiamine biosynthesis, Thi1 seems to have a secondary function of limiting DNA damage, particularly in mitochondria [68]. Increased thiamine in response to biotic [69, 70] and abiotic stress is associated with stress tolerance and is purportedly linked to the oxidative stress response pathway [71, 72]. Interestingly, a ^{15}N proteome turnover study of barley (*Hordeum vulgare*) identified Thi1 as the most turned over protein during the light period in leaves [73]. Characterising the function of Thi1 and thiamine in heat-stress should be a prioritised area of future research considering the frequency in which it is identified in heat stress proteome studies and the limited knowledge about its exact role in mitigating heat stress. A good starting point would be to separating the importance of Thi1 independent of thiamine biosynthesis and quantifying thiamine levels in heat stressed tissue.

7.5 Protein-Protein Interactions

There is growing evidence that many of the proteins most responsive to heat are associated with one another (Fig. 7.3). For example, it is intriguing that sHSP18.1 prevents thermal aggregation at temperatures ranging from 34 to 45 °C of the heat sensitive malate dehydrogenase (MDH) and glyceraldehyde phosphate dehydrogenase (GADPH) [29]. The prevention of aggregation of MDH and GADPH is particularly interesting considering they are two of the other commonly reported proteins that differ in abundance in response to heat, as reported in Table 7.1. Oxygen evolving enhancer protein and ATP synthase are two proteins most commonly affected by heat, and as previously mentioned; sHSP26 from Maize is shown to interact with these two proteins during heat exposure. Furthermore, a pull down experiment of rice sHSP16.9 identified interaction with ATP synthase and GADPH [38]. HSP70 and Cpn60 are complementary to one another and highly represented in the heat-stress literature. Both HSP70 and Cpn60 interact with

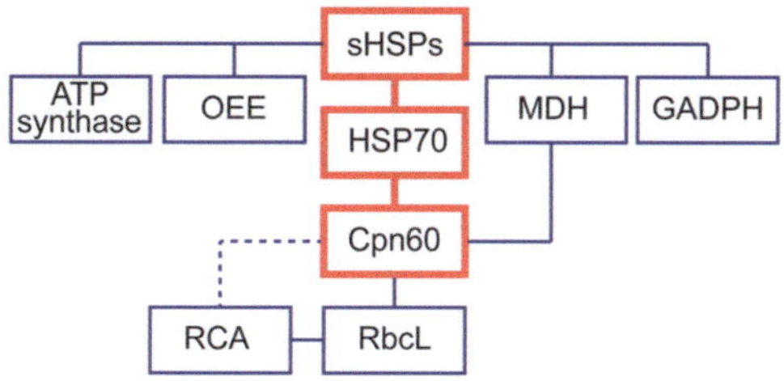

Fig. 7.3 A schematic diagram with connecting lines showing the established interactions between the most commonly reported proteins that respond to heat in crops: small heat shock proteins (sHSPs), heat shock protein 70 (HSP70), chaperonin 60 (Cpn60), ATP synthase, oxygen-evolving enhancer proteins (OEE), malate dehydrogenase (MDH), glyceraldehyde-3-phosphate dehydrogenase (GADPH), Rubisco large subunit (RbcL) and Rubisco activase (RCA). The core HSPs of the heat stress response are outlined with *thicker text boxes* and *connecting lines*. The *dashed line* represents a circumspect interaction

sHSP. Moreover, Cpn60 facilitates the folding of the Rubisco large subunit [74]. There may also be an association between Cpn60 and Rubisco activase during heat stress based on co-elution of the proteins during affinity chromatography [75]. Induction of sHSP, HSP70 and Cpn60 across species and tissue indicates that the HSP response system is critical for crop heat tolerance. This is not surprising considering aggregation of proteins are particularly detrimental to the functioning of the cell, not only because the aggregating protein is rendered non-functional, but also due to interference of other proteins and cellular processes coming into contact with protein aggregates [76].

Piecing together how these *core* heat-responsive proteins and pathways directly interact in the context of heat stress should be a priority of proteomic studies over the coming years and could dramatically increase our knowledge of how plants respond and ultimately are able to tolerate thermal stress.

7.6 Reproductive Heat Susceptibility

Heat can have a detrimental impact on crop yield at both the vegetative and reproductive life stages. Inhibition of vegetative growth due to heat stress can lead to reduced biomass at anthesis and subsequent reduction in yield as grain filling is reliant on mobilisation of photoassimilates acquired by vegetative tissue. For example, wheat stem reserves of non-structural carbohydrates correlate with grain filling, irrespective of high temperature, so increased stem reserves acquired during vegetative growth lead to greater yield when wheat is heat stressed during grain filling [77]. Heat-stress during the reproductive phase can reduce yield directly through inhibition of fruit or grain development. The reproductive tissue of ovaries, anthers and pollen in many species including the cereals wheat, barley and rice, are susceptible to high temperature [9]. The uninucleate stage of pollen development is particularly susceptible to high temperature [78]. Pollen susceptibility is evident in that increasing day temperatures from 28 to 36 °C resulted in pollen abortion of between 70 and 100 % across a diverse range of eight species, despite all the species being endemic to hot environments [79]. In rice, a temperature of 36 °C relative to 5 °C cooler ambient temperature (i.e. 30 °C) caused 48 % spikelet sterility averaged across 10 cultivars [80]. Putting this in perspective, the temperature optimum of net photosynthesis of rice, where assimilation of carbohydrates is at a maximum, occurs at 30 °C and has minimal decline until temperature is greater than 40 °C [81, 82]. Clearly, from a food security perspective, reproductive rather than vegetative susceptibility to heat is more important, when one considers that the threshold for thermal damage is lower and reproductive success equates directly to yield. A qualification may be that the most susceptible reproductive processes such as anther development and pollen microspore biosynthesis occurs over days to weeks while vegetative growth occurs over a longer period of months, increasing the probability of more frequent exposure to heat of the latter, although for many crops reproduction occurs at a warmer part of the year. Nevertheless, it is surprising that there are limited proteomic studies focusing on heat exposure of reproductive organs like anthers and pollen grains. One such study by Jagadish et al. [30] looked at the anther proteome comparing rice cultivars differing in spikelet fertility. Two cultivars, N22 and IR64, with superior fertility rates to a third highly susceptible cultivar, Moroberekan, had a significant increase in only two identified proteins, a 24 kDa sHSP and another 19 kDa stress-responsive protein. The susceptible Moroberekan had a significant increase in an iron deficiency protein with heat. Reduced pollen development for this cultivar was therefore attributed to a lack of iron, which is needed for pollen development. Notably, six of the 13 proteins responding to heat had unknown function [30] and therefore improved characterisation of plant reproductive genes is required to supplement future proteomic studies of reproductive tissue.

Yield is not only dependent on fertilisation rates but fruit or seed mass, which is susceptible to heat. Indeed, there is a positive correlation between starch synthesis and starch synthase activity in wheat grain and activity of starch synthase is severely reduced above 30 °C [83]. Starch synthase in the endosperm and water-soluble fraction of wheat grain increases in abundance with exposure to 34 °C during grain filling [31, 84]. The heat dependent reduction in starch synthase activity may therefore be compensated for by heat dependent induction of starch synthase.

Further exploration of the differences and similarities in the proteome response to heat between developmental stages and tissues will provide insight into what proteins are essential for a systemic response to heat stress and what proteins are localised to a certain life-stage or tissue.

7.7 Future Perspectives

The proteomic study of crops to heat stress is enabling an understanding of the biochemical mechanisms of heat tolerance. The most notable discovery so far is the ubiquity of small chaperone proteins in the heat-stress response. sHSP increase in abundance in leaf, grain and reproductive organs of many crop species and across many temperature treatments. Understanding the exact mechanism and interaction partners of sHSP should be a priority of future proteomic research. Other proteins responsive to heat include HSP70 and Cpn60, Rubisco and Rubisco activase, ATP synthase and oxygen-evolving enhancer protein. Elucidating the temperature-dependent function of these commonly occurring heat responsive proteins and how they interact with one another is needed. Identifying post-translational modification and thermally stable homologues of these heat-responsive proteins may provide a mechanism for improving the tolerance of crop species to heat by introgressing desirable germplasm either through conventional breeding or transgenic modification. One observation of the heat stress proteome literature is that experimental treatments sit within a narrow band of temperature and exposure times. This should be expanded to cover progressive heat application, sustained over longer time frames, providing greater coverage of the natural temperature regimes crops experience. This will enable temperature acclimation and identification of proteins responsible for acclimation. Finally, a narrow focus on vegetative leaf tissue and cereal grain should be expanded to cover other important tissue types, particularly anthers, ovaries and pollen gametes, considering the overt heat-susceptibility of these reproductive tissues to high temperature.

References

1. Lobell DB, Schlenker W, Costa-Roberts J (2011) Climate trends and global crop production since 1980. Science 333:616–620
2. Peng S, Huang J, Sheehy JE, Laza RC, Visperas RM, Zhong X et al (2004) Rice yields decline with higher night temperature from global warming. Proc Natl Acad Sci USA 101:9971–9975
3. Welch JR, Vincent JR, Auffhammer M, Moya PF, Dobermann A, Dawe D (2010) Rice yields in tropical/subtropical Asia exhibit large but opposing sensitivities to minimum and maximum temperatures. Proc Natl Acad Sci 107:14562–14567
4. Hamerlynck EP, Huxman TE, Loik ME, Smith SD (2000) Effects of extreme high temperature, drought and elevated CO_2 on photosynthesis of the Mojave Desert evergreen shrub, *Larrea tridentata*. Plant Ecol 148:183–193
5. Mooney HA, Björkman O, Collatz GJ (1978) Photosynthetic acclimation to temperature in the desert shrub, *Larrea divaricata*: Carbon dioxide exchange characteristics of intact leaves. Plant Physiol 61:406–410
6. Garrett TY, Huynh C-V, North GB (2010) Root contraction helps protect the "living rock" cactus *Ariocarpus fissuratus* from lethal high temperatures when growing in rocky soil. Am J Bot 97:1951–1960
7. Wahid A, Gelani S, Ashraf M, Foolad MR (2007) Heat tolerance in plants: an overview. Environ Exp Bot 61:199–223

8. Way D, Yamori W (2014) Thermal acclimation of photosynthesis: on the importance of adjusting our definitions and accounting for thermal acclimation of respiration. Photosynth Res 119:89–100
9. Barnabás B, Jäger K, Fehér A (2008) The effect of drought and heat stress on reproductive processes in cereals. Plant Cell Environ 31:11–38
10. Berry J, Bjorkman O (1980) Photosynthetic response and adaptation to temperature in higher plants. Annu Rev Plant Physiol 31:491–543
11. Atkin OK, Tjoelker MG (2003) Thermal acclimation and the dynamic response of plant respiration to temperature. Trends Plant Sci 8:343–351
12. Vinocur B, Altman A (2005) Recent advances in engineering plant tolerance to abiotic stress: achievements and limitations. Curr Opin Biotechnol 16:123–132
13. Mittler R (2006) Abiotic stress, the field environment and stress combination. Trends Plant Sci 11:15–19
14. Ort DR, Long SP (2014) Limits on yields in the corn belt. Science 344:484–485
15. Park S-Y, Peterson FC, Mosquna A, Yao J, Volkman BF, Cutler SR (2015) Agrochemical control of plant water use using engineered abscisic acid receptors. Nature 520:545–548
16. Field CB, Jackson RB, Mooney HA (1995) Stomatal responses to increased CO_2: implications from the plant to the global scale. Plant Cell Environ 18:1214–1225
17. Loveys BR, Egerton JJG, Ball MC (2006) Higher daytime leaf temperatures contribute to lower freeze tolerance under elevated CO_2. Plant Cell Environ 29:1077–1086
18. Guy CL (1990) Cold acclimation and freezing stress tolerance: role of protein metabolism. Annu Rev Plant Physiol Plant Mol Biol 41:187–223
19. Neilson KA, Gammulla CG, Mirzaei M, Imin N, Haynes PA (2010) Proteomic analysis of temperature stress in plants. Proteomics 10:828–845
20. Ashoub A, Baeumlisberger M, Neupaertl M, Karas M, Brüggemann W (2015) Characterization of common and distinctive adjustments of wild barley leaf proteome under drought acclimation, heat stress and their combination. Plant Mol Biol 87:459–471
21. Yamori W, Hikosaka K, Way D (2014) Temperature response of photosynthesis in C_3, C_4, and CAM plants: temperature acclimation and temperature adaptation. Photosynth Res 119:101–117
22. Hong S-W, Vierling E (2000) Mutants of *Arabidopsis thaliana* defective in the acquisition of tolerance to high temperature stress. Proc Natl Acad Sci 97:4392–4397
23. Queitsch C, Hong S-W, Vierling E, Lindquist S (2000) Heat shock protein 101 plays a crucial role in thermotolerance in Arabidopsis. Plant Cell 12:479–492
24. Skylas DJ, Cordwell SJ, Hains PG, Larsen MR, Basseal DJ, Walsh BJ et al (2002) Heat shock of wheat during grain filling: proteins associated with heat-tolerance. J Cereal Sci 35:175–188
25. Wang X, Dinler BS, Vignjevic M, Jacobsen S, Wollenweber B (2015) Physiological and proteome studies of responses to heat stress during grain filling in contrasting wheat cultivars. Plant Sci 230:33–50
26. Lee D-G, Ahsan N, Lee S-H, Kang KY, Bahk JD, Lee I-J et al (2007) A proteomic approach in analyzing heat-responsive proteins in rice leaves. Proteomics 7:3369–3383
27. Sun W, Van Montagu M, Verbruggen N (2002) Small heat shock proteins and stress tolerance in plants. Biochim Biophys Acta (BBA) Gene Struct Expr 1577:1–9
28. Lee GJ, Pokala N, Vierling E (1995) Structure and in vitro molecular chaperone activity of cytosolic small heat shock proteins from pea. J Biol Chem 270:10432–10438
29. Lee GJ, Roseman AM, Saibil HR, Vierling E (1997) A small heat shock protein stably binds heat-denatured model substrates and can maintain a substrate in a folding-competent state. EMBO J 16:659–671
30. Jagadish SVK, Muthurajan R, Oane R, Wheeler TR, Heuer S, Bennett J et al (2010) Physiological and proteomic approaches to address heat tolerance during anthesis in rice (*Oryza sativa* L.). J Exp Bot 61:143–156
31. Majoul T, Bancel E, Triboï E, Ben Hamida J, Branlard G (2004) Proteomic analysis of the effect of heat stress on hexaploid wheat grain: characterization of heat-responsive proteins from non-prolamins fraction. Proteomics 4:505–513

32. Sule A, Vanrobaeys F, Hajos G, Van Beeumen J, Devreese B (2004) Proteomic analysis of small heat shock protein isoforms in barley shoots. Phytochemistry 65:1853–1863
33. Chauhan H, Khurana N, Nijhavan A, Khurana JP, Khurana P (2012) The wheat chloroplastic small heat shock protein (sHSP26) is involved in seed maturation and germination and imparts tolerance to heat stress. Plant Cell Environ 35:1912–1931
34. Lee GJ, Vierling E (2000) A small heat shock protein cooperates with heat shock protein 70 systems to reactivate a heat-denatured protein. Plant Physiol 122:189–198
35. Hartl FU (1996) Molecular chaperones in cellular protein folding. Nature 381:571–580
36. Priya S, Sharma SK, Goloubinoff P (2013) Molecular chaperones as enzymes that catalytically unfold misfolded polypeptides. FEBS Lett 587:1981–1987
37. Katiyar-Agarwal S, Agarwal M, Grover A (2003) Heat-tolerant basmati rice engineered by over-expression of hsp101. Plant Mol Biol 51:677–686
38. Chen X, Lin S, Liu Q, Huang J, Zhang W, Lin J et al (2014) Expression and interaction of small heat shock proteins (sHsps) in rice in response to heat stress. Biochim Biophys Acta (BBA) Proteins Proteom 1844:818–828
39. Murakami T, Matsuba S, Funatsuki H, Kawaguchi K, Saruyama H, Tanida M et al (2004) Over-expression of a small heat shock protein, sHSP17.7, confers both heat tolerance and UV-B resistance to rice plants. Mol Breed 13:165–175
40. Hu X, Yang Y, Gong F, Zhang D, Zhang L, Wu L et al (2015) Protein sHSP26 improves chloroplast performance under heat stress by interacting with specific chloroplast proteins in maize (*Zea mays*). J Proteom 115:81–92
41. Fragkostefanakis S, Röth S, Schleiff E, Scharf K-D (2015) Prospects of engineering thermotolerance in crops through modulation of heat stress transcription factor and heat shock protein networks. Plant Cell Environ 38:1881–1895
42. Vierling E (1991) The roles of heat shock proteins in plants. Annu Rev Plant Physiol Plant Mol Biol 42:579–620
43. Salvucci ME, Anderson JC (1987) Factors affecting the activation state and the level of total activity of ribulose bisphosphate carboxylase in tobacco protoplasts. Plant Physiol 85:66–71
44. Salvucci ME, Osteryoung KW, Crafts-Brandner SJ, Vierling E (2001) Exceptional sensitivity of Rubisco activase to thermal denaturation in vitro and in vivo. Plant Physiol 127:1053–1064
45. Crafts-Brandner SJ, Salvucci ME (2000) Rubisco activase constrains the photosynthetic potential of leaves at high temperature and CO_2. Proc Natl Acad Sci USA 97:13430–13435
46. Salvucci ME, Crafts-Brandner SJ (2004) Inhibition of photosynthesis by heat stress: the activation state of Rubisco as a limiting factor in photosynthesis. Physiol Plant 120:179–186
47. Carmo-Silva E, Scales JC, Madgwick PJ, Parry MJ (2015) Optimizing Rubisco and its regulation for greater resource use efficiency. Plant, Cell Environ 38:1817–1832
48. Zhang N, Portis AR (1999) Mechanism of light regulation of Rubisco: a specific role for the larger Rubisco activase isoform involving reductive activation by thioredoxin-f. Proc Natl Acad Sci 96:9438–9443
49. Crafts-Brandner SJ, Van De Loo FJ, Salvucci ME (1997) The two forms of ribulose-1,5-bisphosphate carboxylase/oxygenase activase differ in sensitivity to elevated temperature. Plant Physiol 114:439–444
50. Wang D, Li XF, Zhou ZJ, Feng XP, Yang WJ, Jiang DA (2010) Two Rubisco activase isoforms may play different roles in photosynthetic heat acclimation in the rice plant. Physiol Plant 139:55–67
51. Scafaro AP, Haynes PA, Atwell BJ (2010) Physiological and molecular changes in *Oryza meridionalis* Ng., a heat-tolerant species of wild rice. J Exp Bot 61:191–202
52. Parry MAJ, Andralojc PJ, Scales JC, Salvucci ME, Carmo-Silva AE, Alonso H et al (2013) Rubisco activity and regulation as targets for crop improvement. J Exp Bot 64:717–730
53. Yamori W, Suzuki K, Noguchi KO, Nakai M, Terashima I (2006) Effects of Rubisco kinetics and Rubisco activation state on the temperature dependence of the photosynthetic rate in spinach leaves from contrasting growth temperatures. Plant Cell Environ 29:1659–1670
54. Chen X, Zhang W, Zhang B, Zhou J, Wang Y, Yang Q et al (2011) Phosphoproteins regulated by heat stress in rice leaves. Proteome Sci 9:37

55. Bose A, Ghosh B (1995) Effect of heat stress on ribulose 1,5-bisphosphate carboxylase in rice. Phytochemistry 38:1115–1118
56. Boyer PD (1997) The ATP synthase—a splendid molecular machine. Annu Rev Biochem 66:717–749
57. Buchert F, Schober Y, Römpp A, Richter ML, Forreiter C (2012) Reactive oxygen species affect ATP hydrolysis by targeting a highly conserved amino acid cluster in the thylakoid ATP synthase γ subunit. Biochim Biophys Acta (BBA) Bioenerg 1817:2038–2048
58. Sweetlove LJ, Heazlewood JL, Herald V, Holtzapffel R, Day DA, Leaver CJ et al (2002) The impact of oxidative stress on Arabidopsis mitochondria. Plant J 32:891–904
59. Tezara W, Mitchell VJ, Driscoll SD, Lawlor DW (1999) Water stress inhibits plant photosynthesis by decreasing coupling factor and ATP. Nature 401:914–917
60. Lefebvre-Legendre L, Vaillier J, Benabdelhak H, Velours J, Slonimski PP, Di Rago J-P (2001) Identification of a nuclear gene (FMC1) required for the assembly/stability of yeast mitochondrial F1-ATPase in heat stress conditions. J Biol Chem 276:6789–6796
61. Allakhverdiev S, Kreslavski V, Klimov V, Los D, Carpentier R, Mohanty P (2008) Heat stress: an overview of molecular responses in photosynthesis. Photosynth Res 98:541–550
62. Takahashi S, Murata N (2008) How do environmental stresses accelerate photoinhibition? Trends Plant Sci 13:178–182
63. Takahashi S, Badger MR (2011) Photoprotection in plants: a new light on photosystem II damage. Trends Plant Sci 16:53–60
64. Semin BK, Davletshina LN, Ivanov II, Seibert M, Rubin AB (2012) Rapid degradation of the tetrameric Mn cluster in illuminated, PsbO-depleted photosystem II preparations. Biochemistry (Moscow) 77:152–156
65. Popelkova H, Commet A, Kuntzleman T, Yocum CF (2008) Inorganic cofactor stabilization and retention: the unique functions of the two PsbO subunits of eukaryotic photosystem II. Biochemistry 47:12593–12600
66. Enami I, Kitamura M, Tomo T, Isokawa Y, Ohta H, Katoh S (1994) Is the primary cause of thermal inactivation of oxygen evolution in spinach PS II membranes release of the extrinsic 33 kDa protein or of Mn? Biochim Biophys Acta (BBA) Bioenerg 1186:52–58
67. Yang EJ, Oh YA, Lee ES, Park AR, Cho SK, Yoo YJ et al (2003) Oxygen-evolving enhancer protein 2 is phosphorylated by glycine-rich protein 3/wall-associated kinase 1 in Arabidopsis. Biochem Biophys Res Commun 305:862–868
68. Machado CR, Costa De Oliveira RL, Boiteux S, Praekelt UM, Meacock PA, Menck CFM (1996) *Thi1*, a thiamine biosynthetic gene in *Arabidopsis thaliana*, complements bacterial defects in DNA repair. Plant Mol Biol 31:585–593
69. Ahn I-P, Kim S, Lee Y-H (2005) Vitamin B_1 functions as an activator of plant disease resistance. Plant Physiol 138:1505–1515
70. Wang G, Ding X, Yuan M, Qiu D, Li X, Xu C et al (2006) Dual function of rice *OsDR8* gene in disease resistance and thiamine accumulation. Plant Mol Biol 60:437–449
71. Rapala-Kozik M, Kowalska E, Ostrowska K (2008) Modulation of thiamine metabolism in *Zea mays* seedlings under conditions of abiotic stress. J Exp Bot 59:4133–4143
72. Tunc-Ozdemir M, Miller G, Song L, Kim J, Sodek A, Koussevitzky S et al (2009) Thiamin confers enhanced tolerance to oxidative stress in Arabidopsis. Plant Physiol 151:421–432
73. Nelson CJ, Alexova R, Jacoby RP, Millar AH (2014) Proteins with high turnover rate in barley leaves estimated by proteome analysis combined with in planta isotope labeling. Plant Physiol 166:91–108
74. Gutteridge S, Gatenby AA (1995) Rubisco synthesis, assembly, mechanism, and regulation. Plant Cell 7:809–819
75. Salvucci ME (2008) Association of Rubisco activase with chaperonin-60b: a possible mechanism for protecting photosynthesis during heat stress. J Exp Bot 59:1923–1933
76. Tyedmers J, Mogk A, Bukau B (2010) Cellular strategies for controlling protein aggregation. Nat Rev Mol Cell Biol 11:777–788
77. Blum A, Sinmena B, Mayer J, Golan G, Shpiler L (1994) Stem Reserve mobilisation supports wheat-grain filling under heat stress. Funct Plant Biol 21:771–781

78. Harsant J, Pavlovic L, Chiu G, Sultmanis S, Sage TL (2013) High temperature stress and its effect on pollen development and morphological components of harvest index in the C3 model grass *Brachypodium distachyon*. J Exp Bot 64:2971–2983
79. Sage TL, Bagha S, Lundsgaard-Nielsen V, Branch HA, Sultmanis S, Sage RF (2015) The effect of high temperature stress on male and female reproduction in plants. Field Crops Res 182:30–42
80. Prasad PVV, Boote KJ, Allen LH Jr, Sheehy JE, Thomas JMG (2006) Species, ecotype and cultivar differences in spikelet fertility and harvest index of rice in response to high temperature stress. Field Crops Res 95:398–411
81. Makino A, Sage RF (2007) Temperature response of photosynthesis in transgenic rice transformed with 'sense' or 'antisense' *rbcS*. Plant Cell Physiol 48:1472–1483
82. Scafaro AP, Yamori W, Carmo-Silva AE, Salvucci ME, von Caemmerer S, Atwell BJ (2012) Rubisco activity is associated with photosynthetic thermotolerance in a wild rice (*Oryza meridionalis*). Physiol Plant 146:99–109
83. Keeling PL, Bacon PJ, Holt DC (1993) Elevated temperature reduces starch deposition in wheat endosperm by reducing the activity of soluble starch synthase. Planta 191:342–348
84. Majoul T, Bancel E, Triboï E, Ben Hamida J, Branlard G (2003) Proteomic analysis of the effect of heat stress on hexaploid wheat grain: characterization of heat-responsive proteins from total endosperm. Proteomics 3:175–183
85. Li W, Wei Z, Qiao Z, Wu Z, Cheng L, Wang Y (2013) Proteomics analysis of alfalfa response to heat stress. PLoS One 8:e82725
86. Rocco M, Arena S, Renzone G, Scippa GS, Lomaglio T, Verrillo F et al (2013) Proteomic analysis of temperature stress-responsive proteins in *Arabidopsis thaliana* rosette leaves. Mol BioSyst 9:1257–1267
87. Zhang M, Li G, Huang W, Bi T, Chen G, Tang Z et al (2010) Proteomic study of *Carissa spinarum* in response to combined heat and drought stress. Proteomics 10:3117–3129
88. Han F, Chen H, Li X-J, Yang M-F, Liu G-S, Shen S-H (2009) A comparative proteomic analysis of rice seedlings under various high-temperature stresses. Biochim Biophys Acta (BBA) 1794:1625–1634
89. Ferreira S, Hjerno K, Larsen M, Wingsle G, Larsen P, Fey S et al (2006) Proteome profiling of *Populus euphratica* Oliv. upon heat stress. Ann Bot 98:361–377
90. Rollins JA, Habte E, Templer SE, Colby T, Schmidt J, Von Korff M (2013) Leaf proteome alterations in the context of physiological and morphological responses to drought and heat stress in barley (*Hordeum vulgare* L.). J Exp Bot 64:3201–3212
91. Yang Y, Chen J, Liu Q, Ben C, Todd CD, Shi J et al (2012) Comparative proteomic analysis of the thermotolerant plant *Portulaca oleracea* acclimation to combined high temperature and humidity stress. J Proteome Res 11:3605–3623
92. Xu C, Huang B (2008) Root proteomic responses to heat stress in two Agrostis grass species contrasting in heat tolerance. J Exp Bot 59:4183–4194

Chapter 8
Proteomics Approach for Identification of Nutrient Deficiency Related Proteins in Crop Plants

Mohammad-Zaman Nouri, Mohammad-Reza Ghaffari, Hamid Sobhanian and Mohammad-Reza Hajirezaei

Abstract Nutrients play an essential role in growth and development of plants and their deficiencies trigger diverse responses from architecture to genes and proteins expression. Deficiency of macro and micronutrients significantly alters uptake and transport in plant cells. Proteins are main player in compensation of the deficiency and according to the plant type, a wide range of proteins are recognized as nutrient deficiency responsive proteins. In this chapter, proteins involve in nutrients uptake and transport are classified and their roles in homeostasis, signaling and interactions under nutrients deficiency described in plant cells. Further, a comprehensive review of proteome studies in various plants under macro and micronutrients deficiency and protein overexpression to improve nutrient use efficiency are summarized.

Keywords Metabolism · Nutrient deficiency · Plant · Proteomics

---

M.-Z. Nouri (✉)
Rice Research Institute of Iran, Mazandaran Branch, Agricultural Research, Education and Extension Organization (AREEO), Amol 46191-91951, Iran
e-mail: m.nouri@areo.ir

M.-R. Ghaffari
Systems Biology Department, Agricultural Biotechnology Research Institute of Iran, Agricultural Research, Education and Extension Organization (AREEO), Karaj 31535-1897, Iran

M.-R. Hajirezaei
Department of Physiology and Cell Biology, Molecular Plant Nutrition, Institute of Plant Genetics and Crop Plant Research, Corrensstrasse 3, 06466 Gatersleben, Germany

H. Sobhanian
Department of Biology, Payame Noor Universtiy, PO Box 19395-3697, Tehran, Iran

G.H. Salekdeh (ed.), *Agricultural Proteomics Volume 2*,
DOI 10.1007/978-3-319-43278-6_8

8.1 Introduction

Plant growth is determined by accumulating of biomass, which is the result of the ultimate expression of metabolism, cell proliferation with a concomitant accumulation of biomaterials mainly through polymer biosynthesis and cell expansion driven by water, micro and macronutrient uptake [1, 2]. Plants need at least 14 well-established mineral elements for their physiological functions and productivity. The macronutrients nitrogen (N), phosphorus (P), potassium (K), calcium (Ca), magnesium (Mg) and sulphur (S) constitute between 1000 and 15,000 $\mu g\ g^{-1}$ plant dry weight. However, the micronutrients chlorine (Cl), boron (B), iron (Fe), manganese (Mn), copper (Cu), zinc (Zn), nickel (Ni) and molybdenum (Mo) are found in relatively small amounts in plants [3].

The nutrients C, N, H, and S, the constituents of amino acids, proteins, enzymes and nucleic acids, are taken up from the soil or atmosphere as CO_2, HCO_3^-, H_2O, O_2, NO_3^-, NH_4^+, N_2, SO_4^{2+}, and SO_2^- ions and play major roles in enzymatic or assimilation by oxidation-reduction reactions. However, the nutrients K, Na, Ca, Mg, Mn and Cl play a role in maintaining the osmotic potential, the optimal conformation of enzymes (enzyme activation), bridging of reaction partners, balancing anions, controlling membrane permeability and electrochemical potentials. Furthermore, Fe, Cu, Zn and Mo as ion or chelates enable electron transport by valence change. The nutrients P, B, Si in form of inorganic chemical are generally obtained from the soil and maintain electrochemical potential, electron transport and energy transfer reactions [4].

The first step of nutrient accumulation in plants is their uptake by roots which is principally dependent on plant demand, environmental factors such as water content, pH, redox potential, abundance of organic matter, and microorganisms in soils. Besides, chemical and physical properties of the soil have a major effect on the availability of the micronutrients [5]. For instance, nutrient mobility of Fe and P is restricted within roots by the water conditions [6]. Furthermore, acquisition of Fe, as one of the most generous metal in the earth's crust, is very limited under well-aerated calcareous or alkaline soils and the availability of Zn is markedly reduced in calcareous soils, particularly under arid or semi-arid conditions [5, 7]. There are two different and more efficient strategies for nutrient uptake modulated by the root system architecture comprising of primary roots, lateral roots and root hair development and modulation of transport activity [8]. It has been reported that while primary root elongation was inhibited under phosphorus starvation, lateral roots formation was enhanced in Arabidopsis (Arabidopsis thaliana L.) [9]. Such changes in root morphology cause an increase in root surface area and alleviate P stress in plants.

Any morphological changes in plant architecture and cellular alterations to improve nutrient uptake is directly controlled by the expression of related genes and proteins. Once plants encounter a specific nutrient deficiency, coordinated gene activation for an efficient uptake, transport and nutrient usage is activated. Depending on the source of the nutrient, mechanism of uptake and subsequently type of the regulated genes and proteins can be different. In addition, nutrients may have synergistic or antagonistic interactions which not only affects the level of uptake, but also alters gene expression and protein function [10]. Besides genomic approaches for gene identification, proteomics is a valuable approach to identify the proteins crucial for nutrient uptake in plants. In this chapter, nutrient metabolism, homeostasis and signaling in plants are described. Further, a summary for identified proteins responding to nutrient deficiency in various plants is presented.

8.2 Metabolism of Macronutrients in Plants Under Deficient Conditions

Nutritional status in plant is strongly impaired by environmental factors. Any reduction in availability, uptake, transport and utilization of micro and macroelements leads to a significant reduction in the quantity and quality of crop yield due to scarcity of resources. In the following section, metabolism and importance of macronutrients will be described.

8.2.1 Nitrogen Metabolism in Plants Under Deficient Conditions

Nitrogen is one of the most important macronutrient that functions as a signal to regulate many biological processes, from metabolism to resource allocation, growth, and development [11]. The plant systems cope with the heterogeneity and dynamic variations of nitrate and ammonium concentrations ranging from lower than 100 μM to higher than 10 mM in soil solutions with their high and low affinity transporters to produce 1 kg of dry plant biomass per 20–50 g of N taken in most non-legume plants [12]. Besides, the N-containing compounds of amino acids, peptides in forms of di- and tri-peptides, and proteins are other major nitrogenous resources [13]. Nitrate as a key source of nitrogen is taken up by membrane bound transporters like nitrate transporter 1/peptide transporter (NRT1/PTR), NRT2, chloride channel (CLC), and slow anion channel-associated 1 homolog 3 (SLAC1/SLAH) [14]. These proteins are profoundly involved in nitrate uptake, allocation, and storage in most wild and crop species or adaptably act in nitrate sensing, plant development, pathogen defense, and/or stress response [14].

Ammonium in turn, which stimulates root growth, root branching and lateral root elongation is taken up by ammonium transporters. It further assimilated into amino acids via the GS/glutamine-2-oxoglutarate aminotransferase cycle and acts as central precursor of nucleic acids, proteins and other organic molecules, as well as a product of their catabolism and localized nitrogen supply. During N-deficiency, plants phenotypically develop more extensive root and lateral root growth while shoot growth is restricted because of leaf senescence [15, 16]. It also results in the accumulation of primary metabolites in the leaf, particularly starch and secondary metabolites such as phenylpropanoids, anthocyanins and flavonoids (e.g. rutin and ferulic acid) [17].

While growth is limited by N deficiency with the channeling of available N into essential metabolic processes and defense compounds, plants show a range of adaptive responses by maintaining a broader range of metabolic C:N ratio composition. At the early stages of N limitation, the adaptation mechanism is more systematic than selective to triage constituents. For instance, the levels of many minor amino acids and major amino acids Gln, Glu and Ser are even higher in plants grown on low N than on high N while Asp, Asn and Ala are lower in low-N conditions, indicating a restricted transfer of amino groups from Glu to Asp and Ala, and from Ser to other amino acids. In the same way, there is depletion in organic acid pools in plants grown on low N, reflecting a low rate of nitrate assimilation in the low-N grown plants. Furthermore, there are evidences that a few hours after nitrate resupply respiration rises, levels of amino acids of Gln, Glu, Asp, Ala, Asn, Arg and His increase, and the TCA cycle intermediates malate and fumarate accumulate [18]. Further nitrate resupply strongly induced the genes involved in nitrogen assimilation like nitrate reductase and production of reducing equivalents [18]. These evidences indicate that many of the metabolic changes observed in nutrient-deficient plants are stress-related symptoms that can be avoided if the plant adjusts its growth rate to the nutrient assimilation rate.

8.2.2 *Metabolism of Phosphorus in Plants Under Deficient Conditions*

Phosphorus is an essential macronutrient that is required for making up of inherited material, nucleic acids, and biosynthesis of cell membrane phospholipids. It also takes part in cellular signaling cascades by acting as mediator of signal transduction or serving as a reservoir of energy in cells. Phosphorus is also required for photosynthesis and respiration processes [19]. Plant takes Pi up by low-Pi-inducible high-affinity and constitutive low-affinity Pi phosphate transporters. These transporters are expressed mainly in the epidermal cells and root hair systems when the plant is exposed to Pi limitation. Phosphate transporters are not only involved in Pi uptake, but also in delivering Pi to leaves, stems, cotyledons, pollen grains, seeds, flowers, and potato tubers [20].

Plants exhibit several physiological characteristics to Pi deficiency including primary root growth restriction, massive lateral and hairy root production, increase in root to shoot ratio, inhibition of meristem and cell elongation activity [21]. Accumulation of anthocyanin pigment is a biochemical response to P deficiency which commonly appears in leaf to protect chloroplasts and nucleic acids from the intense or ultraviolet light [22]. There are strong evidences that organic acids exudation increases in response to Pi deficiency [23]. This is a major trait in breeding crops with improved phosphate acquisition efficiency which contributes to high-phosphate utilization efficiency.

Phytohormones play important role in root system architecture under Pi deficiency. Upon Pi starvation, auxin and the associated polar auxin transport mechanism trigger lateral root formation and modulation of root system. However, cytokinin, which is involved in regulation of root architecture modulation seems to negatively regulates Pi starvation responsive genes [24]. Furthermore, ethylene is involved in primary root elongation and root hair formation in seedlings grown in Pi-limited medium. Detailed omics analyses have recently shown that Pi significantly decreases the metabolic levels of carbohydrate metabolism like glucose, pyruvate and chlorophyll, and genes related to carbon metabolism in rice (Oryza sativa L.) [25].

When the availability of Pi is limited in soil, plant systems can undergo a range of adaptive biochemical responses. It includes the acquisition of more Pi from the environment, which is improved by exudation of organic acids and phosphatases into the soil or mobilization of Pi within the body of the plant. The physiological responses are also changing in root systems, morphology and architecture and establishment of a symbiotic association with arbuscular mycorrhizal fungi to gain Pi through a mutualistic interaction.

8.2.3 Potassium Metabolism in Plants Under Deficient Conditions

Potassium in the form of free ion (K^+) is one of the most important mineral nutrient elements, which plays a crucial role in many plant physiological processes like maintenance of cytoplasmic pH homeostasis, maintenance of electrochemical gradients across membranes, inorganic anions and metabolites transport and signal transduction [26]. It activates numerous enzymes of central metabolism involved in energy metabolism, protein synthesis, and solute transport. Pyruvate kinase, which catalyzes the transfer of a phosphate group from phosphoenolpyruvate to ADP, yielding one molecule of pyruvate and one molecule of ATP is particularly sensitive to cytoplasmic K and its activity in root cells can be rapidly inhibited under K deficiency [18].

Potassium is absorbed from the soil against the K^+ concentration gradient and its translocation from roots to shoots or from source to sink is also mediated by K^+ transporters and channels [27]. Early investigations revealed a dual affinity of high and low K^+ mechanisms of which high-affinity K^+ uptake mechanism is mediated primarily by K^+ transporters at low external K^+ (below 0.2 mM). However, in high external K^+ (above 0.3 mM) the primary mechanism is mediated by K^+ channels. K^+ deficient plants phenotypically show a decrease in lateral root and shoot formation at seedling stage in rice [28]. Plants also show a root-hair elongated phenotype with a decreased lateral root length and lateral root number in Arabidopsis [29]. To adapt to K^+ deficiency, plants activate a complex series of signaling and network adaptive events at both physiological and morphological levels. Phosphorylation and dephosphorylation of K^+ channel proteins at the posttranslational level is one adaptive mechanism to enhance K^+ uptake from the environment and translocation in plant cells along with mobilization of K^+ ions from vacuoles.

Upon K^+ deficiency, plant cells exhibit long and short physiological and biochemical responses of several signal components, including membrane potential, reactive oxygen species (ROS), phytohormones (ethylene and auxin) and metabolic changes [30]. Regarding metabolites, K^+ deficient plants accumulate more basic or neutral amino acids and slightly increase total amino acid and protein content in their roots and shoots. However, a strong decrease of pyruvate and organic acids, specifically is occurring in the roots [18]. Changes in the activity of enzyme involved in sugar metabolism like acid invertase, glucokinase and fructokinase and of enzymes of glycolysis and TCA cycle like glycerol aldehyde phosphate dehydrogenase (GAPDH) and malic dehydrogenase and of enzymes involved in nitrogen assimilation such as nitrate reductase, glutamine synthase (GS), glutamine dehydrogenase (GDH) and ferredoxin-dependent glutamine:2-oxoglutarate aminotransferase (Fd-GOGAT) in roots are known as an adaptive mechanism against K^+ deficiency.

8.3 Macronutrient Homeostasis and Signaling in Plants

Sessile and terrestrial plants must directly encounter changes of environment, therefore, they developed complex strategies to survive and to cope with different stresses [31]. These are the integration of local and systemic sensing and signaling for maintenance of cellular nutrient homeostasis at different stages of growth. Upon nutrient fluctuations, roots perceive the signal in extracellular nutrient levels and transfer it to the shoot via the xylem. The signal subsequently transmits to the shoot apices and roots via the phloem to adjust developmental processes and nutrient uptake [21]. Furthermore, long-distance transport signals in the phloem from source leaves in the shoot to sink tissues in the root including root apices and lateral root primordium comes into play upon nutrient deficiency. At the cellular level, a range of signals including changes in cytosolic Ca^{2+}, pH and potential of membrane, ROS, nitric oxide and different hormones including auxin, gibberellins (GA), cytokinins (CKs), absicic acid (ABA), ethylene, strigolactones (SLs) and brassinosteroids are involved in development and changing root architecture to improve the efficiency of nutrient uptake [31] (Fig. 8.1).

8.3.1 Nitrate Sensing and Signaling in Plants

Plants sense nitrate fluctuations quickly through several possible scenarios. While the external nitrate may be sensed most probably by a membrane-bound protein, the intracellular nitrate level senses either in the cytosol or in other cellular compartments such as vacuole. Furthermore, nitrate-transporting/metabolizing proteins, are known as a nitrate flux sensors [32]. It is hypothesized that nitrate sensing is mediated by membrane nitrate transporters [33]. At least two transporters, NRT1.1 and NRT2.1, are involved in nitrate sensing in Arabidopsis [34]. There is a tight

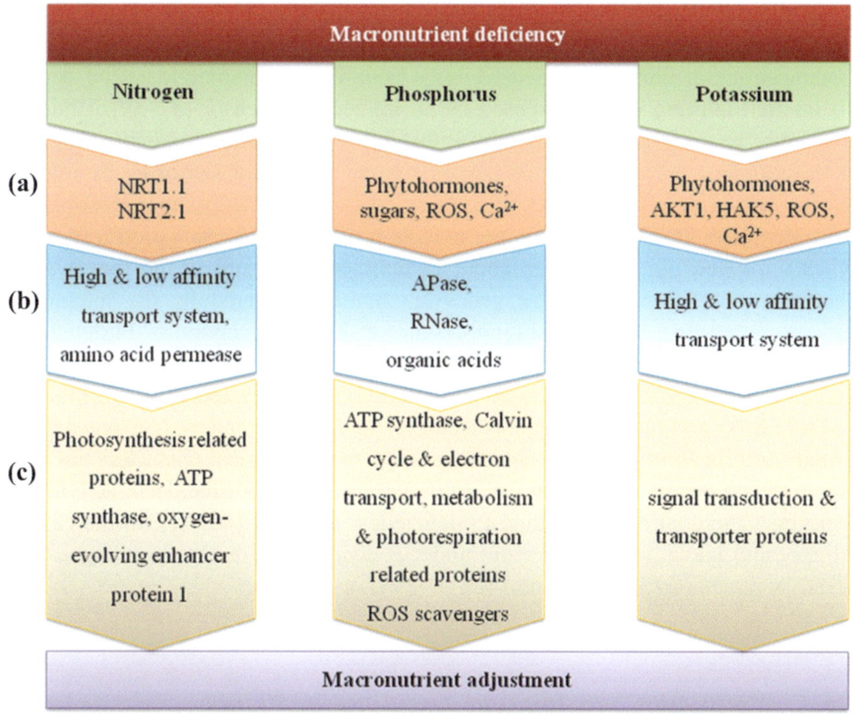

Fig. 8.1 Schematic presentation of the processes occurring upon macronutrient deficiency from sensing to protein regulation in plants. Deficiency of nutrients is sensed by specific sensors (*a*), which activates mechanisms to improve uptake and transport (*b*). Regulation of proteins in response to nutrient deficiency (*c*) is an adaptive mechanism to adjust plants with the stress conditions

connection between NO_3^- and CKs signaling in plants in which NO_3^- induces both CKs biosynthesis in roots and translocation to the shoot [35]. A significant induction of adenosine phosphate-isopentenyltransferase3, a CKs biosynthesis gene, by NO_3^- is reported in the roots. [11].

8.3.2 Sensing and Signaling of Phosphorus in Plants

Plants have evolved complex strategies and adaptive mechanisms for uptake, remobilization and recycling of Pi to sustain Pi homeostasis. Local Pi sensing and signaling triggers adjustments in root system architecture for increasing of Pi absorption, while the systemic, or long-distance signaling pathways act to adjust Pi uptake, mobilization and redistribution [36]. Pi deficiency senses locally through roots where phytohormones are important signaling components. Auxin, ethylene, CKs, ABA, GA, SLs, along with sugars, miRNAs and Ca^{2+} have all been shown to be involved in Pi local and systemic sensing and signaling pathways. There are two ways by which plants are supposed to sense Pi status around rhizosphere: (i) external Pi concentration changes is sensed by a root cell membrane-localized sensor, (ii) internal nutrient levels is sensed by an intracellular sensor [37–39]. Under Pi deficiency, an initial signal or stimulus (most likely Pi concentration in the apoplasm of the root tip) perceives by a plasma-membrane-localized sensor or by internal sensors in root tip cells. To date, none of external or internal Pi-stress sensors have been identified. Investigations on the induction of phosphate starvation responsive genes improved our knowledge about Pi deficiency sensing in the root system. Phosphate starvation responsive genes are mainly triggered by the internal Pi amount, rather than by external (apoplasmic) Pi level. Perception of Pi deficiency signal by the root sensing system causes the activation of downstream adaptive signaling pathways to generate both cell-autonomous and systemic signals [39, 40].

Ca^{2+}, ROS and inositol polyphosphates as universal secondary messengers play a major role in Pi sensing and signaling. Some morphological characteristics such as leaf development, time regulation of flowering, and shoot meristem activity are also under the control of systemic Pi deficiency signaling [40, 41]. In the case of primary and lateral root architecture under P deprivation, auxin signalling is supposed as an adaptive mechanism [42].

8.3.3 Potassium Sensing and Signaling in Plants

At cellular level, cytosolic K^+ concentrations are maintained about 100 mmol L^{-1}, whereas vacuolar K^+ concentrations are variable depending on the external K^+ concentrations and plant situation. K^+ influx into the vacuole and efflux from it controls cytosolic K^+ homeostasis. This cation is transported actively or passively from the soil to the plant cell and under K^+-deficiency, ROS, Ca^{2+} and phytohormones are involved as regulatory signals [27]. Transport of external K^+ through the plasma membrane, and its compartmentation within the plant are mediated by diverse K^+ channels and transporters. Sensing of K^+ is a main step for regulation of K^+ homeostasis. AKT1, a Shaker-type inward-rectifying K^+ channel as the most putative candidate K^+ sensor, plays a crucial role in K^+ efflux under starvation conditions. Some other channels, such as high-affinity K^+ transporter $_5$ (HAK_5), guard cell outwardly-rectifying K^+ channel, and K^+ uptake permease 4, have also been suggested as K^+ sensors but their sensor activities have not been confirmed in plants [43].

Plant roots steadily monitor exogenous K^+ concentrations in soil and sense K^+ availability. When plants sense K^+ deficiency, a short-term deficiency response is switched on and the high-affinity K^+ uptake system is activated within a few hours [27]. The high-affinity K^+ uptake mechanism in plants is mediated by the high-affinity channels such as HAK5 and AKT1 in Arabidopsis. ROS can activate these channels within 6 h from deficiency detection [44]. Calcium is another player in K^+-deficiency signaling cascade and has a major role in K^+/Na^+ homeostasis. Cytoplasmic Ca^{2+} is a secondary signal and mediates the downstream transcriptional and post-translational response to the external K^+ availability [27]. By activation of AKT1 via phosphorylation through CIPK23, the cytosolic Ca^{2+} amounts are immediately elevated under K^+ deficiency [45, 46].

8.4 Interactions Among Nutrients in Plant Cells

Interactions between essential nutrients, both in soil and within plant cells, lead to deficiency or toxicity of some minerals and consequently slow down the plant growth and crop yield. Fluctuations of a nutrient affect plant processes and uptake of other nutrients. Therefore, maximum biomass and nutrient accumulation is associated with optimal uptake of both macro and micronutrients. On the other hand, individual pair of nutrients can have temporal and specific synergistic or antagonistic interactions [10]. Investigations of antagonistic or synergistic interactions proved the ability of one particular nutrient to decrease or to increase the uptake of other nutrients.

Not only nutrient deficiency, but also excessive concentration of nutrients may negatively affect growth and productivity of crops. The negative effect of P, Cu, Zn and Mn toxicity could be considerably eliminated by increasing of Ca, Mg and Fe, as well as N, K, Mo, Co and B in substrate [47]. Increase in a nutrient such as Zn, may inhibit uptake of other nutrients, i.e. Pi. In wheat, when efficiency of Pi uptake is high, Zn uptake may decrease resulting in a reduction of Zn levels in grains grown in soils with low Pi. It was reported that Pi mobilization to reproduction system is blocked, because of the antagonistic effect of Zn triggered in roots [48]. In general, foliar application of Zn singular or in combination with its soil application decreases the Pi uptake in shoots at booting stage and in grains [49]. The evidence of coordination of homeostasis among different nutrients showed that Pi starvation down regulates the expression of an N recycling protein, RING-type ubiquitin E3 ligase and nitrogen limitation adaptation in Arabidopsis [50].

8.5 Molecular Approaches to Improve Nutrient Uptake in Crops

Plants adopt complex mechanisms for an efficient uptake of nutrients from the soil. To enhance nutrient acquisition, plants activate developmental programs to alter the root system architecture. This is an adaptive mechanism when the distribution of nutrients in the soil is not homogenous [51]. Lateral root development under N starvation is one of the adaptive mechanisms. Root N uptake relies at the molecular level on transporters. Three forms of N including NO_3^-, NH_4^+, and urea are taken up by members of the nitrate transporter, ammonium transporter and urea transporter, respectively [52]. High and Low affinity transport systems (HATS and LATS) are involved in the process of N uptake. These systems are predominant for soil nitrate concentration below and higher than 1 mM, respectively. Plant uses the mechanism of N remobilization to improve N efficiency. This mechanism is under the control of four proteases including FtsH, a chloroplastic protease, aspartic protease, proteasome β subunit A1 and SAG12 (cysteine protease) which are strongly induced during leaf senescence [53].

Upon P starvation, plants alter root growth and architecture by development of lateral roots and by increasing root hair number and length. In this respect, plants secrete APase, RNase and organic acids into the rhizosphere to improve Pi availability in the soil [54]. APase as a marker gene improves mobilization of immobile phosphate present in the soil or inside plant cells [55]. The mechanisms of plant response to P starvation are highly dependent on the plant type and on the tolerance to low-P conditions. It was reported that maize low-P-tolerant mutant 99038 could accumulate and secrete more citrate and higher proportion of sucrose in the total soluble sugars as well as higher proliferation in the root meristem [56]. Bucher (2007) reviewed the molecular and biochemical mechanisms involved in Pi transport, emphasizing on the role of root hair cells and mycorrhizal cortical cells. He showed that the Pi uptake kinetics in plants is not similar at low Pi concentrations compared to high Pi concentrations [57].

Because of complex soil dynamics, the availability of potassium in the plant is highly variable. Potassium deficiency not only triggers developmental responses in roots, but also activates various types of channels and transporters for potassium acquisition [58]. Two main transport components, including high- and low-affinity transport system are involved in K^+ uptake. Transcriptome analysis of Arabidopsis root transporters under K^+ deficiency indicated the majority of K^+ transporters is regulated post transcriptionally rather than at the transcript level [59].

8.5.1 Macronutrient Deficiency Related Proteins and Metabolic Changes in Crops

Once plants encounter deficiencies in the macronutrients such as N, P and K, a considerable change occurs in primary metabolism, carbohydrate concentrations, biomass allocation between shoot and root and subsequently plant morphology [60]. According to current knowledge, sugars and starch are accumulated in the leaves under N deficiency and this may cause a reduction of photosynthesis. Reduction of Pi and triose phosphate, which are crucial for the regulation of photosynthesis and starch metabolism can be considered as the reason in photosynthesis inhibition under P starvation [61]. Expression of genes and proteins related to photosynthesis and sucrose synthesis is also highly affected by macronutrient starvation [16, 60]. Deficiency of P can restrict ATP synthesis, which subsequently causes deactivation of Ribulose-1,5-bisphosphate carboxylase/oxygenase (RuBisCO) and accumulation of ribulose-1,5-bisphosphate [62]. Taken together, photosynthesis and carbon metabolism are the main part of a plant cell mechanism that is highly affected by macronutrient deficiency.

Transcriptome and metabolite data have indicated that N deficiency stimulates the expression of genes involved in amino acid catabolism, protein degradation, autophagy and ubiquitin-proteasome. However, the genes encoding phenyl-propanoid metabolism pathways or those involved in photosynthetic proteins and biosynthetic pathways that require N-containing metabolites tend to be down-regulated [63, 64]. While deprivation of N in Arabidopsis repressed the genes involved in photosynthesis apparatus and chlorophyll synthesis, it induced the expression of genes involved in the biosynthesis of anthocyanin and phenyl-propanoid pathways as well as protein degradation [16, 65]. At the proteome level, down regulation of RuBisCO large subunit, RuBisCO activase and ferredoxin NADP reductase in ramie (*Boehmeria nivea*) leaf; changes in the expression of ATP synthase in rice root [66] and oxygen-evolving enhancer protein 1 and RuBisCO LSU in the wheat shoot have been reported [67].

In response to P starvation, a coordinated change in the expression of auxin-inducible genes [42]; genes involved in the remodeling of membrane lipids, Pi acquisition and carbohydrate flux [68], P uptake related genes [69] and P transport and recycling-related genes were already reported in different plants [70]. Gene expression analysis of potato leaves identified proteins involved in lipid, protein, and carbohydrate metabolism like novel patatin like protein under Pi deficiency [71]. At the proteome level, down regulation of the proteins involved in Calvin cycle and the electron transport system and CO_2 enrichment [72], down regulation of energy production, photorespiration, ROS scavengers and defense related proteins [54] and proteins involved in phytohormone biosynthesis, and signal transduction were already reported in plants in response to P deficiency [73].

Despite the importance of K in plant physiological and developmental processes, few studies have been performed to analyze the regulated genes and proteins under K deficiency compared to other macronutrients. Transcriptome analysis of rice roots under K starvation revealed that high affinity K^+ transporters have been changed compared to P or N deficiency. Proteome analysis in Arabidopsis under K-deficiency indicated that different proteins involved in signal transduction and transport processes were regulated [58, 74].

A summary of already identified proteins responding to N, P and K deficiencies is illustrated in Table 8.1. Rice and Arabidopsis were the most studied plants and two-dimensional polyacrylamide gel electrophoresis (2-DE) for protein separation followed by MALDI TOF-TOF MS for protein identification was the major proteomics technique. Although there were no correlations among the plant and sample type as well as the type of the nutrient deficiency, the identified proteins were functionally similar. Metabolism related proteins were mostly common in the regulated proteins under macronutrient deficiency in various plants. Proteins related to photosynthesis can be classified as the second group of the regulated proteins. Defense/stress, signal transduction and transporter proteins are also highly affected by nutrient deficiency (Table 8.1). In general, macronutrient starvation impairs plant metabolism as well as photosynthesis apparatus which usually lead to a dramatic reduction in the yield of crop plants.

Table 8.1 Macronutrient deficiency related proteins in plants

Type of nutrient	Plant species	Organ	Proteomics technique	No. of identified proteins (up/down regulation)	Functional classification of the regulated proteins	Reference
Nitrogen	*Oryza sativa* L.	Root	2-DE/MALDI TOF MS	11	Metabolism; stress response	[66]
	Hordeum vulgare L.	Root and shoot	2-DE/MALDI TOF-TOF MS	*Long-term stress* Root: 37 (14/23) Shoot: 35 (19/16) *Short-term stress* Root: 30 (17/13) Shoot: 7 (1/6)		[67]
	Boehmeria nivea	Leaf	2-DE/MALDI TOF-TOF MS	32 (10/22)	Photosynthesis; protein destination and storage; metabolism; disease/defence; signal transduction	[108]
	Zea mays L.	Leaf	2-DE/LC-MS/MS	40 (14/26)	Signalling; stress/defence; proteolysis; transcription	[64]
	Gossypium hirsutum L.	Fiber	2-DE/MALDI TOF-TOF MS	49 (28/21)	Metabolism; cell wall synthesis and transportation; wall modification; cytoskeleton formation; cell response/signaling; redox homeostasis	[14]
	Triticum aestivum L.	Leaf	2-DE/MALDI TOF MS	25	Photosynthesis; glycolysis; nitrogen metabolism; sulphur metabolism; defence	[109]
	Arabidopsis thaliana L.	Seedling	2-DE/MALDI TOF-TOF MS	170 (70/100)	Nitrogen and protein metabolism; photosynthesis; cytoskeleton; redox homeostasis; signal transduction	[14]

(continued)

Table 8.1 (continued)

Type of nutrient	Plant species	Organ	Proteomics technique	No. of identified proteins (up/down regulation)	Functional classification of the regulated proteins	Reference
Phosphorus	*Zea mays* L.	Root	2-DE/MALDI TOF MS	106	Phytohormone biosynthesis; carbon and energy metabolism; protein synthesis; signal transduction; cell cycle; cellular organization; defense; secondary metabolism	[73]
	Arabidopsis thaliana L.	Suspension cell	2-DE/MALDI TOF MS	24 (18/6)	Cell wall modification; proteolysis; pathogen responses; and ROS metabolism	[110]
	Arabidopsis thaliana L.	Root	iTRAQ/LC-MS/MS	356	Lipid and glucose metabolism; transporter	[68]
	Oryza sativa L.	Root	2-DE/MALDI TOF-TOF MS	32	ROS scavengers; citric acid cycle; signal transduction; defense	[111]
	Zea mays L.	Leaf	2-DE/MALDI TOF-TOF MS	100 (46/54)	Photosynthesis; carbohydrate and energy metabolism; signal transduction; protein synthesis; cell rescue; defense	[21]
	Glycine max L.	Nodule	2-DE/MALDI TOF-TOF MS	44 (17/27)	Metabolism; transcription and signaling; stress response; transporter	[95]
	Brassica napus L.	Root and leaf	MALDI TOF-TOF MS	*Root:* 21 (11/10) *Leaf:* 15 (9/6)	Defense and stress response; carbohydrate and energy metabolism; signaling and regulation; protein process; biogenesis and cellular component	[112]
	Oryza sativa L.	Root	MALDI TOF-MS	21(15/6)	Metabolism; oxidation/reduction; nucleotide related protein	[113]
	Oryza sativa L.	Root	2-DE/MALDI TOF MS	10 (2/8)	Metabolism; defense/stress response	[54]

(continued)

Table 8.1 (continued)

Type of nutrient	Plant species	Organ	Proteomics technique	No. of identified proteins (up/down regulation)	Functional classification of the regulated proteins	Reference
	Solanum lycopersicum L.	Leaf	2-DE/MALDI TOF MS	46 (31/15)	Defense; transcription and translation processes; carbohydrate and energy metabolism; transporter; biosynthesis process; DNA binding	[114]
	Arabidopsis thaliana L.	Seedling	2-DE/MALDI TOF MS	24 (18/6)	Phytohormone biosynthesis or signaling; carbon and energy metabolism; signal transduction	[74]
	Oryza sativa L.	Root	2-DE/MALDI TOF MS	5 (1/4)	Metabolism; defense	[54]
	Hordeum vulgare L.	Leaf	TMT/LC-MS/MS	110 (76/34)	Defense; transcription; signal transduction; energy and protein synthesis	[115]

8.5.2 Microelement Deficiency Related Proteins in Crops

A survey of the plant proteome analyses under nutrient deficiency indicates that most studies are focused on the identification of proteins involved in N, P, and Fe deficiency and information about the responsive proteins to other essential nutrient deficiency remains scarce [75]. Several novel microelement deficiency-responsive proteins are illustrated in Table 8.2. In crop plants, especially those grown on calcareous soils, Fe deficiency is a major nutritional disorder causes a decrease in growth, yield and quality. In Arabidopsis roots, 4454 proteins were identified and 2882 were reliably quantified under Fe deficiency. Out of all identified proteins, 101 were related to stress defense, C metabolism and N metabolism [46]. Further, oxidative stress related proteins were differentially expressed under Fe deficiency due to a change of redox homeostasis and the induction of ROS formation via Fenton reactions. Many proteins involved in oxidative stress, such as peroxidases and catalase are Fe-containing proteins. An increase in the expression of superoxide dismutases (CuZnSOD and MnSOD), monodehydro ascorbate reductase, peroxidase and a decrease in the expression of catalase was reported [76].

Germin-Like-Protein, a defense related protein localized in the plasmodesmata, was up regulated under Fe^{2+} deficiency. This protein has a role in regulating primary root growth via controlling phloem-mediated distribution of resources between primary and lateral root meristems [77]. Carbon metabolism, especially glycolysis related proteins shows the most consistent changes in Fe^{2+} deficiency. Several proteins belonging to this pathway were up-regulated in different plant species, showing the fact that controlling of carbon metabolism that occurs under Fe^{2+} deficiency is strongly conserved among plant species. Proteomic results are in agreement with transcriptomic and biochemical studies in which up-regulation in glycolysis and TCA cycle enzymes upon Fe^{2+} deficiency was reported [78].

Fe^{2+} deficiency has also an impact on N metabolism by upregulation of glutamine synthetase and S-adenosylmethionine synthase1 (SAM1) and down regulation of SAM2. Enzymes involved in ammonia release (urease and dihydrolipoamide dehydrogenase), assimilation (GLN1 and GLN2), and amino group transfer (alanine aminotransferase and aspartate aminotransferase) were up regulated. Fe^{2+} deficiency decreased photosynthesis rate by the reduction of granal number and stromal lamellae, down regulation of thylakoid membrane components like proteins, electron carriers and lipids, electron transfer proteins, including light harvesting components and the core of the PSI and PSII complexes and cytochrome b_6/f as well as RuBisCO as leaf C fixation protein [76, 79].

Boron (B), a member of the metalloid group, is an essential micronutrient for all vascular plants. Deficiency of B negatively affects various biochemical and physiological processes [80]. Proteome analysis of rape seed (*Brassica napus* L.) under B deficiency resulted in the identification of proteins related to carbohydrate and energy metabolism, cell wall, stress response, signaling and regulation, protein process, fatty acid, amino acid and nucleic acid metabolism [75]. Further, glycolysis and TCA cycle are impaired in rape seed under B deficiency. Down regulation of

Table 8.2 Micronutrient deficiency related proteins in plants

Type of nutrient	Plant species	Organ	Proteomics technique	No. of identified proteins (Up/Down regulation)	Functional classification of the regulated proteins	References
Iron	*Arabidopsis thaliana*	Leaf	2-DE/MALDI LIFT TOF–TOF	140 (27/7)	Photosynthesis; electron transport	[99]
	Malus pumila	Leaf	Tandem mass spectrometry	98 (12/33)	Photosynthesis	[100]
	Pisum sativum	Root	2-DE, IEF/MALDI-TOF-TOF	nd (2/13)	Stress defense; C metabolism	[101]
	Solanum lycopersicum	Root	2-DE/MALDI-TOF and nanoLC-MS/MS	nd (25/4)	Stress defense; C metabolism	[102]
	Cucumis sativus	Root	2-DE/LC-ESI-MS/MS	44 (21/23)	C metabolism; nitrogen related proteins; stress defence; cell wall	[103]
	Medicago Truncatula	Root	2-DE/MALDI-TOF-MS, LIFT TOF/TOF	69 (24/27)	Glycolysis and TCA cycle	[104]
	Arabidopsis thaliana	Root	iTRAQ/LC-MS/MS	101	Transport; oxidative stress; respiration	[46]
	Arabidopsis thaliana	Root	iTRAQ Labeling/MS/MS Analysis	4454 (59/42)	Stress defense; C metabolism; N metabolism	[68]
	Beta vulgaris	Root	2-DE, IEF/MALDI-MS	148 (16/6)	Glycolysis and TCA cycle	[105]
Boron	*Brassica napus*	Root	MALDI-TOF/TOF MS.	800 (35/10)	C metabolism; stress response; signaling and regulation	[75]
Copper	*Brassica napus*	Leaf	2–DE, ESI LC–MS/MS	693 (36/12)	Stress responses; energy processes; ATP synthase	[106]
Zinc	*Lactuca sativa*	Leaf	Tandem MS Analysis	122	Photosynthesis; metabolism; membrane transport; cell wall metabolism	[107]

malate dehydrogenase involved in aerobic metabolism and up regulation of pyruvate decarboxylase, the main enzyme in the fermentation process, 6-phosphogluconate dehydrogenase, a rate-limiting enzyme of the pentose phosphate pathway, SAMS3, lignin biosynthesis related proteins were reported to respond to B deficiency [81]. Annexin-like protein was down-regulated sharply under B deficiency and then increased during the recovery stage, suggesting that this protein has a role in protection of plant from oxidative damage [82–84].

8.6 Root Nodules and Nutrient Deficiency

Nitrogen and ammonium obtained from the soil solution are a major nitrogen resource for most leguminous and non-leguminous plants. However, leguminous have obtained an added option to fix N_2 by symbiotic bacterial called rhizobia in their root nodules. Nodule formation is the most key event for the establishment of a mutualistic interaction in which atmospheric dinitrogen is fixed by bacteria inside of nodules to produce ammonia and dicarboxylic acids. The initiation of root nodule formation is started by differentiation in root cortical cells resetting by rhizobial infection to produce pre-nodule whose function is an obligatory step of intracellular infection. Nodule primordium occurring in dividing cells in the pericycle and in front of the xylem pole will grow and develop into the true nodule [85, 86]. There are two morphologically distinct types of nodules comprising of indeterminate and determinate nodules. In indeterminate type, the nodules zones such as nodule meristem, the rhizobial infection zone, and the nitrogen fixation zone are well established like those in *Medicago truncatula* and *Pisum sativum*. However in *Lotus japonicus* and *Glycine max* the nodule type is determinate nodule in which the nodule zone is not clearly distinguished [87]. Besides, nodules commonly have been improved with more peripheral vascular bundles to exchange nutrients between roots and nodules.

Thought in plants relying on dinitrogen fixation, P has more impact on nodulation and N_2 fixation than its function on plant growth and development. Nutrient deficiency may limit N_2 fixation through its direct effects on nodule and mass number of rhizobia and symbiosome formation [88, 89]. For instance, under Pi limitation, the mass number of nodule was restricted and led to a lower symbiotic dinitrogen fixation capacity or specific nitrogenase activity of nodules in common bean (Phaseolus vulgaris L.) and *Alnus incana* [89, 90].

In the process of rhizobial symbioses, Pi is an essential nutrient to convert N_2 into NH_4 usable for N nutrition in legumes. Low-P availability in the soil reduces N_2 fixation and subsequently slows down the growth of legumes. Addition of P fertilizer increased N content in shoots and roots [91, 92]. Phytases are a group of enzymes that are able to hydrolyze a variety of inositol phosphates and have some functions in nodule of legume. Effect of phytase in the nodule of common bean using two contrasting recombinant inbred lines for N_2 fixation under P deficiency has been studied. Under P deficiency, phytase transcripts were significantly increased in the P-efficient line leading to improve the efficiency in the use of the rhizobial symbiosis by 34 % [93]. A study on common bean inoculated with *Rhizobium tropici* CIAT899 grown under P-deficient and P-sufficient conditions indicated that nodulation and nitrogenase activity were significantly reduced in plants grown in P-deficient conditions [94].

Proteome analysis of soybean nodules was performed under phosphate starvation in which 44 phosphate-starvation responsive proteins were identified (Table 8.1). Since N fixation is an energy-consuming process, reduction in P availability inhibited nodule growth, number and mass as well as nodule nitrogenase activity. It clearly indicates that P has a major role not only in morphology, but also in the activity of nodules. Interestingly, almost 40 % of the identified Pi-starvation responsive proteins in nodules were from bacteroids. Down regulation of proteins involved in carbon metabolism and transport indicated that metabolism of carbon and exchange of nutrient was inhibited in bacterial cells [95].

8.7 Overexpression of Proteins to Improve Nutrient Use Efficiency

In recent decades, application of synthetic N significantly increased and it caused considerable negative impacts on the environment. Therefore, demand for an efficient use of N in crop plants is increasing [96]. Attempts at improving N use efficiency in crop plants using transgenic approaches are mainly focused on altering N uptake and assimilation as well as N remobilization and regulation [52]. A review on the engineering N use efficient crops indicated that the predicted proteins did not improve N use efficiency in a specific crop. For instance, overexpression of nitrate reductase in cereal crops and glutamine synthetase in maize did not affect plant N use efficiency. However, overexpression of genes involved in primary N metabolism, amino acid biosynthesis, carbon-nitrogen balance, photosynthesis, carbon metabolism and transcription factors could improve N use efficiency in crops [96].

Genetic engineering assists in the breeding of crops to improve P acquisition efficiency and thus reduces inputs of P fertilizer for optimal growth [57]. Use of P transporter proteins and expression of PHT1 gene to improve P uptake was already reported. Mitsukawa et al. [97] isolated a high-affinity phosphate transporter gene from Arabidopsis. Overexpression of this gene at high levels in tobacco-cultured cells improved the rate of P uptake. Biomass production of the transgenic cells increased when the supply of phosphate was limited [97]. By contrast, overexpression of the same gene in barley did not improve the rate of P uptake [98]. Therefore, other mechanisms such as post-translational modifications, Pi availability and plant type may affect the Pi transport activity. Protein phosphatases were also cloned and characterized for the ability to improve P efficiency. Overexpression of two protein phosphatase genes, PvPS2:1 and PvP2:2 in bean improved root growth and P accumulation [50]. The target for production of transgenic plants tolerant to low P conditions has mainly focused on the genes encoding transcription factors, PAPs, Pi transporters, protein kinases and the genes involved in organic acid production [69]. Improvement in P efficiency and uptake in several transgenic crops indicates the potential of this technique to produce plants with more efficiency in P acquisition.

8.8 Conclusions

Plants represent a diverse response to nutrient deficiency from changes in root architecture to expression of responsible genes and proteins to improve nutrient uptake and transportation. Because of synergistic or antagonistic interactions among nutrients, a complex network of protein regulation is involved in plant cell to adjust uptake and to balance nutrient content (Fig. 8.1). This chapter represented an overview of the mechanisms of homeostasis, signaling, uptake and transport of macro and micronutrients in plant cells. Responsible proteins to deficiency of macro and micronutrients were also classified.

Deficiency of macronutrients highly affects photosynthesis and carbon metabolism. Therefore, it is reasonable that most of the regulated proteins in response to macronutrient deficiency are metabolism and photosynthesis related proteins. According to the review of literatures, most of the proteome analysis to identify nutrient deficiency related proteins is performed using 2-DE for protein separation followed by MALDI TOF-TOF MS for protein identification in the model plants of rice and Arabidopsis. The main challenge after the identification of the nutrient starvation responsive proteins is how to use this information in the improvement of nutrient uptake in plants. Although generation of plants with the overexpressed proteins to improve nutrient use efficiency is widely reported, however, more efforts should be made to have crops with the improved nutrient use efficiency in commercial levels.

References

1. Anastasiou E, Lenhard M (2007) Growing up to one's standard. Curr Opin Plant Biol 10:63–69
2. Smith AM, Stitt M (2007) Coordination of carbon supply and plant growth. Plant Cell Environ 30:1126–1149
3. Prinzenberg AE, Barbier H, Salt DE, Stich B, Reymond M (2010) Relationships between growth, growth response to nutrient supply, and ion content using a recombinant inbred line population in Arabidopsis. Plant Physiol 154:1361–1371
4. Mengel K, Kosegarten H, Kirkby EA, Appel T (2001) Principles of plant nutrition. Springer, Berlin

5. Marschner H (2011) Marschner's mineral nutrition of higher plants. Academic Press, New York
6. Jobbágy EG, Jackson RB (2001) The distribution of soil nutrients with depth: global patterns and the imprint of plants. Biogeochemistry 53:51–77
7. Broadley MR, White PJ, Hammond JP, Zelko I, Lux A (2007) Zinc in plants. New Phytol 173:677–702
8. Aibara I, Miwa K (2014) Strategies for optimization of mineral nutrient transport in plants: multilevel regulation of nutrient-dependent dynamics of root architecture and transporter activity. Plant Cell Physiol 55:2027–2036
9. Williamson LC, Ribrioux SP, Fitter AH, Leyser HO (2001) Phosphate availability regulates root system architecture in Arabidopsis. Plant Physiol 126:875–882
10. Rietra R, Heinen M, Dimkpa C, Bindraban P (2015) Effects of nutrient antagonism and synergism on fertilizer use efficiency. VFRC Report 2015/5. Virtual fertilizer research center, Washington, D.C., p 42
11. Bouguyon E, Gojon A, Nacry P (2012) Nitrate sensing and signaling in plants. In: Seminars in cell & developmental biology. Elsevier, Amsterdam, pp 648–654
12. Xu G, Fan X, Miller AJ (2012) Plant nitrogen assimilation and use efficiency. Annu Rev Plant Biol 63:153–182
13. Miller AJ, Fan X, Orsel M, Smith SJ, Wells DM (2007) Nitrate transport and signalling. J Exp Bot 58:2297–2306
14. Wang Y-Y, Hsu P-K, Tsay Y-F (2012) Uptake, allocation and signaling of nitrate. Trends Plant Sci 17:458–467
15. Comadira G, Rasool B, Karpinska B, Morris J, Verrall SR, Hedley PE et al (2015) Nitrogen deficiency in barley (*Hordeum vulgare*) seedlings induces molecular and metabolic adjustments that trigger aphid resistance. J Exp Bot 66:3639–3655
16. Scheible W-R, Morcuende R, Czechowski T, Fritz C, Osuna D, Palacios-Rojas N et al (2004) Genome-wide reprogramming of primary and secondary metabolism, protein synthesis, cellular growth processes, and the regulatory infrastructure of Arabidopsis in response to nitrogen. Plant Physiol 136:2483–2499
17. Foyer CH, Noctor G (2002) Photosynthetic nitrogen assimilation: inter-pathway control and signaling. In: Photosynthetic nitrogen assimilation and associated carbon and respiratory metabolism. Springer, Berlin, pp 1–22
18. Amtmann A, Armengaud P (2009) Effects of N, P, K and S on metabolism: new knowledge gained from multi-level analysis. Curr Opin Plant Biol 12:275–283
19. Ha S, Tran L-S (2014) Understanding plant responses to phosphorus starvation for improvement of plant tolerance to phosphorus deficiency by biotechnological approaches. Crit Rev Biotechnol 34:16–30
20. López-Arredondo DL, Leyva-González MA, González-Morales SI, López-Bucio J, Herrera-Estrella L (2014) Phosphate nutrition: improving low-phosphate tolerance in crops. Annu Rev Plant Biol 65:95–123
21. Zhang Z, Liao H, Lucas WJ (2014) Molecular mechanisms underlying phosphate sensing, signaling, and adaptation in plants. J Integr Plant Biol 56:192–220
22. Bustos R, Castrillo G, Linhares F, Puga MI, Rubio V, Pérez-Pérez J et al (2010) A central regulatory system largely controls transcriptional activation and repression responses to phosphate starvation in Arabidopsis. PLoS Genet 6:e1001102
23. Ryan P, Delhaize E, Jones D (2001) Function and mechanism of organic anion exudation from plant roots. Annu Rev Plant Biol 52:527–560
24. Sato A, Miura K (2011) Root architecture remodeling induced by phosphate starvation. Plant Signal Behav 6:1122–1126
25. Park MR, Baek S-H, Benildo G, Yun SJ, Hasenstein KH (2012) Transcriptome profiling characterizes phosphate deficiency effects on carbohydrate metabolism in rice leaves. J Plant Physiol 169:193–205
26. Amtmann A, Troufflard S, Armengaud P (2008) The effect of potassium nutrition on pest and disease resistance in plants. Physiol Plant 133:682–691

27. Wang Y, Wu W-H (2013) Potassium transport and signaling in higher plants. Annu Rev Plant Biol 64:451–476
28. Ma T-L, Wu W-H, Wang Y (2012) Transcriptome analysis of rice root responses to potassium deficiency. BMC Plant Biol 12:161
29. Szczerba MW, Britto DT, Ali SA, Balkos KD, Kronzucker HJ (2008) NH^{4+}-stimulated and-inhibited components of K^+ transport in rice (*Oryza sativa* L.). J Exp Bot 59:3415–3423
30. Coskun D, Britto DT, Kronzucker HJ (2015) The nitrogen-potassium intersection: membranes, metabolism, and mechanism. Plant Cell Environ doi:10.1111/pce.12671
31. Kochian LV, Lucas WJ (2014) Plant mineral nutrient sensing and signaling. J Integr Plant Biol 56:190–191
32. Hanson J, Smeekens S (2009) Sugar perception and signaling—an update. Curr Opin Plant Biol 12:562–567
33. Gojon A, Krouk G, Perrine-Walker F, Laugier E (2011) Nitrate transceptor (s) in plants. J Exp Bot 62:2299–2308
34. Liu KH, Tsay YF (2003) Switching between the two action modes of the dual-affinity nitrate transporter CHL1 by phosphorylation. EMBO J 22:1005–1013
35. Kiba T, Kudo T, Kojima M, Sakakibara H (2011) Hormonal control of nitrogen acquisition: roles of auxin, abscisic acid, and cytokinin. J Exp Bot 62:1399–1409
36. Linkohr BI, Williamson LC, Fitter AH, Leyser H (2002) Nitrate and phosphate availability and distribution have different effects on root system architecture of Arabidopsis. Plant J 29:751–760
37. Forde B, Lorenzo H (2001) The nutritional control of root development. Plant Soil 232:51–68
38. Nagarajan VK, Smith AP (2012) Ethylene's role in phosphate starvation signaling: more than just a root growth regulator. Plant Cell Physiol 53:277–286
39. Chiou T-J, Lin S-I (2011) Signaling network in sensing phosphate availability in plants. Annu Rev Plant Biol 62:185–206
40. Lucas WJ, Groover A, Patrick JW, Sperry J, Yoshida A, López-Millán A-F et al. (2013) The plant vascular system: evolution, development and functions| NOVA. The University of Newcastle's Digital Repository
41. Lough TJ, Lucas WJ (2006) Integrative plant biology: role of phloem long-distance macromolecular trafficking. Annu Rev Plant Biol 57:203–232
42. Al-Ghazi Y, Muller B, Pinloche S, Tranbarger T, Nacry P, Rossignol M et al (2003) Temporal responses of Arabidopsis root architecture to phosphate starvation: evidence for the involvement of auxin signalling. Plant Cell Environ 26:1053–1066
43. Grabov A (2007) Plant KT/KUP/HAK potassium transporters: single family–multiple functions. Ann Bot 99:1035–1041
44. Shin R, Schachtman DP (2004) Hydrogen peroxide mediates plant root cell response to nutrient deprivation. Proc Natl Acad Sci USA 101:8827–8832
45. Hirsch RE, Lewis BD, Spalding EP, Sussman MR (1998) A role for the AKT1 potassium channel in plant nutrition. Science 280:918–921
46. Lan W-Z, Lee S-C, Che Y-F, Jiang Y-Q, Luan S (2011) Mechanistic analysis of AKT1 regulation by the CBL–CIPK–PP2CA interactions. Mol Plant 4:527–536
47. Rinkis GY, Ramane H, Paegle G, Kunitskaya T (1989) An optimization system and diagnostic methods of plant mineral nutrition. Zinatne, Riga
48. Khan A, Zende G (1977) The site for Zn–P interactions in plants. Plant Soil 46:259–262
49. Khattak SG, Dominy PJ, Ahmad W (2015) Assessment of ZN interaction with nutrient cations in alkaline soil and its effect on plant growth. J Plant Nutr 38:1110–1120
50. Liang CY, Chen ZJ, Yao ZF, Tian J, Liao H (2012) Characterization of two putative protein phosphatase genes and their involvement in phosphorus efficiency in *Phaseolus vulgaris.* J Integr Plant Biol 54:400–411
51. López-Bucio J, Cruz-RamíRez A, Herrera-Estrella L (2003) The role of nutrient availability in regulating root architecture. Curr Opin Plant Biol 6:280–287
52. Fischer JJ, Beatty PH, Good AG, Muench DG (2013) Manipulation of microRNA expression to improve nitrogen use efficiency. Plant Sci 210:70–81

53. Christophe S, Jean-Christophe A, Annabelle L, Alain O, Marion P, Anne-Sophie V (2011) Plant N fluxes and modulation by nitrogen, heat and water stresses: a review based on comparison of legumes and non legume plants. Abiotic Stress in Plants–Mechanisms and Adaptations. Intech Open Access Publisher, Rijeka, pp 79–118
54. Kim SG, Wang Y, Lee CH, Mun BG, Kim PJ, Lee SY et al (2011) A comparative proteomics survey of proteins responsive to phosphorous starvation in roots of hydroponically-grown rice seedlings. J Korean Soc Appl Biol Chem 54:667–677
55. Nanamori M, Shinano T, Wasaki J, Yamamura T, Rao IM, Osaki M (2004) Low phosphorus tolerance mechanisms: phosphorus recycling and photosynthate partitioning in the tropical forage grass, Brachiaria hybrid cultivar Mulato compared with rice. Plant Cell Physiol 45:460–469
56. Li K, Xu C, Li Z, Zhang K, Yang A, Zhang J (2008) Comparative proteome analyses of phosphorus responses in maize (*Zea mays* L.) roots of wild-type and a low-P-tolerant mutant reveal root characteristics associated with phosphorus efficiency. Plant J 55:927–939
57. Bucher M (2007) Functional biology of plant phosphate uptake at root and mycorrhiza interfaces. New Phytol 173:11–26
58. Ashley M, Grant M, Grabov A (2006) Plant responses to potassium deficiencies: a role for potassium transport proteins. J Exp Bot 57:425–436
59. Maathuis FJ, Filatov V, Herzyk P, Krijger GC, Axelsen BK, Chen S et al (2003) Transcriptome analysis of root transporters reveals participation of multiple gene families in the response to cation stress. Plant J 35:675–692
60. Hermans C, Hammond JP, White PJ, Verbruggen N (2006) How do plants respond to nutrient shortage by biomass allocation? Trends Plant Sci 11:610–617
61. Lee WC (2013) Identification of phosphorus efficient potato cultivars. University of Florida
62. De Groot CC, Van Den Boogaard R, Marcelis LF, Harbinson J, Lambers H (2003) Contrasting effects of N and P deprivation on the regulation of photosynthesis in tomato plants in relation to feedback limitation. J Exp Bot 54:1957–1967
63. Schlüter U, Mascher M, Colmsee C, Scholz U, Bräutigam A, Fahnenstich H et al (2012) Maize source leaf adaptation to nitrogen deficiency affects not only nitrogen and carbon metabolism but also control of phosphate homeostasis. Plant Physiol 160:1384–1406
64. Amiour N, Imbaud S, Clément G, Agier N, Zivy M, Valot B et al (2012) The use of metabolomics integrated with transcriptomic and proteomic studies for identifying key steps involved in the control of nitrogen metabolism in crops such as maize. J Exp Bot 63:5017–5033
65. Vidal EA, Gutiérrez RA (2008) A systems view of nitrogen nutrient and metabolite responses in Arabidopsis. Curr Opin Plant Biol 11:521–529
66. Hakeem KR, Mir BA, Qureshi MI, Ahmad A, Iqbal M (2013) Physiological studies and proteomic analysis for differentially expressed proteins and their possible role in the root of N-efficient rice (*Oryza sativa* L.). Mol Breed 32:785–798
67. Møller AL, Pedas P, Andersen B, Svensson B, Schjoerring JK, Finnie C (2011) Responses of barley root and shoot proteomes to long-term nitrogen deficiency, short-term nitrogen starvation and ammonium. Plant Cell Environ 34:2024–2037
68. Lan P, Li W, Schmidt W (2012) Complementary proteome and transcriptome profiling in phosphate-deficient Arabidopsis roots reveals multiple levels of gene regulation. Mol Cell Proteom 11:1156–1166
69. Elanchezhian R, Krishnapriya V, Pandey R, Rao AS, Abrol YP (2015) Physiological and molecular approaches for improving phosphorus uptake efficiency of crops. Curr Sci 108:1271
70. Calderón-Vázquez C, Sawers RJ, Herrera-Estrella L (2011) Phosphate deprivation in maize: genetics and genomics. Plant Physiol 156:1067–1077
71. Hammond JP, Broadley MR, Bowen HC, Spracklen WP, Hayden RM, White PJ (2011) Gene expression changes in phosphorus deficient potato (*Solanum tuberosum* L.) leaves and the potential for diagnostic gene expression markers. PLoS One 6: e24606. doi:10.1371/journal.pone.0024606

72. Zhang K, Liu H, Tao P, Chen H (2014) Comparative proteomic analyses provide new insights into low phosphorus stress responses in maize leaves. PLoS One 9: e98215. doi:10.1371/journal.pone.0098215
73. Li K, Xu C, Zhang K, Yang A, Zhang J (2007) Proteomic analysis of roots growth and metabolic changes under phosphorus deficit in maize (*Zea mays* L.) plants. Proteomics 7:1501–1512
74. Kang JG, Pyo YJ, Cho JW, Cho MH (2004) Comparative proteome analysis of differentially expressed proteins induced by K^+ deficiency in *Arabidopsis thaliana*. Proteomics 4:3549–3559
75. Wang Z, Wang Z, Chen S, Shi L, Xu F (2011) Proteomics reveals the adaptability mechanism of Brassica napus to short-term boron deprivation. Plant Soil 347:195–210
76. López-Millán A-F, Grusak MA, Abadía A, Abadía J (2013) Iron deficiency in plants: an insight from proteomic approaches. Front in Plant Sci 4:254 doi:10.3389/fpls.2013.00254
77. Ham B-K, Li G, Kang B-H, Zeng F, Lucas WJ (2012) Overexpression of Arabidopsis plasmodesmata germin-like proteins disrupts root growth and development. Plant Cell 24:3630–3648
78. Zocchi G (2006) Metabolic changes in iron-stressed dicotyledonous plants. In: Iron nutrition in plants and rhizospheric microorganisms. Springer, pp 359–370
79. Morales F, Abadía A, Abadía J (1990) Characterization of the xanthophyll cycle and other photosynthetic pigment changes induced by iron deficiency in sugar beet (*Beta vulgaris* L.). Plant Physiol 94:607–613
80. Bolaños L, Lukaszewski K, Bonilla I, Blevins D (2004) Why boron? Plant Physiol Biochem 42:907–912
81. Shen B, Li C, Tarczynski MC (2002) High free-methionine and decreased lignin content result from a mutation in the Arabidopsis S-adenosyl-L-methionine synthetase 3 gene. Plant J 29:371–380
82. Mohr G, Lambowitz AM (2003) Putative proteins related to group II intron reverse transcriptase/maturases are encoded by nuclear genes in higher plants. Nucleic Acids Res 31:647–652
83. Ishimizu T, Sasaki A, Okutani S, Maeda M, Yamagishi M, Hase S (2004) Endo-β-mannosidase, a plant enzyme acting on N-Glycan purification, molecular cloning, and characterization. J Biol Chem 279:38555–38562
84. Rawsthorne S (2002) Carbon flux and fatty acid synthesis in plants. Prog Lipid Res 41:182–196
85. Callaham D, Torrey JG (1977) Prenodule formation and primary nodule development in roots of Comptonia (Myricaceae). Can J Bot 55:2306–2318
86. Péret B, Swarup R, Jansen L, Devos G, Auguy F, Collin M et al (2007) Auxin influx activity is associated with Frankia infection during actinorhizal nodule formation in *Casuarina glauca*. Plant Physiol 144:1852–1862
87. Higashi S, Kushiyama K, Abe M (1986) Electron microscopic observations of infection threads in driselase treated nodules of *Astragalus sinicus*. Can J Microbiol 32:947–952
88. Israel DW (1987) Investigation of the role of phosphorus in symbiotic dinitrogen fixation. Plant Physiol 84:835–840
89. Gentili F, Wall LG, Huss-Danell K (2006) Effects of phosphorus and nitrogen on nodulation are seen already at the stage of early cortical cell divisions in *Alnus incana*. Ann Bot 98:309–315
90. Kouas S, Labidi N, Debez A, Abdelly C (2005) Effect of P on nodule formation and N fixation in bean. Agron Sustain Dev 25:389–393
91. Cabeza RA, Liese R, Lingner A, Von Stieglitz I, Neumann J, Salinas-Riester G et al (2014) RNA-seq transcriptome profiling reveals that *Medicago truncatula* nodules acclimate N2 fixation before emerging P deficiency reaches the nodules. J Exp Bot 65:6035–6048
92. Robson A, O'hara G, Abbott L (1981) Involvement of phosphorus in nitrogen fixation by subterranean clover (*Trifolium subterraneum* L.). Funct Plant Biol 8:427–436

93. Lazali M, Zaman-Allah M, Amenc L, Ounane G, Abadie J, Drevon J-J (2013) A phytase gene is overexpressed in root nodules cortex of *Phaseolus vulgaris*–rhizobia symbiosis under phosphorus deficiency. Planta 238:317–324
94. Hernández G, Valdés-López O, Ramírez M, Goffard N, Weiller G, Aparicio-Fabre R et al (2009) Global changes in the transcript and metabolic profiles during symbiotic nitrogen fixation in phosphorus-stressed common bean plants. Plant Physiol 151:1221–1238
95. Chen Z, Cui Q, Liang C, Sun L, Tian J, Liao H (2011) Identification of differentially expressed proteins in soybean nodules under phosphorus deficiency through proteomic analysis. Proteomics 11:4648–4659
96. Mcallister CH, Beatty PH, Good AG (2012) Engineering nitrogen use efficient crop plants: the current status. Plant Biotechnol J 10:1011–1025
97. Mitsukawa N, Okumura S, Shirano Y, Sato S, Kato T, Harashima S et al (1997) Overexpression of an *Arabidopsis thaliana* high-affinity phosphate transporter gene in tobacco cultured cells enhances cell growth under phosphate-limited conditions. Proc Natl Acad Sci 94:7098–7102
98. Rae AL, Jarmey JM, Mudge SR, Smith FW (2004) Over-expression of a high-affinity phosphate transporter in transgenic barley plants does not enhance phosphate uptake rates. Funct Plant Biol 31:141–148
99. Andaluz S, Lopez-Millan AF, De Las RJ, Aro EM, Abadıia J, Abadia A (2006) Proteomic profiles of thylakoid membranes and changes in response to iron deficiency. Photosynth Res 89:141–155
100. Laganowsky A, Gomez S M, Whitelegge JP, Nishio JN (2009) Hydroponics on a chip: analysis of the Fe deficient Arabidopsis thylakoid membrane proteome. J Proteomics 72:397–415
101. Meisrimler CN, Planchon S, Renaut J, Sergeant K, Luthje S (2011) Alteration of plasma membrane-bound redox systems of iron deficient pea roots by chitosan. J Proteomics 74:1437–1449
102. Brumbarova T, Matros A, Mock H P, Bauer P (2008) A proteomic study showing differential regulation of stress, redox regulation and peroxidase proteins by iron supply and the transcription factor FER. Plant J 54:321–334
103. Donnini S, Prinsi B, Negri AS, Vigani G Espen L, Zocchi G (2010) Proteomic characterization of iron deficiency responses in Cucumis sativus L. roots. BMC Plant Biol 10:268–282
104. Rodriguez-Celma J, Lattanzio G, Grusak MA, Abadia A, Abadia J, Lopez-Millan AF (2011) Root responses of Medicago truncatula plants grown in two different iron deficiency conditions: changes in root protein profile and riboflavin biosynthesis. J Proteome Res 10:2590–2601
105. Rellan-Alvarez R, Andaluz S, Rodriguez-Celma J,Wohlgemuth G, Zocchi G, alvarez-Fernandez 1A, Fiehn O, Lopez-Millan AF, Abadia J (2010) Changes in the proteomic and metabolic profiles of Beta vulgaris root tips in response to iron deficiency and resupply. BMC Plant Biol 10:120–134
106. Billard V, Ourry A, Maillard A, Garnica M, Coquet L, Jouenne T, Cruz F, Garcia-Mina J M, Yvin J C, Etienne P (2015) Copper-Deficiency in Brassica napus Induces Copper Remobilization, Molybdenum Accumulation and Modification of the Expression of Chloroplastic Proteins. Plos One. doi:10.1371/journal.pone.0109889
107. Lucini L, Bernardo L (2015) Comparison of proteome response to saline and zinc stress in lettuce. Front Plant Sci doi:10.3389/fpls.2015.00240
108. Deng G, Liu LJ, Zhong XY et al (2014) Comparative proteome analysis of the response of ramie under N, P and K deficiency. Planta 239:1175–1186
109. Chandna R, Ahmad A (2015) Nitrogen stress-induced alterations in the leaf proteome of two wheat varieties grown at different nitrogen levels. Physiol Mol Biol Plants 21:19-33
110. Tran HT, Plaxton WC (2008) Proteomic analysis of alterations in the secretome of Arabidopsis thaliana suspension cells subjected to nutritional phosphate deficiency. Proteomics 8:4317-4326

111. Torabi S, Wissuwa M, Heidari M et al (2009) A comparative proteome approach to decipher the mechanism of rice adaptation to phosphorous deficiency. Proteomics 9:159-170
112. Chen S, Ding G, Wang Z et al (2015) Proteomic and comparative genomic analysis reveals adaptability of Brassica napus to phosphorus-deficient stress. J Proteomics 117:106-119
113. Fukuda T, Saito A, Wasaki J et al (2007) Metabolic alterations proposed by proteome in rice roots grown under low P and high Al concentration under low pH. Plant Science 172:1157-1165
114. Muneer S, Jeong BR (2015) Pr oteomic analysis provides new insights in phosphorus homeostasis subjected to Pi (Inorganic Phosphate) starvation in tomato plants (Solanum lycopersicum L.). PLoS One 10:e0134103. doi: 10.1371/journal.pone.0134103
115. Zeng J, He X, Quan X, Cai S, Han Y, Nadira UA et al. (2015) Identification of the proteins associated with low potassium tolerance in cultivated and Tibetan wild barley. J Proteomics 126:1–11

Chapter 9
Plant Response to Bacterial Pathogens: A Proteomics View

Sheldon Lawrence II, Jennifer Parker and Sixue Chen

Abstract Plants rely on their innate immunity comprised of pattern-triggered immunity (PTI) and effector-triggered immunity (ETI) for defense against pathogens. The evolution of this immune response has resulted in a highly effective system of defense that is able to resist potential attack by pathogens. Bacterial pathogens are major threats to crop production and result in vast losses in revenue each year. Thus, better understanding of the interactions between plants and pathogenic bacteria is a promising avenue for the improvement of crop productivity and agriculture sustainability. Proteomic technologies provide a unique angle to study the intricate interactions between plants and pathogens. Approaches for proteomic analysis can not only lead to the identification of proteins, but also provide quantification and characterization of post-translational modification (PTM). Here we highlight the current knowledge of plant innate immunity in response to bacterial pathogens. We also discuss interesting plant proteomic responses, as well as address the exciting areas of secretome and PTM proteomics as they closely relate to plant-bacteria interactions.

Keywords Proteomics · Bacterial pathogen · Secretome · Post-translational modification

---

S. Lawrence II · J. Parker · S. Chen (✉)
Plant Molecular and Cellular Biology, University of Florida, Gainesville, FL, USA
e-mail: schen@ufl.edu

S. Lawrence II · J. Parker · S. Chen
Department of Biology, University of Florida, Gainesville, FL, USA

S. Lawrence II · J. Parker · S. Chen
Genetics Institute, University of Florida, Gainesville, FL, USA

S. Chen
Interdisciplinary Center for Biotechnology Research, University of Florida, Gainesville, FL, USA

G.H. Salekdeh (ed.), *Agricultural Proteomics Volume 2*,
DOI 10.1007/978-3-319-43278-6_9

9.1 Introduction

Plants are in constant interaction with microbes. Most interactions are not harmful, however, those with pathogenic microbes can lead to diseases that affect plants ability to thrive and reproduce. The diseases caused by the phytopathogens are widespread and often results in a significant decrease in crop yield and economic loss, thus threatening the global food security. It is estimated that over 10 % of the world's agriculture crops are lost to plant-pathogens annually with much greater losses occur during times of epidemics [1]. As the world's population continues to grow, land available for farming decreases and pathogens continue to evolve resistance to plant defenses, better understanding of plant-pathogen interactions is critical to the development of more useful strategies of plant protection against pathogens and increase crop productivity and global food security [2]. Among the many types of phytopathogens, which include fungi, viruses and oomycetes, disease causing bacteria are major threats to crop production [3] and will be the focus of this review.

Bacteria are single-celled prokaryotic organisms found in vast numbers in nearly every place on Earth. They are classified into two major groups, Gram negative and gram positive, based on their chemical composition and cell wall structure [4]. The interactions between bacteria and plants are often beneficial for the plants and/or the bacteria. Such is the case for nitrogen-fixing bacteria in the roots of certain legumes and other plant growth-promoting bacteria as well as those involved in the decomposition of plant remains [5]. Beneficial interactions are common and pose agronomic importance, and plant microbiome research has started to gain a lot of attraction [6]. However, in many cases bacteria can be harmful to plants causing disease and result in major economic loss and increase famine due to decrease in crop production [1]. Most disease causing plant bacterial pathogens belong to the following genera: *Pantoea, Burkholderia, Acidovorax, Clavibacter, Streptomyces, Spiroplasma, and Phytoplasma, Pseudomonas, Ralstonia, Xanthomona, Pectobacterium, Agrobacterium, Xylella, Erwinia, Dickeya* [7]. The top ten scientific and economically important bacterial pathogens fall in the genera of the latter eight and cause diseases such as bacterial speck of tomato, potato brown rot, leaf blight of rice, and fire blight of apple and others that lead to significant economic loss of agriculture crops [8]. These pathogens often use sophisticated molecular strategies to cause plant disease, and plants must overcome them in a biological race to compete and survive.

Currently, an environmental friendly method of battling diseases of agriculture crops involves the development of crop cultivars with enhanced resistance to those pathogenic bacteria that normally cause disease [9]. This requires knowledge of molecular interactions involved in promoting or suppressing disease. Given that proteins play a large role in the interactions between plant and pathogen, proteomics is a logical tool for investigating and elucidating molecular mechanisms that lead to disease. In recent years there has been significant progress made to better understanding plant-bacterial interactions with the use of proteomics technologies and approaches. The use of these approaches has enabled the identification of and changes in proteins involved in plant-bacteria communication during infection. Here we first highlight the current knowledge of plant innate immunity in response to bacterial challenge. Next, we discuss plant proteomic responses to bacteria pathogens with special attention to agriculture crops. Lastly, we address the expanding areas of secretome and post-translational modification (PTM) proteomics as they closely relate to plant-pathogen interactions.

9.2 Current Knowledge of Plant Innate Immune Response to Bacteria

9.2.1 Pathogen Associated Molecular Patterns (PAMPs)

The interaction between plant and bacterial pathogens involves a multifaceted process mediated by both plant and pathogen derived molecules that included proteins, sugars and lipopolysaccharides [10]. To cause disease, pathogens must successfully enter and colonize inside the host. The cell wall and cuticle of plants act as natural physical barriers to prevent the invasion of pathogens. However, these natural defenses are not always effective at preventing invading pathogens. Successful resistance comes from the plants dual level innate immune system of active defense responses [11]. The first level of plant active defense involves the recognition of microbial molecules or elicitors by the plant. These molecules or elicitors are known as pathogen-associated molecular patterns (PAMPs) and are evolutionarily conserved components of microbes/pathogens [12]. Research throughout the years has generated a list of known PAMPs from various bacteria and their induced responses. Common examples of PAMPs include flagellin, elongation factor Tu (EF-Tu), cold shock proteins (CPS), and lipopolysaccharides (LPS) and are all characteristic of gram negative bacteria [13]. Although not all PAMPs are identical across bacteria, conserved protein regions have been shown to provide an avenue for studying PAMP response mechanisms in plants. The most extensively studied PAMP is the small peptide flg22 derived from bacterial flagellin. Flagellin is a protein component of flagella, the structure that provides mobility to certain cells including many gram negative bacteria. Both the N and C terminal regions of this protein are evolutionarily conserved across a number of bacterial species [14], making flagellin useful when studying plant defense against bacterial pathogens. In an early study, purified flagella from *P. tabaci* was shown to initiate downstream responses to the bacterial PAMP in both tomato and Arabidopsis [15]. Early work provided multiple examples of pathogens capable of evading the plant recognition system due to mutations within flg22 epitope, which further demonstrated the importance of flagellin in plant-bacterial interactions [15–17]. In addition to flagellin, several other PAMPs (e.g., EF-Tu and LPS) have been identified and shown to induce downstream responses to bacterial PAMPs in different systems [18–22].

9.2.2 PAMP Triggered Immunity (PTI)

In response to PAMPs, plants have evolved mechanisms to detect conserved bacteria components. In plants, PAMPs are recognized by pattern recognition receptors (PRRs). These receptors are localized on the plant cell membrane and are either leucine-rich receptor-like kinases (LRR-RLK) or receptor-like proteins (RLP) [23]. Like PAMPs, several PPRs have been characterized and include Flagellin Sensitive 2 (FLS2), the flagellin receptor and EF-Tu receptor (EFR), the bacterial elongation factor EF-Tu receptor. Recognition of PAMPs by plant PRRs leads to the activation of the first level of plant innate immune response (Fig. 9.1). This PAMP activated immune response and subsequent changes within the plant are often referred to as basal defense response, general elicitor response and PAMP triggered immunity (PTI), each describing the same level of response that PAMP recognition induces within the plant. This form of immunity is characterized by an assortment of molecular and genetic changes, which include alkalinization, changes in ion flux, increase in reactive oxygen species (ROS), and activation of mitogen-activated protein kinase (MAPK) cascades (Fig. 9.1) [24–26]. PTI can be considered a

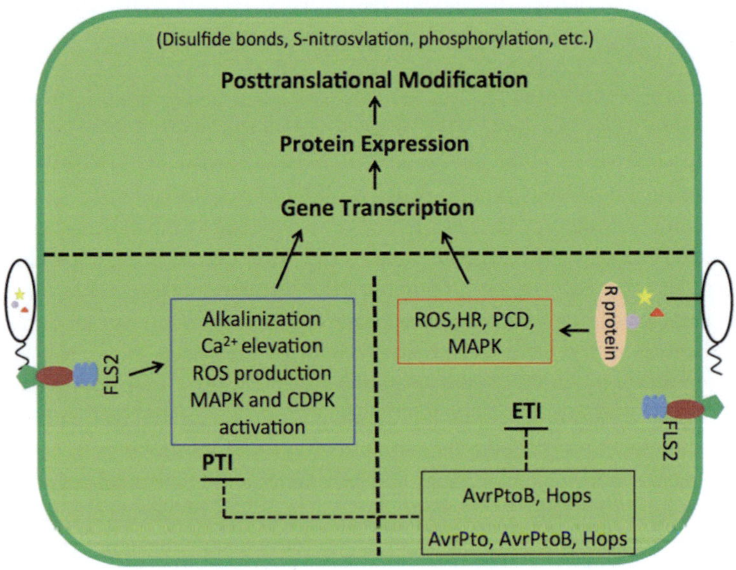

Fig. 9.1 Diagram showing interaction between plants and pathogen infection. Plants respond to PAMPs such as flg22 through PRRs and induce PTI mechanisms including alkylation, Ca^{2+} elevation, ROS production, MAPK, and CDPKs activation. These response mechanisms are inhibited by T3Es such as AvrPto, AvrPtoB, and Hops. ETI response mechanisms are activated through identification of T3Es and production of ROS, HR, PCD, and activation of MAPKs. ETI is also regulated through T3Es such as AvrPtoB and Hops. Both early and late response mechanisms have been shown to initiate changes in gene transcription, translation and PTMs such as disulfide bond formation, S-nitrosylation and phosphorylation

"priming" mechanism of plants defense response to bacteria, so that if a plant is challenged by a secondary infection after initial PTI has been allowed to develop, bacterial growth is often restricted in host plants and eliminated in non-host plants [27]. Over the last decade, there has been much research that has led to our understanding of PTI signaling cascades. Research here has involved mutations of known receptors such as FLS2 and EFR in order to examine phenotype of PRR mutants as well as elucidate signaling components of PAMP induced PTI and pathogen development [19, 27–29].

9.2.3 Effector Triggered Immunity (ETI)

As previously described, PTI is a basal defense response utilized by the plant to prevent bacterial growth. In order to successfully infiltrate the apoplast and thrive, bacteria must bypass this basal defense. Bacteria have therefore evolved mechanisms to overcome different components of PTI by delivering 'effector' proteins into the plant. A mechanism in which many pathogenic bacteria such as certain species from the genera *Pseudomonas, Ralstonia, Xanthomona, Erwinia* accomplish delivery of effectors is through the use of a type III secretion system (TTSS). Utilizing this system, bacteria are able to bypass the plant cell wall and plasma membrane to inject effectors directly into the cell (Fig. 9.1) and successfully suppress PTI. Plants are often able to perceive this and mount a subsequent response. The second level of plant innate immune response involves the recognition and interaction of these specific effectors known as avirulence (Avr) proteins or type three effectors (T3E) by resistance proteins (R proteins) within the plant and is referred to as effector-triggered immunity (ETI). An adaptive zig-zag model has been proposed and is representative of the changes in the infection and defense processes [11]. Successful recognition of avr proteins, like AvrPto, AvrPtoB, and Hops from *Pseudomonas syringae* by plant R proteins leads to a localized cell death (hypersensitive response), the expression of pathogenesis related (PR) genes, changes in the proteome and systemic acquired resistance (SAR) in order to kill the pathogen, pathogen infected cells and help prevent later infections [30, 31]. A number of studies have highlighted the role and provided support to the importance of several factors involved in ETI [32–35]. Although ETI response is described as being faster and more robust when compared to PTI, both PTI and ETI act together to effectively defend plants against pathogens and common to them both is an increase in reactive oxygen species (ROS), production of nitric oxide (NO), and changes in PTM events in order to prevent pathogen infection or inhibition of its growth.

9.3 Proteomic Responses to Bacterial Pathogens

Despite the knowledge of plant innate immunity, many components of PTI and ETI in response to bacteria pathogens remain largely unknown. As previously stated, proteins play an undeniable role in plant-pathogen interactions, and proteomics has become a valuable resource for better understanding these interactions. It is known that information obtained from genomics and transcriptomics does not always correlate to protein changes [36]. Providing insights to protein localization, enzymatic complexes, protein-protein interactions and PTMs, proteomics allow a more direct view of cellular processes and activities that occur during specific plant-pathogen interactions. Advances in proteomic technologies such as two-dimensional gel electrophoresis (2-DE), two-dimensional difference gel electrophoresis (2D-DIGE), and gel free methods such as isotope-coded affinity tag (ICAT), isobaric tags for relative and absolute quantitation (iTRAQ), cysteine reactive tandem mass tags (cysTMT), and stable isotope labeling by amino acids in cell culture (SILAC) in tandem with liquid chromatography (LC)-mass spectrometry (MS) have made it possible for the large scale analysis of proteomic changes during plant-pathogen interactions under various treatment conditions and times (Fig. 9.2). Results of proteomic work aid in filling knowledge gaps that arose during the genomic and transcriptomic area. With the great improvements in these technologies within the past 20 years there has been an increase in the amount proteomic literature in the area of plant-bacteria interactions (Table 9.1). Much of the research here highlights changes in protein modification and expression levels induced by treatment with bacterial pathogens or elicitors and relies on the use of MS approaches to provide qualitative and quantitative information. In the area of plant-bacteria proteomics, the symbiotic interaction between nitrogen fixing bacteria and legumes has long been the most well studied [37]. In more recent years, however, research has been increasingly focused on the proteomes of plant during their interactions with bacterial pathogens.

9.3.1 General Proteomic Studies

A number of proteomics studies using non-crop model systems have contributed to what is known regarding plant–pathogen interactions. An important system is that of the model plant *Arabidopsis thaliana*, which has been invaluable to the understanding of molecular events leading to the progression of disease or successful pathogen defense in crops [38]. Arabidopsis along with bacterial pathogens and bacterial elicitors has been used to identify specific proteins that respond to bacterial pathogens. Using bacterial elicitors rather than the pathogen itself to treat the host plants or cell suspension cultures simplifies the system, making it easier to study rapid changing in the proteome. Therefore, some plant defense studies have utilized Flg22, chitin or other bacterial elicitors to induce defense responses [39, 40]. However, the

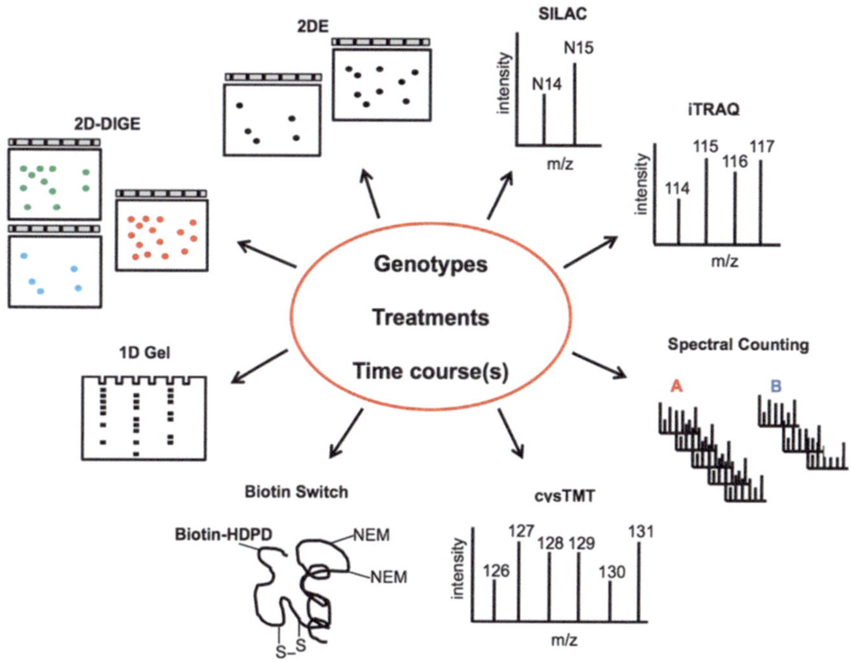

Fig. 9.2 Analytical approaches used in plant pathogen interaction proteomics. 1D SDS-PAGE, 2-DE, and 2DIGE are gel methods used to identify global protein changes and PTMs. SILAC is a gel free method that uses the changes in mass due to the metabolic incorporation of amino acid isotopes to examine quantitative changes between samples. iTRAQ and cysTMT are isobaric tag labeling systems that are used to label protein extracts when examining global protein changes. Labeling with cysTMT and biotin-HDPD is used to identify cysteine redox changes due to different treatment. Spectral counting is a label-free method capable of examining changes in protein levels and PTMs

use of bacterial pathogens themselves provides a more complete system and often reveals more information about the interactions and will be the focus here.

As a lab model system, the pathogenic bacterium *Pseudomonas syringae* pv. *tomato* (*Pst*) strain DC3000 has often been used to study plant response to bacteria pathogens. *Pst* is a gram negative bacterium that causes bacterial speck disease in *A. thaliana* and *Solanum lycopersicum* (tomato). Because of the necrotic lesion that forms on the plant after infection, plants and their fruits become less valuable. Jones and colleagues analyzed PAMP, TTSS, and ETI proteomic changes in Arabidopsis when challenged with different near isogenic lines of (*Pst*) DC3000 [41, 42]. The study reported defense-related PR-9 antioxidant proteins and metabolic enzymes. Two groups of proteins (glutathione S-transferase (GST) and peroxiredoxin (Prx) changed in response to treatment with *Pst* and its variants, *hrpA* DC3000 and *Pst* (*avrRPM1*) [41]. These two groups of antioxidant enzymes are suggested to play important roles in the regulation of redox conditions during pathogen infection. Further studies using early time points, protein extracts from the chloroplast,

Table 9.1 Proteomic studies of plant response to bacterial pathogens

Organism	Pathogen/elicitor	Proteomic techniques	Identified proteins	Main findings	References
*A. thaliana/*suspension cells	Flg22 or chitin fragments	2-DE, nESI-MS/MS	AtPhos43, and related proteins	AtPhos43 phosphorylation is rapid and dependent on FLS2	[39]
*A. thaliana/*suspension-cultured cells	Flg22	Labelling with 32P, SAX-IMAC, MALDI MS, nanoLC-MS/MS	299 phosphopeptides; ~200 phosphoproteins	Large-scale ID of phosphorylation sites; previously unknown phosphorylation sites on proton ATPase isoforms	[40, 67]
*A. thaliana/*leaves	*Pseudomonas syringae* pv. *tomato* DC3000, DC3000 *avrRpm1*, DC3000 *hrpA*	2-DE, LC-MS/MS	53 proteins, two groups GSTs and Prxs analyzed	glutathione S-transferases and peroxiredoxins, enzyme changes	[41]
*A. thaliana/*leaves	*P. syringae*	iTRAQTM	5 proteins	Phosphoproteome changes of dehydrin, co-chaperone, HSP, RuBisCO, plastid-associated protein	[42]
*A. thaliana/*leaves	*P. syringae* pv. *tomato* DC3000, DC3000 *avrRpm1*, DC3000 *hrpA*	2-DE, LC- MS/MS	36 total soluble, 8 chloroplast, 13 mitochondrial proteins	Identified significant changes to defense-related antioxidants and metabolic enzymes involved in ETI and PTI	[43]
A. thaliana/ suspension cells	*P. syringae pv. tomato* DC3000, DC3000 *avrRpm1*, DC3000 *hrpA*	iTRAQ coupled with LC-MS/MS and Q-TOF-MS	45 differentially expressed proteins	Proteins involved in metabolism, redox regulation, defense, and cell wall maintenance	[60]
A. thaliana/ suspension cells	Flg22	iTRAQ, SAX-IMAC, nanoLC-MS/MS	11 proteins	Phosphorylated defense response proteins: NADPH oxidase, RBOHD, and ABC transporter, pleiotropic drug resistance 8, and penetration 3	[67]
*A. thaliana/*suspension cells	Flg22 or xylanase	$^{4}N/^{15}N$- labelling, SCX-TiO_2, LC-MS/MS	76 proteins, 98 phosphopeptides	Phosphorylated changes in unidentified membrane associated proteins and known defense proteins	[66]

(continued)

Table 9.1 (continued)

Organism	Pathogen/elicitor	Proteomic techniques	Identified proteins	Main findings	References
A. thaliana and *M. sativa*/roots	*P. syringae pv. tomato* DC3000 or *Sinorhizobium meliloti* strain Rm1021	2-DE, nanoLC-MS/MS	~100 secreted proteins	Seven plant proteins, such as hydrolases, peptidases, and peroxidases were increased in *M. sativa–S. meliloti* interaction; Four bacterial proteins increased	[56]
A. thaliana/leaves	*P. syringae pv. tomato* DC3000, DC3000 *AvrB*	2-DE, immunoblotting, MALDI-TOF- TOF	16 proteins	Identification of S-nitrosylated proteins involved in metabolism and defense signaling	[76]
Oryza sativa/transgenic and/or inoculated	*Xanthomonas oryzae* pv. oryzae	2-DE, MS/MS Protein sequencer	10 proteins	Identification of differential expressed proteins related to energy, metabolism, and defense	[51]
O. sativa/suspension cell cultures	*X. oryzae* pv. oryzae	2-DE, MS/MS	20 proteins	Identification of plant proteins involved in early response to bacterial blight	[52]
O. sativa/leaves	*X. campestris* pv. Oryzicola	2-DE, MALDI-TOF MS	32 proteins	Identification of proteins involved in disease resistance signal transduction, pathogenesis, and regulation of cell metabolism	[53]
O. longistaminata/ leaves	*X. oryzae* pv. oryzae	2-DE coupled with MALDI-TOF MS	29 proteins	Proteins for tolerance against the disease: r40c1, cyclin-dependent kinase C, Ent-isokaur-15-ene synthase, glutathione-dependent dehydroascorbate reductase 1 (GSH- DHAR1), and germin-like protein	[54]
O. sativa	*X. oryzae*	2-DE/MudPIT and MALDI-TOF/MS or nESI-LC-MS/MS	109 proteins	Differential enriched proteins involved in pathogenicity, protease/peptidase, and host defense *in planta*, and in vitro	[61]
Lycopersicon hirsutum	*Clavibacter michiganensis* subp. *michiganensis*	2-DE, ESI-MS/MS	47 proteins	Revealed distinct mechanisms conferred by loci Rcm 2.0 and Rcm 5.1, proteins involved in defense, stress responses and cell signaling	[47]

(continued)

Table 9.1 (continued)

Organism	Pathogen/elicitor	Proteomic techniques	Identified proteins	Main findings	References
Solanum lycopersicum	*P. solanacearum*	2-DE, Protein sequence	15 proteins	Proteins involved in protein destination and storage (60 kDa chaperonin, PDI, heat shock protein), protein synthesis, metabolism (e.g., RuBisCO activase, glycine dehydrogenase), and defense (calgranulin, AMA)	[49]
S. lycopersicum	*P. syringae* pv. *tomato* DC3000	iTRAQ, LC-MS/MS	477 proteins	Proteins involved in oxidation-reduction, response to stress, defense signaling, cell wall organization and hormone signaling	[50]
S. lycopersicum	*P. syringae* pv. *tomato* DC3000	CysTMT LC-MS/MS	90 proteins	Potential redox-regulated proteins in carbohydrate/energy metabolism, Cys biosynthesis, sucrose and brassinosteroid, cell wall biogenesis, polysaccharide/starch biosynthesis, cuticle development, lipid metabolism, proteolysis, TCA cycle, protein targeting to vacuole, and oxidation–reduction	[71]
Lotus japonicas	Rhizobial lipochitooligosaccharides nodulation factors (NF)/ Flg22	^{33}P-orthophosphate labeling, 2-DE immunoblotting, kinase assay	13 proteins	Phosphorylation of unique and shared proteins during symbiotic and defense responses	[68]

mitochondria, and total soluble protein and 2-DE and MS/MS, 57 proteins with differences in expression levels were identified [42]. These proteins were determined to be involved in biological processes such as defense (GST), transcription (squamosa promoter binding protein-like 14), protein stability (cyclophilin) and metabolism (glyceraldehydes 3-phosphate dehydrogenase). These results suggest PAMP responses could be attributed to regulation of glycolysis, tricarboxylic acid (TCA) cycle, and antioxidation activities. However, these processes were also altered by the presence of virulent T3E and ETI. PAMP-induced changes at the protein level were reflected by 2D gel spot density changes between mock and *hrpA* DC3000 treatment, a TTSS mutant used to examine PAMP responses and changes induced by T3Es. This revealed a total of 30 proteins identified as having changes in spot density in response to PAMPs. Jones et al. [42, 43] additionally analyzed changes in the phosphoproteome of Arabidopsis during *Pst* infection. Phosphoprotein enrichment followed by iTRAQ tagging was employed and proteins that were differentially phosphorylated in soluble *A. thaliana* leaf extracts were identified. Phosphoproteome changes of 4 proteins (dehydrin, co-chaperone, heat shock protein, plastid-associated protein) and ribulose-1,5-bisphosphate carboxylase/oxygenase large subunit (RuBisCO LSU) was determined [43]. This first study utilizing iTRAQ to study plant-pathogen interactions highlighted the reproducibility and utility as well as problems often associated with the quantitative analysis of changes in a complex phosphoproteome. In an approach to decrease high abundant proteins in Arabidopsis leaves after *Pst* infection combinatorial hexapeptide ligand libraries (CPLL) were used. The high abundant proteins from infected leaves shown by 2-DE were reduced to a low level and less abundant protein were enriched. Mass spectrometric analysis then led to the identification of 312 bacterial proteins from the infected leaf tissue [44].

The plasma membrane (PM) is where detection of PAMPs occurs and signal transduction is initiated during pathogen response. Using transgenic Arabidopsis with dexamethasone (Dex) inducible AvrRpt2 expression (*GVG-AvrRpt2*), ETI was induced and proteomic changes observed. Using SDS-PAGE and MS/MS, a total of 2336 proteins were identified in PM enriched samples, with 423 proteins showing significant changes in expression [45]. A total of 235 proteins with increased levels were involved in processes such as camalexin biosynthesis (CYP71B15, CYP71A12, CYP71A13), membrane trafficking (syntaxin of plants (SYP) 122, soluble N-ethylmaleimide-sensitive factor adaptor protein 33), protein phosphorylation (Pep1 receptor 1, wall-associated kinase 1, cysteine-rich receptor kinase), and methionine metabolism (S-adenosylmethionine synthase, 1-aminocyclopropane-1-carboxylate oxidase). Proteins with decrease levels, a total of 188, were involved in biological processes that included glucosinolate metabolism (e.g., CYP83A1 and CYP83B1), membrane transport (auxin and Ca^{2+} transporters), and photosynthesis (thylakoid membrane proteins). This information provided insight into the specific ETI regulated changes occurring at the PM. Virulent *Pst* and avirulent *Pst* (avrRpt2) treatment of Arabidopsis was used to investigate the function of time in the development of the response mechanisms [46]. Utilizing 2-DE and MS/MS, the authors identified 800 proteins, with 147 showing significant changes. After the virulent infection, 794 spots were present at 4 hai, 795 present at 8 hai, and 772 present at 24 hai. Avirulent infection resulted in observation of 808 spots present at 4 hai, 810 at 8 hai, and 739 at 24 hai. Proteins were categorized as being shared between the virulent and avirulent response mechanism, or unique to one or the other response. A total of 23 proteins that were not previously identified as being involved in defense were identified, e.g., RAS GTP-binding nuclear protein and NUDIX (nucleoside diphosphates linked to moiety X) hydrolase homolog. Results of this study provide information on proteomic changes occurring not only between virulent and avirulent responses, but also provide a snapshot of the proteomic changes occurring at different time points during the infection.

Although proteomic research using Arabidopsis is extensive and has improved what is known about plant-pathogen communication (additional studies with the model plant will be described in other sections), studies using other systems and various pathogens has been helpful in studying specific plant responses to bacterial infection. Proteomic work utilizing the wild tomato species *L. hirsutum* revealed proteins that are regulated in response to *Clavibacter michiganensis* subp. *michiganensis* infection. This bacterial pathogen is responsible for bacterial canker disease in tomato, leading to leaf necrosis, leaf wilt, cankers on the stem and plant death. Two partially resistant lines that contained the quantitative trait loci Rcm2.0 and Rcm 5.1 which control resistance to the disease and a susceptible line were compared using 2-DE and MS [47]. Analysis identified 26 differentially regulated proteins, of which three superoxide dismutase (SOD) enzymes and nine other enzymes directly related to plant defense were identified. The identification of these enzymes indicates the importance of stress related proteins including those related to oxidative stress in response to pathogenic infection and revealed distinct mechanisms conferred by two loci: Rcm 2.0 and Rcm 5.1. In a study to detect genes that control tomato (*L. esculentum*) bacterial wilt infected with *Ralstonia solanacearum* [48] reported that expression of the caffeoyl coenzyme A (CoA) 3-O-methyltransferase gene was down-regulated in seedlings of different susceptible cultivars of tomato that were inoculated with the bacteria. The new discovery of this down-regulated gene in infected tomato suggests its role in tomato response to stress such as that caused by bacterial infection. Further, proteomic research in the tomato system reported the differential levels of proteins in response to the necrotrophic bacteria *Pseudomonas solanacearum* [49]. Using 2-DE and Edman sequencing, the comparison of protein fold changes between bacterial wit-sensitive and wit-resistant cultivars revealed nine proteins that were highly expressed in resistant cultivars. The proteins were found to be related to plant defense, protein storage and protein trafficking. As well, the apical membrane antigen (AMA) was found to be increased in the susceptible cultivars. This provided support that a previously unidentified protein has a role in tomato resistance to this bacteria pathogen. Additionally, Parker and colleagues recently examined plant pathogen interactions between tomato and *Pst* DC3000 and identified 2369 proteins amongst two genotypes (*PtoR* and *prf3*) and two time points (4 hai and 24 hai) [50]. Of the proteins, 477 showed significant changes in levels. Proteins with significant changes in *PtoR* 4 hai (225) included proteins involved in cell wall organization and reduction of oxidative stress. *PtoR* 24 hai had the most proteins (325) with significant changes when compared amongst genotypes and time points. Those proteins were identified as being involved in oxidation-reduction, response to stress, and signaling. When 164 proteins identified in *prf3* 4 hai were examined, it was observed that although there were fewer proteins identified in this genotype and treatment when compared to both time points in *PtoR*, more proteins were identified when compared to *prf3* 24 hai. It was also noted that *prf3* 4 hai and *PtoR* 24 hai had shared biological processes such as immune response, response to biotic stimulus, and hormone metabolic process, suggesting defense response processes may be induced early after infection of susceptible genotypes, but repressed in *prf3* 24 hai. Biological processes associated with abiotic stress and oxidation-reduction were observed in *prf3* 24 hai. This genotype and time point had the fewest number of proteins (128) with significant changes in levels, but the most proteins with a decrease in expression that paralleled the decreased defense response and onset of the diseased phenotype.

Additional proteomic studies employed other plant species as pathosystems to studying plant response to pathogens. In rice bacterial blight, caused by *Xanthomonas oryzae* pv. oryzae (Xoo) is one of the most severe diseases of this crop, leading to major crop loss during server epidemics. In work by Mahmood and colleagues [51], the defense-related antioxidants PR-9 and PR-5 in rice were determined to be induced in response to treatment with Xoo. Analysis revealed four defense-related proteins PR-5, Probenazole-inducible protein (PBZ1), SOD and Prx were induced for both compatible and incompatible *X. oryzae* pv. Rapid induction and higher expression of PR-5 proteins and PBZ1 were seen in incompatible interactions and in the presence of the plant hormone jasmonic acid (JA). As well the work here presented PR-5 (Thaumatin like protein) as a candidate to use in plant biotechnology against bacterial blight in rice. In an effort to study the early defense responses involved in the rice receptor kinase Xa21 mediated resistance proteins from rice, plasma membrane from suspension cell cultures inoculated with both compatible and incompatible Xoo race strains were analyzed [52]. Twenty proteins were found differentially regulated in cultures that were induced by the pathogen at 12 and 24 h post-inoculation. Of the twenty proteins that have potential function in rice defense, eight were plasma membrane associated and two were non plasma membrane associated.

Li and colleagues identified proteins responsive to *X. campestris* pv. *Oryzicola,* the causative agent of bacterial leaf streak (BLS) another major disease in rice. Using 2-DE coupled with matrix assisted laser desorption ionization (MALDI)-time of flight (TOF) MS analysis, 32 increased proteins that are potentially involved in disease resistance signal transduction, pathogenesis, and regulation of cell metabolism were identified. In addition, seven gene transcripts were shown to be increased after bacterial infection [53]. This helped to elucidate the molecular mechanisms underlying BLS disease. Recently in a comparative proteomic analysis of total foliar protein isolated from infected rice leaves of susceptible Pusa Basmati 1 (PB1) and resistant *Oryza longistaminata* genotypes, proteins belonging to a large number of biological and molecular functions potentially involved in *Xoo* infection as well as candidate genes conferring tolerance against bacterial blight were identified [54]. Using 2-DE coupled with MALDI-TOF MS, 29 protein spots encoding unique proteins from both the genotypes were identified. Among the proteins those related to biotic and abiotic stress were induce during infection, which suggests that both pathways are activated during infection. The identified candidate genes for tolerance against the disease include putative r40c1, cyclin-dependent kinase C, Ent-isokaur-15-ene synthase and glutathione-dependent dehydroascorbate reductase 1 (GSH-DHAR1), which were induced, and the germin-like protein which was induced only in the resistant genotype.

9.3.2 Secretome Studies

Pathogenesis depends on the ability of the pathogen to manipulate the plant metabolism and to inhibit plant immunity, which depends to a large degree on the plant's capacity to recognize pathogen elicitors. The first interaction between plant and pathogens occurs in the apoplast, thus analyzing the changes of apoplastic proteins through a proteomics approach is important to the understanding of the components of signal perception and signal transduction during pathogen attack. The extracellular secreted proteins in the apoplast at a given time are known as the secretome. Despite its importance, the secretome during plant-bacteria interactions remains poorly characterized compared to the intracellular proteome [38]. The dearth of literature in this area most likely results from the difficulty involved in obtaining apoplastic material without damaging the plant cell and the lack of better methods of preventing contamination of cytoplasmic proteins [55]. In a study of the root secretomes during the interaction between *Medicago sativa* (alfalfa) and the bacterial symbiont *Sinorhizobium meliloti* and between *A. thaliana* and the bacterial pathogen *P. syringae*, bacteria were shown to change the proteins they secreted during infection depending on the identity of the host plant [56]. Using a vacuum infiltration (VIC) method to isolate the apoplastic proteins from the root exudates, more than 100 proteins were identified as differentially expressed during the different plant-bacteria interactions. Of the identified proteins seven plant proteins, such as hydrolases, peptidases, and peroxidases were increased in *M. sativa-S. meliloti* interaction. In addition, four bacterial proteins increased during *S. meliloti/alfalfa* interaction and nine plant defense-related proteins increased during *P. syringae* DC3000/Arabidopsis compatible interaction. This study uncovered a specific, protein level cross-talk between roots and two bacteria pathogens. Several studies have used suspension cell cultures in secretome research [57–59]. Secreted proteins can be easily separated from suspended cells in suspension cell cultures without disrupting the cell, thus preventing potential contamination by cytoplasmic proteins. In an experiment to examine proteomic changes in the secretome after pathogen infection, *A. thaliana* suspension cell cultures were treated with different strains of *Pst.* Using sodium dodecyl sulfate-polyacrylamide gel electrophoresis (SDS-PAGE) and iTRAQ, the researchers identified changes in 45 proteins, which include glucan endo-1,3-β-glucosidase 7, Prx 53, strictosidine synthase, and EF-2 [60]. Nine proteins, that contained signaling peptides, were observed to be commonly suppressed after treatment with each strain, suggesting that their levels may be regulated by PAMP detection and response mechanisms. Additionally, eight proteins were identified as regulated specifically by T3Es. Gene for gene resistance induced by *Pst* (*avrRpm1*) caused accumulation of 12 proteins. The role that T3Es and gene for gene resistance play in the regulation of proteins induced by PAMP response was also examined. *Pst* (*hrpA*$^{-}$) induced three proteins, and one was suppressed by T3Es. However, one protein induced by PAMP was suppressed by T3Es, but the levels increased again after gene for gene resistance was introduced through *Pst* (*avrRPM1*). A total of six proteins without a signaling peptide were suppressed by PAMP treatment. The same proteins were identified as being induced after T3E treatment and had reduced levels after gene for gene resistance was induced. Regulation of protein levels by T3Es suggests a role in successful pathogen infection.

Apoplastic proteins were also analyzed during bacterial infection of rice [61]. Analysis of Xoo (compatible race K3)-secreted proteins, isolated from its in vitro culture and *in planta* infected rice leaves using 2-DE coupled with MS, 109 proteins were identified. Only six of the identified proteins were secreted from rice, indicating that the percentage of bacterial-secreted proteins is much higher than its host rice. The identified proteins secreted from *X. oryzae* in vitro and *in planta*, belonged to multiple biological and molecular functions. Proteins involved in metabolic and nutrient uptake activities were common to both in vitro and *in planta* secretomes. However proteins involved in pathogenicity, protease/peptidase, and host defense were highly enriched *in planta*, but not detected in vitro. Information obtained from this work furthers the knowledge of rice bacteria blight disease.

9.3.3 PTM Proteomics

PTMs are known to control many physiological processes by affecting protein structure, activity, and stability. To date hundreds of PTMs have been described, however, only a few have been analyzed using large-scale proteomic techniques. Proteins can undergo different PTMs such as phosphorylation and redox-based regulation such as nitrosylation. Constitutive activation of defense response mechanisms often results in reduced plant growth and yield. The ability to turn on and off a "molecular switch" through PTMs such as phosphorylation and nitrosylation allows plants to respond quickly and efficiently to environmental challenges (Fig. 9.1). It should be noted that pathogens also have the ability to modify proteins. Modifications to the proteins involved in plant-pathogen interaction often suppress or initiate their activities [62]. For example, it is known that AvrPto-induced HR is mediated through autophosphorylation of Pto and that mutations rendering the protein incapable of autophosphorylation results in decreased ETI [63]. It is also known that phosphorylation on T199A of Pto P+1 loop is required for the interaction with Pto and AvrPtoB to occur [30]. In addition, PTMs can take place on T3Es themselves, thus affecting their ability to modulate plant defense responses. For example, AvrPto is phosphorylated by a 30-40 kDa kinase on serine 149 (S149). Treatment of *prf3* with the S149A mutant resulted in decreased bacterial growth. Also, treatment of PtoR with S149A showed a decrease in bacteria, showing that S149A is important for the avirulence responses initiated by ETI [64]. HopAI1 is another example of T3E capable of direct T3E-triggered modifications. In Arabidopsis, this protein interacts with MPK3/MPK6 and through de-phosphorylation inactivates the kinases involved in PTI and leads to dampening of PTI activated MAPK signaling cascades [65].

The study of phosphorylation in early response to flg22 and xylanase has been performed in Arabidopsis [66]. Phosphorylated peptides in the PM of Arabidopsis cell culture were examined using SILAC and titanium dioxide (TiO_2) enrichment along with LC-MS/MS. Kinase activity was found to be the highest at 5–10 min after flg22 or xylanase treatment and decreased after 30 min. At 10 min after treatment, the researchers identified 472 phosphorylated proteins, and a total of 76 of the proteins with 98 peptides were differentially phosphorylated. The phosphorylated proteins include CDPKs, RLKs, respiratory burst oxidase homolog D (RBOHD), and vesicle trafficking associated SYP121 [66]. These phosphorylated residues and their corresponding proteins provided previously unknown information into the membrane phosphorylation events occurring after PAMP treatment. In a separate study, Nuhse and co-workers utilized iTRAQ along with immobilized metal ion affinity chromatography (IMAC) enrichment to identify phosphoproteins involved in flg22 response in Arabidopsis [67]. Identified were 11 peptides with PTMs and significant fold changes. Known defense response proteins NADPH oxidase, RBOHD, and an adenosine triphosphate (ATP) binding cassette transporter, pleiotropic drug resistance8, and penetration3 were shown to be phosphorylated. Further characterization of the phosphorylation sites of RBOHD, S343, and S347 revealed that phosphorylation of the residues is necessary for RBOHD mediated ROS production. Additionally, proteins annotated as being involved in membrane trafficking and ubiquitination such as dynamin, RING-H2 protein, and Arabidopsis toxicos en levadura 6 (ATL6) were also identified as being phosphorylated. These findings help to elucidate regulatory mechanisms of plant innate immune responses via PTMs.

Jones and co-workers analyzed changes in the phosphoproteome of Arabidopsis during *Pst* infection. Phosphoprotein enrichment followed by iTRAQ tagging was employed and proteins that were differentially phosphorylated in soluble *A. thaliana* leaf extracts were identified. Examining proteins with changes during PAMPs response, dehydrin, a protein involved in water stress, and a putative p23 co-chaperone were identified as having a decrease in phosphorylation. Additionally, a plastid-lipid associated protein associated with transportation, a heat shock protein, and a proton-dependent oligopeptide transporter were observed to have an increase in phosphorylation [43]. In addition, a recent study comparing *Lotus japonicas* roots treated with a nodulation factor and a pathogen elicitor flg22 revealed differential phosphorylation of shared and unique proteins during symbiotic and defense responses [68].

As previously stated, ROS and NO are involved in the signaling mechanisms that characterize PTI, HR and PCD. Their actions are usually carried out through the PTMs of proteins via cysteine residues that act as versatile sulfur switches. The role of redox-based PTMs in plant-pathogen interaction is currently an area of interest. Like other PTMs, the formation of molecular disulfide bonds often affects protein properties. Preliminary redox proteomics was performed to examine pathogen responsive proteins in Arabidopsis [41]. The researchers observed that GST (F6, F7 and F8) had a shift in protein PI, possibly due to redox modifications. Another study with Arabidopsis suspension cells treated with 5 mM H_2O_2 revealed redox induced modifications using two thiol reactive tags, 5-(iodoacetamido) fluorescein (IAF) a fluorescent probe used alongside 2D-DIGE as well as N-(biotinoyl)-N′-(iodoacetyl)-ethylenediamine (BIAM) [69]. A total of 84 potentially redox responsive proteins were identified and they were involved in processes such as metabolism, antioxidation, translation, and protein folding. Parker and co-workers [70] have shown that labeling using the isobaric tags can provide multiplex high throughput analysis of redox responsive cysteines and proteins. When mock treated and *Pst* treated tomato samples were compared, a change in the reporter ion spectra can be compared between labeled samples showing quantitative cysteine PTM changes. This method allowed researchers to quantify redox changes and map the modified cysteines across six samples. The recently developed cysTMT tags have been used to examine potentially redox-regulated proteins [71]. Here cysTMT labeling helped to identify similarities and differences of protein redox modifications in tomato resistant (*PtoR*) and susceptible (*prf3*) genotypes in response to *Pst* infection. A total of 4348 proteins were identified by LC-MS/MS, 90 of which were identified to be potentially redox-regulated. The 90 potential redox-regulated proteins fell into diverse categories including carbohydrate and energy metabolism, biosynthesis of cysteine, sucrose and brassinosteroid, cell wall biogenesis, polysaccharide/starch biosynthesis, cuticle development, lipid metabolism, proteolysis, tricarboxylic acid cycle, protein targeting to vacuole, and oxidation–reduction.

Although several different redox-based PTMs are known to occur in plants (Fig. 9.3), protein nitrosylation is considered as one of the key mechanism regulating protein function [72]. Nitrosylation is a PTM in which a nitric oxide radical oxidizes a free thiol group as well as tyrosine, thus NO produced during plant-pathogen interactions could exert their signaling action through nitrosylation of specific proteins [73, 74]. To examine potentially S-nitrosylated proteins, Arabidopsis suspension cells were treated with NO donors S-nitrosoglutathione (GSNO) and sodium nitroprusside, and leaves were exposed to NO gas [75]. A total of 63 proteins from the GSNO treatment and 52 proteins from the NO treatment were identified as being S-nitrosylated. The proteins were found to be involved in stress, redox regulation, signaling, cytoskeleton structure, metabolism, and photosynthesis and included proteins such as SOD (copper (Cu) and zinc (Zn)), a glutathione peroxidase, elongation factor 1α-chain, actin 2 and 7, fructose

1,6-bisphosphate aldolase, and Rieske Fe-S protein. This study provided information on potential targets of S-nitrosylation as well as the sites where the modification is likely to occur.

Differential thiol nitrosylation during HR was observed in Arabidopsis after avirulent *Pst* (*avrB*) treatment at 0 hai, 4 hai, and 8 hai [76]. A total of 18 differentially modified proteins were identified. For example, RuBisCO in photosynthesis was identified along with an allene oxide cyclase involved in JA signaling. The other

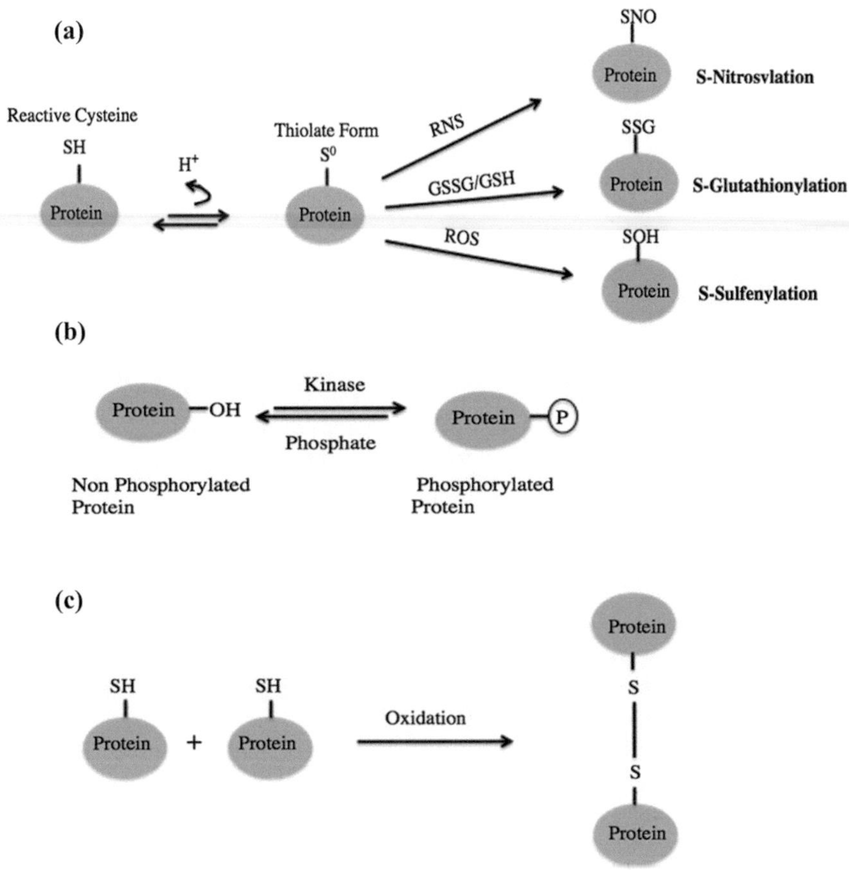

Fig. 9.3 PTM proteomics methods used in plant bacterial pathogen interactions. **a** A reactive cysteine (SH) with a low pka can readily lose a hydrogen ion leading to the formation of a highly reactive thiol S^0, which can react with reactive nitrogen species (RNS) to become S-nitrosylated (SNO), Glutathione disulfide (GSSG)/reduced Glutathione (GSH) to become S-Glutathionylated (SSG) or reactive oxygen species (ROS) to become S-Sulfenylated. These oxidative modifications are reversible. (*Note* Further oxidation of sulfenic acid to sulfinic acid (RSO_2H) and sulfonic acid (RSO_3H) is thought to be generally irreversible.) **b** During protein phosphorylation, phosphate moieties are transferred by protein kinases to serine, threonine or tyrosine residues of proteins. This reaction can be reverse by protein phosphatases that hydrolyze phosphate moieties. **c** Depiction of the formation of a disulfide bond as a result of oxidation of two sulfhydryl groups

processes observed included metabolism with the identification of glyceraldehydes-3-phosphate dehydrogenase and redox regulation including a germin-like protein. In addition, the S-nitrosylated sites on the proteins were mapped. A change in the nitroproteome of Arabidopsis through tyrosine nitrosylation is also evident during HR [77]. With peroxynitrate ($ONOO^-$) treatment, eight proteins were identified as being nitrosylated including the 33 kDa oxygen evolving protein, RuBisCO, and glutamine synthetase 2, proteins involved in photosynthesis, Calvin cycle and glycolysis, and nitrate assimilation. A time course of treatment with avirulent *Pst* (*avrB*) revealed that nitrosylation peaked from 4 hai to 8 hai. The results here indicate that tyrosine nitrosylation may play a role in pathogen response.

9.4 Conclusions and Perspectives

The study of plant-pathogen interactions is a broad and important area of plant biology research. PTI and ETI have been examined extensively in Arabidopsis and to a lesser extent in crop plants. Bacterial responses have been characterized and modeled in order to describe not only how infection occurs, but how bacteria respond to plant defense mechanisms. Progress has been made in the analysis of individual genes and proteins as well as transcriptomic changes however, plant-bacteria proteomics has just touched the surface and secretome as well as PTM proteomic changes due to pathogen infection are still in their infancy. The information gathered from global proteomics and PTM proteomics allows researchers to observe processes underlying pathogen infection and plant defense responses. The knowledge will allow for the connection between molecular networks to be made and further characterized in order to gain a more dynamic perspective on pathogen response mechanisms. Over the past years, proteomics research has markedly contributed to the knowledge of the changes in proteins and associated biological functions in response to pathogen infection however further progress is required. For example, PTMs go beyond phosphorylation, S-nitrosylation, and disulfide bond formation, and those areas have yet to be described in the context of plant pathogen interactions. The area of proteomics and the technology used to characterize global protein changes including PTMs are rapidly advancing, and with those advancements, researchers are able to examine regulatory mechanisms beyond protein level changes. With the fast advancement in high resolution separation, high sensitivity, and versatile mass spectrometers, whole proteome coverage of plants and bacteria is in sight. Large-scale multiplexing (e.g., Neucode [78]), characterization of low abundance proteins, protein-protein interaction, and PTM crosstalk are new directions of proteomics that will allow construction of molecular networks underlying plant pathogen interactions and achievement of the ultimate goal of determining critical nodes and edges in the networks in order to achieve rational engineering/breeding of crops for enhanced yield, quality and defense.

References

1. Strange RN, Scott PR (2005) Plant disease: a threat to global food security. Annu Rev Phytopathol 43:83–116
2. Agrawal GK, Sarkar A, Righetti PG, Pedreschi R, Carpentier S, Wang T et al (2013) A decade of plant proteomics and mass spectrometry: translation of technical advancements to food security and safety issues. Mass Spectrom Rev 32:335–365
3. Mehta A, Brasileiro AC, Souza DS, Romano E, Campos MA, Grossi-De-Sá MF et al (2008) Plant-pathogen interactions: what is proteomics telling us? FEBS J 275:3731–3746
4. Gotō M (1992) Fundamentals of bacterial plant pathology. Academic Press, San Diego
5. Cheng Z, Mcconkey BJ, Glick BR (2010) Proteomic studies of plant–bacterial interactions. Soil Biol Biochem 42:1673–1684
6. Rosier A, Bishnoi U, Lakshmanan V, Sherrier DJ, Bais HP (2016) A perspective on inter-kingdom signaling in plant-beneficial microbe interactions. Plant Mol Biol 90:537–548
7. Burkholder WH (1948) Bacteria as plant pathogens. Annu Rev Microbiol 2 1:389–412
8. Mansfield J, Genin S, Magori S, Citovsky V, Sriariyanum M, Ronald P et al (2012) Top 10 plant pathogenic bacteria in molecular plant pathology. Mol Plant Pathol 13:614–629
9. Rampitsch C, Bykova NV (2012) Proteomics and plant disease: advances in combating a major threat to the global food supply. Proteomics 12:673–690
10. Boyd LA, Ridout C, O'sullivan DM, Leach JE, Leung H (2013) Plant-pathogen interactions: disease resistance in modern agriculture. Trends Genet 29:233–240
11. Jones JDG, Dangl JL (2006) The plant immune system. Nature 444:323–329
12. Dodds PN, Rathjen JP (2010) Plant immunity: towards an integrated view of plant-pathogen interactions. Nat Rev Genet 11:539–548
13. Nürnberger T, Brunner F, Kemmerling B, Piater L (2004) Innate immunity in plants and animals: striking similarities and obvious differences. Immunol Rev 198:249–266
14. Gómez-Gómez L, Boller T (2002) Flagellin perception: a paradigm for innate immunity. Trends Plant Sci 7:251–256
15. Felix G, Duran JD, Volko S, Boller T (1999) Plants have a sensitive perception system for the most conserved domain of bacterial flagellin. Plant J 18:265–276
16. Sun W, Dunning FM, Pfund C, Weingarten R, Bent AF (2006) Within-species flagellin polymorphism in *Xanthomonas campestris* pv. campestris and its impact on elicitation of Arabidopsis FLAGELLIN SENSING2-dependent defenses. Plant Cell 18:764–779
17. Pfund C, Tans-Kersten J, Dunning F, Alonso J, Ecker J, Allen C et al (2004) Flagellin is not a major defense elicitor in *Ralstonia solanacearum* cells or extracts applied to *Arabidopsis thaliana*. Mol Plant Microbe Interact 17:696–706
18. Dow M, Newman MA, Von Roepenack E (2000) The induction and modulation of plant defense responses by bacterial lipopolysaccharides. Annu Rev Phytopathol 38:241–261
19. Zipfel C, Kunze G, Chinchilla D, Caniard A, Jones JDG, Boller T et al (2006) Perception of the bacterial PAMP EF-Tu by the receptor EFR restricts Agrobacterium-mediated transformation. Cell 125:749–760
20. Nguyen HP, Chakravarthy S, Velásquez AC, Mclane HL, Zeng L, Nakayashiki H et al (2010) Methods to study PAMP-triggered immunity using tomato and *Nicotiana benthamiana*. Mol Plant Microbe Interact 23:991–999
21. Newman MA, Von Roepenack-Lahaye E, Parr A, Daniels MJ, Dow JM (2002) Prior exposure to lipopolysaccharide potentiates expression of plant defenses in response to bacteria. Plant J 29:487–495
22. Felix G, Boller T (2003) Molecular sensing of bacteria in plants. The highly conserved RNA-binding motif RNP-1 of bacterial cold shock proteins is recognized as an elicitor signal in tobacco. J Biol Chem 278:6201–6208
23. Göhre V, Spallek T, Häweker H, Mersmann S, Mentzel T, Boller T et al (2008) Plant pattern-recognition receptor FLS2 is directed for degradation by the bacterial ubiquitin ligase AvrPtoB. Curr Biol 18:1824–1832

24. Hammond-Kosack KE, Jones JD (1996) Resistance gene-dependent plant defense responses. Plant Cell 8:1773–1791
25. Lecourieux D, Ranjeva R, Pugin A (2006) Calcium in plant defence-signalling pathways. New Phytol 171:249–269
26. Jeworutzki E, Roelfsema MR, Anschütz U, Krol E, Elzenga JT, Felix G et al (2010) Early signaling through the Arabidopsis pattern recognition receptors FLS2 and EFR involves Ca-associated opening of plasma membrane anion channels. Plant J 62:367–378
27. Zipfel C, Robatzek S, Navarro L, Oakeley EJ, Jones JDG, Felix G et al (2004) Bacterial disease resistance in Arabidopsis through flagellin perception. Nature 428:764–767
28. Gómez-Gómez L, Felix G, Boller T (1999) A single locus determines sensitivity to bacterial flagellin in *Arabidopsis thaliana*. Plant J 18:277–284
29. Chinchilla D, Bauer Z, Regenass M, Boller T, Felix G (2006) The Arabidopsis receptor kinase FLS2 binds flg22 and determines the specificity of flagellin perception. Plant Cell 18:465–476
30. Oh CS, Martin GB (2011) Effector-triggered immunity mediated by the Pto kinase. Trends Plant Sci 16:132–140
31. Mysore KS, Crasta OR, Tuori RP, Folkerts O, Swirsky PB, Martin GB (2002) Comprehensive transcript profiling of Pto- and Prf-mediated host defense responses to infection by *Pseudomonas syringae* pv. tomato. Plant J 32:299–315
32. Fu ZQ, Dong X (2013) Systemic acquired resistance: turning local infection into global defense. Annu Rev Plant Biol 64:839–863
33. Delledonne M, Xia Y, Dixon RA, Lamb C (1998) Nitric oxide functions as a signal in plant disease resistance. Nature 394:585–588
34. Martin GB, Brommonschenkel SH, Chunwongse J, Frary A, Ganal MW, Spivey R et al (1993) Map-based cloning of a protein kinase gene conferring disease resistance in tomato. Science 262:1432–1436
35. Gutierrez JR, Balmuth AL, Ntoukakis V, Mucyn TS, Gimenez-Ibanez S, Jones AM et al (2010) Prf immune complexes of tomato are oligomeric and contain multiple Pto-like kinases that diversify effector recognition. Plant J 61:507–518
36. Ning K, Fermin D, Nesvizhskii AI (2012) Comparative analysis of different label-free mass spectrometry based protein abundance estimates and their correlation with RNA-Seq gene expression data. J Proteome Res 11:2261–2271
37. Jorrín-Novo JV, Maldonado AM, Echevarría-Zomeño S, Valledor L, Castillejo MA, Curto M et al (2009) Plant proteomics update (2007–2008): Second-generation proteomic techniques, an appropriate experimental design, and data analysis to fulfill MIAPE standards, increase plant proteome coverage and expand biological knowledge. J Proteomics 72:285–314
38. Nishimura MT, Dangl JL (2010) Arabidopsis and the plant immune system. Plant J 61:1053–1066
39. Peck SC, Nühse TS, Hess D, Iglesias A, Meins F, Boller T (2001) Directed proteomics identifies a plant-specific protein rapidly phosphorylated in response to bacterial and fungal elicitors. Plant Cell 13:1467–1475
40. Nühse TS, Stensballe A, Jensen ON, Peck SC (2003) Large-scale analysis of in vivo phosphorylated membrane proteins by immobilized metal ion affinity chromatography and mass spectrometry. Mol Cell Proteomics 2:1234–1243
41. Jones AM, Thomas V, Truman B, Lilley K, Mansfield J, Grant M (2004) Specific changes in the Arabidopsis proteome in response to bacterial challenge: differentiating basal and R-gene mediated resistance. Phytochemistry 65:1805–1816
42. Jones AM, Thomas V, Bennett MH, Mansfield J, Grant M (2006) Modifications to the Arabidopsis defense proteome occur prior to significant transcriptional change in response to inoculation with *Pseudomonas syringae*. Plant Physiol 142:1603–1620
43. Jones AM, Bennett MH, Mansfield JW, Grant M (2006) Analysis of the defence phosphoproteome of *Arabidopsis thaliana* using differential mass tagging. Proteomics 6:4155–4165

44. Fröhlich A, Gaupels F, Sarioglu H, Holzmeister C, Spannagl M, Durner J et al (2012) Looking deep inside: detection of low-abundance proteins in leaf extracts of Arabidopsis and phloem exudates of pumpkin. Plant Physiol 159:902–914
45. Elmore JM, Liu J, Smith B, Phinney B, Coaker G (2012) Quantitative proteomics reveals dynamic changes in the plasma membrane during Arabidopsis immune signaling. Mol Cell Proteomics 11(M111):014555
46. Sghaier-Hammami B, Redondo-Lopez I, Maldonado-Alconada A, Echevarria-Zomeno S, Jorrin-Novo J (2012) A proteomic approach analysing the *Arabidopsis thaliana* response to virulent and avirulent *Pseudomonas syringae* strains. Acta Physiologiae Plantarum 34:905–922
47. Coaker GL, Willard B, Kinter M, Stockinger EJ, Francis DM (2004) Proteomic analysis of resistance mediated by Rcm 2.0 and Rcm 5.1, two loci controlling resistance to bacterial canker of tomato. Mol Plant Microbe Interact 17:1019–1028
48. Miao L, Shou S, Zhu Z, Jiang F, Zai W, Yang Y (2008) Isolation of a novel tomato caffeoyl CoA 3-O-methyltransferase gene following infection with the bacterium Ralstonia solanacearum. J Phytopathol 156:588–596
49. Afroz A, Khan MR, Ahsan N, Komatsu S (2009) Comparative proteomic analysis of bacterial wilt susceptible and resistant tomato cultivars. Peptides 30:1600–1607
50. Parker J, Koh J, Yoo MJ, Zhu N, Feole M, Yi S et al (2013) Quantitative proteomics of tomato defense against *Pseudomonas syringae* infection. Proteomics 13:1934–1946
51. Mahmood T, Jan A, Kakishima M, Komatsu S (2006) Proteomic analysis of bacterial-blight defense-responsive proteins in rice leaf blades. Proteomics 6:6053–6065
52. Chen F, Yuan Y, Li Q, He Z (2007) Proteomic analysis of rice plasma membrane reveals proteins involved in early defense response to bacterial blight. Proteomics 7:1529–1539
53. Li D, Wang L, Teng S, Zhang G, Guo L, Mao Q et al (2012) Proteomics analysis of rice proteins up-regulated in response to bacterial leaf streak disease. Journal of Plant Biology 55:316–324
54. Kumar A, Bimolata W, Kannan M, Kirti PB, Qureshi IA, Ghazi IA (2015) Comparative proteomics reveals differential induction of both biotic and abiotic stress response associated proteins in rice during *Xanthomonas oryzae* pv. oryzae infection. Funct Integr Genomics 15:425–437
55. Delaunois B, Jeandet P, Clément C, Baillieul F, Dorey S, Cordelier S (2014) Uncovering plant-pathogen crosstalk through apoplastic proteomic studies. Front Plant Sci 5:249
56. De-La-Peña C, Lei Z, Watson BS, Sumner LW, Vivanco JM (2008) Root-microbe communication through protein secretion. J Biol Chem 283:25247–25255
57. Cheng F, Blackburn K, Lin Y, Goshe M, Williamson J (2009) Absolute Protein Quantification by LC/MSE for Global Analysis of Salicylic Acid-induced Plant Protein Secretion Responses. J Proteome Res 8:82–93
58. Chivasa S, Simon WJ, Yu XL, Yalpani N, Slabas AR (2005) Pathogen elicitor-induced changes in the maize extracellular matrix proteome. Proteomics 5:4894–4904
59. Jung YH, Jeong SH, Kim SH, Singh R, Lee JE, Cho YS et al (2008) Systematic secretome analyses of rice leaf and seed callus suspension-cultured cells: workflow development and establishment of high-density two-dimensional gel reference maps. J Proteome Res 7:5187–5210
60. Kaffarnik FA, Jones AM, Rathjen JP, Peck SC (2009) Effector proteins of the bacterial pathogen *Pseudomonas syringae* alter the extracellular proteome of the host plant, *Arabidopsis thaliana*. Mol Cell Proteomics 8:145–156
61. Wang Y, Kim S, Wu J, Huh H, Lee S, Rakwal R et al (2013) Secretome analysis of the rice bacterium *Xanthomonas oryzae* (Xoo) using in vitro and in planta systems. Proteomics 13:1901–1912
62. Howden AJ, Huitema E (2012) Effector-triggered post-translational modifications and their role in suppression of plant immunity. Front Plant Sci 3:160
63. Pedley KF, Martin GB (2003) Molecular basis of Pto-mediated resistance to bacterial speck disease in tomato. Annu Rev Phytopathol 41:215–243

64. Anderson JC, Pascuzzi PE, Xiao F, Sessa G, Martin GB (2006) Host-mediated phosphorylation of type III effector AvrPto promotes Pseudomonas virulence and avirulence in tomato. Plant Cell 18:502–514
65. Zhang J, Shao F, Li Y, Cui H, Chen L, Li H et al (2007) A *Pseudomonas syringae* effector inactivates MAPKs to suppress PAMP-induced immunity in plants. Cell Host Microbe 1:175–185
66. Benschop JJ, Mohammed S, O'flaherty M, Heck AJ, Slijper M, Menke FL (2007) Quantitative phosphoproteomics of early elicitor signaling in Arabidopsis. Mol Cell Proteomics 6:1198–1214
67. Nühse TS, Bottrill AR, Jones AM, Peck SC (2007) Quantitative phosphoproteomic analysis of plasma membrane proteins reveals regulatory mechanisms of plant innate immune responses. Plant J 51:931–940
68. Serna-Sanz A, Parniske M, Peck SC (2011) Phosphoproteome analysis of *Lotus japonicus* roots reveals shared and distinct components of symbiosis and defense. Mol Plant Microbe Interact 24:932–937
69. Wang H, Wang S, Lu Y, Alvarez S, Hicks LM, Ge X et al (2012) Proteomic analysis of early-responsive redox-sensitive proteins in Arabidopsis. J Proteome Res 11:412–424
70. Parker J, Zhu N, Zhu M, Chen S (2012) Profiling thiol redox proteome using isotope tagging mass spectrometry. J Vis Exp 61:e3766
71. Balmant KM, Parker J, Yoo MJ, Zhu N, Dufresne C, Chen S (2015) Redox proteomics of tomato in response to *Pseudomonas syringae* infection. Hortic Res 2:15043
72. Lounifi I, Arc E, Molassiotis A, Job D, Rajjou L, Tanou G (2013) Interplay between protein carbonylation and nitrosylation in plants. Proteomics 13:568–578
73. Spoel SH, Loake GJ (2011) Redox-based protein modifications: the missing link in plant immune signalling. Curr Opin Plant Biol 14:358–364
74. Corpas FJ, Del Río LA, Barroso JB (2008) Post-translational modifications mediated by reactive nitrogen species: Nitrosative stress responses or components of signal transduction pathways? Plant Signal Behav 3:301–303
75. Lindermayr C, Saalbach G, Durner J (2005) Proteomic identification of S-nitrosylated proteins in Arabidopsis. Plant Physiol 137:921–930
76. Romero-Puertas MC, Campostrini N, Mattè A, Righetti PG, Perazzolli M, Zolla L et al (2008) Proteomic analysis of S-nitrosylated proteins in *Arabidopsis thaliana* undergoing hypersensitive response. Proteomics 8:1459–1469
77. Cecconi D, Orzetti S, Vandelle E, Rinalducci S, Zolla L, Delledonne M (2009) Protein nitration during defense response in *Arabidopsis thaliana*. Electrophoresis 30:2460–2468
78. Rose CM, Merrill AE, Bailey DJ, Hebert AS, Westphall MS, Coon JJ (2013) Neutron encoded labeling for peptide identification. Anal Chem 85:5129–5137

Chapter 10
Plant Fungus Interaction Proteomics: An Update

Kanika Narula, Pooja R. Aggarwal, Niranjan Chakraborty and Subhra Chakraborty

Abstract Diversity of angiosperm is renowned and mechanism of perception and interaction with different environmental conditions is also variable. Patho-stress response in different plant families varies during the invasion of same or different fungal species. A major puzzle is how interaction and communication could increase fitness in plant at molecular level. Global proteome analysis of plant-pathosystem provides an invaluable resource for the identification of host as well as pathogen proteins involved in disease progression or immunity development. At protein level plant-fungal interaction upsurge the need to understand protein homeostasis and molecular adaptation of building blocks of cell to manifest natural selection for the host. Here, we examine the multilayered facets of interaction between organisms of two diverse kingdoms, namely plant and fungi at protein level based on more than 3000 identified host proteins till date.

Keywords Proteomics · Plant-fungus interaction

10.1 Introduction

Life on the phyllosphere and rhizosphere features specific adaptations and display multipartite relationship between plant and sphere microbes, particularly fungi. Phyllosphere is the aerial sphere of plant whereas rhizosphere encompasses area around the roots. Fungi can be expected to occupy diverse niches influencing plant physiology and cellular processes. Furthermore, there is heterogeneity in their life style following saprophytic, parasitic or mutualistic mode of interactions. Collectively, fungi and fungal-like organisms (FLOs) cause majority of plant diseases [1, 2]. Fungi are capable of attacking a broad range of plant families, including many agricultural crops. Phytopathogenic fungi epitomize high versatility in disease

K. Narula · P.R. Aggarwal · N. Chakraborty · S. Chakraborty (✉)
National Institute of Plant Genome Research, Aruna Asaf Ali Marg,
New Delhi 110067, India
e-mail: subhrac@hotmail.com

G.H. Salekdeh (ed.), *Agricultural Proteomics Volume 2*,
DOI 10.1007/978-3-319-43278-6_10

manifestation in different hosts, plant organs (e.g., leaves, flowers, roots, fruits and seeds) and infection mechanisms which trigger diverse plant defense responses. The recognition of fungal virulence factor *en route* by the host plant profoundly influences the interaction between two organisms [3]. Failure of defense responses is a consequence of hormonal imbalance, including salicylic acid, jasmonic acid, auxin, abscisic acid, and gibberellin [4]. Thus, a paradigm shift in the signaling cascade due to fungal invasion governs the resistance or susceptibility in plants.

The plant sciences and fungal biology are highly integrative and interdisciplinary. Phenomenon of cellular reprogramming in both plant and invading fungus is a result of biological reformation and restoration, immunity, and cell death. These processes are orchestrated in a sophisticated symphony of supramolecular and macromolecular interactions that we have just begun to understand with increasing clarity. A plant-fungal interaction follows the trajectory defined by the structural features and creates a pathway which travels unless perturbed by strong natural selection. Overall, our comprehensive understanding of plant defense response is at a very early stage. Over the ensuing years, completion of few plant and fungal genome projects is considered as dawn of genomics era and also allows the parallel evolution of other "omics", including proteomics to understand the biological circuitry. Proteomics aims at the study of complete set of proteins encoded by the genome at a given point of time and thereby complementing genomics studies [5]. Decoding of gene function using proteomics and proteogenomics involves the extensive use of data mining by informatics analysis to determine the functionality of complicated biological pathways. The availability of proteomic datasets in particular reinforces the potential to predict various biological processes, including immune response which is controlled by intricate regulatory networks.

This overview comprehensively brings together the insights of plant-fungal interaction at proteome level. Delineating the association of inter and intra molecular signaling networks with phenotypes primarily advocates the use of proteomics to identify and assess the patterns of molecular covariance, thereby facilitating conceptualization of phenotypic variation as a mirror icon of the core cellular mechanisms. Thus, proteomic studies may apprehend underlying principles of sphere interaction of fungus with plant and help to develop a deeper understanding of the microbiota and multiple signaling events that dictate plant adaptation under patho-stress.

10.2 Background of Plant-Fungus Interaction

Plants within their ecosystem encounter several environmental cues including patho-stress. The qualitative and quantitative estimation of the molecular patterns responsible for disease tolerance or susceptibility can be estimated at gene, transcript, and/or protein level. Till now hundreds of quantitative trait loci (QTLs) associated with quantitative resistance have been reported for fungal stress in plants, such as blast in rice (*Oryza sativa*) [6], Fusarium wilt in chickpea [7],

Fusarium wilt in watermelon [8], Fusarium head blight in wheat (*Triticum aestivum*) [9] and barley (*Hordeum vulgare*) [10], powdery mildew in wheat [11]. Differential expression analysis of chickpea genes associated with Ascochyta blight has been shown in different chickpea cultivars [12]. Gene expression pattern analysis during chickpea-*Fusarium* interactions has identified several novel regulatory genes specifically showing differential expression, which suggests their putative role in plant immunity during fungal stress [13]. Very recently, genome-wide transcriptome analysis in cotton has elucidated candidate genes in response to fungus *Aspergillus flavous* [14]. In addition, protein qualitative assessment of interaction was reported in The Pathogen-Host Interactions database (PHI-base, http://www.phi-base.org) [15]. This database catalogues experimentally verified pathogenicity, virulence, and effector genes from fungi, oomycetes, and bacterial pathogens, which infect animal, plant, fungus, and insect hosts. Nonetheless, fungal stress in plants induces various post-transcriptional and post-translational changes. Thus, it is imperative to analyze the cellular remodeling at proteome level to fully comprehend the basis of pathogen induced responses in host plant.

Proteomic analyses of plant-fungus interactions has emerged as decisive approach because the conduits used by both plants and fungi rely on stress induced protein synthesis, modification and activity. These encoded building blocks (proteins) provide the global information of biological interactions, know-how of protein partner existence in two counterparts. Interaction proteomics is a powerful approach to understand phenome in great detail, but still has a long road ahead.

10.3 Plant-Fungal Interaction: A Force in Protein Evolution

Co-evolutionary forces reciprocate protein evolution for intimate alliance and inter-kingdom communication especially in plant-fungal interaction resulting in phenotypic plasticity and composite biological processes dictating two counterparts (Fig. 10.1). Fungi and plants comprise a diverse lineage of eukaryotic organisms and estimates suggest that around 1.5 million fungal species and >2.5 million plant species exist in nature. Protein evolution studies highlight conserved sequence blocks, variable length regions and the insertion/deletions (indels). It has provided insight into protein configuration and function [16]. Protein structure and function largely govern plant behavior and evolutionary relationships with pathogenic and symbiotic fungi. Proteome analyses can contribute significantly toward the molecular mechanisms underlying the development of these associations. Earlier structural, functional and phylogenetic importance of eukaryotic protein was assessed using comparative proteome dataset from a taxonomically broad set of metazoa, fungi and viridiplantae. Results yielded 299 substantial (>250aa) universal, single-copy (in paralog only) proteins, from which 901 simple

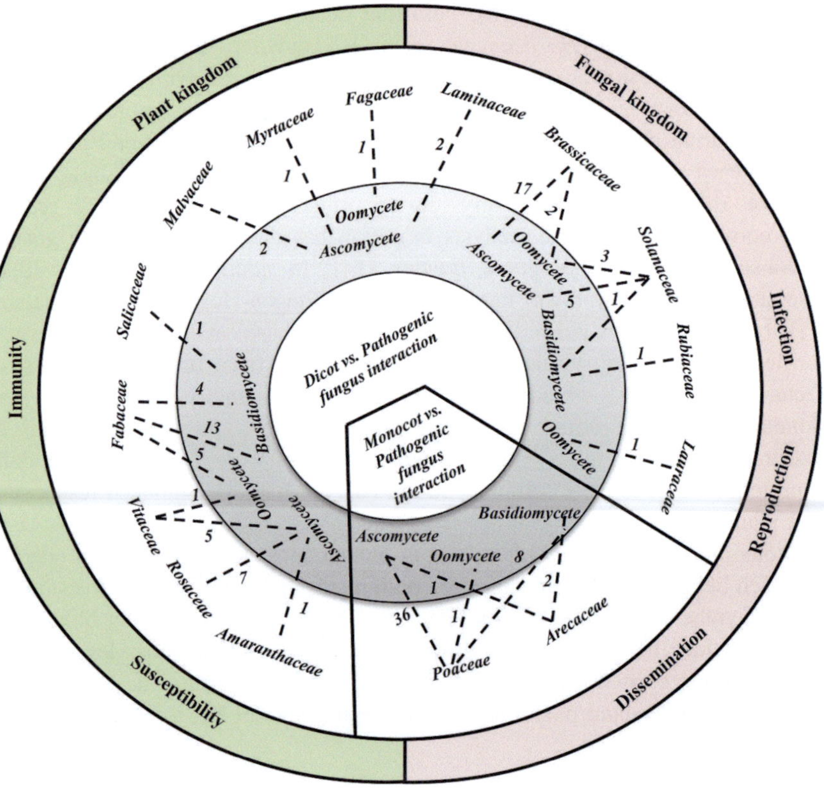

Fig. 10.1 Inter-kingdom crosstalk. Comprehensive study of proteomics analyses in different plant families during fungal pathogen interaction

(present/absent) and 3806 complex (multistate) indels were extracted [17]. As expected the data pinpoints high level of hidden homoplasy (multiple independent origins) for the eukaryotic proteins especially in plants and fungi.

10.4 Assessment of Inter-Kingdom Crosstalk

Cross-Kingdom communication between plants and fungi could be decoded by the elucidation of regulatory networks using proteomic approaches. Essence of protein language in diverse plant-fungi interaction provides a foundation for understanding biologically interesting paradox. In general, plant-derived proteins facilitate resistance to the invasion, while fungus species circumvent this attempt by synthesizing detrimental proteins and the outcome of this molecular battle appears in the form of host tolerance or susceptibility. Interaction outcomes of various pathosystems encompassing diverse classes, family, genus and species with varied phytopathogenic fungi are largely dependent on the extensive array of protein complexes and precise nature of signaling molecules. Protein language perceived by both the partners is thought to be discrete but proteome data depicts that exchange of chemical information varies for different plant species interacting with diverse fungal pathogens. Therefore, the perception between plant and fungi and the functional protein crosstalk may be elaborated based on the taxonomic classification of both the protagonists to better understand the complex nature of protein language.

Decoding fungal stress related proteins across kingdom dramatically and episodically are of utmost importance to understand diversity among the protein complement in plants. The possibility of intra-family comparison of proteins with the use of advanced proteomics techniques are of great value.

10.5 Taxonomical Evaluation of Eudicot Proteomes in Response to Fungal Attack: Family Wide Proteome Study

10.5.1 Brassicaceae-Fungal Pathosystem: Selectivity and Exclusivity of Proteins

A total of nineteen crucifer-fungus interaction proteomes have been reported till date of which eighteen reports focuses mainly on ascomycetes fungi, whereas only two publications involve oomycetes (Table 10.1). Brassicaceae is one the most comprehensively sampled angiosperm family for fungal stress proteome study rivaling larger families such as the grasses and legumes. Multispecies and family wide analytical paradigm showed that eighteen proteomic studies of Brassicaceae members (*Arabidopsis thaliana*, *Brassica carinata*, and *Brassica napus*) involved interaction with ascomycetes fungi (*Alternaria brassicicola, Aspergillus ochraceus, Botrytis cineria, Fusarium moniliforme, Fusarium oxysporum, Fusarium sporotrichioides, Verticillium longisporum, Sclerotinia sclerotiorum, Leptosphaeria maculans*). Of these, ten pathogens followed necrotrophic mode [18–26], whereas seven other showed hemibiotrophic mode of invasion [27–31] (Table 10.1). In two of the studies fungal counterparts were oomycetes (*Albugo candida, Hyaloperonospora parasitica*) showing biotrophic mode of invasion [32, 33]. No report is available for basidomycete-Brassicaceae interaction proteome until now. Proteomic studies were majorly focused on total protein profile patterning and led to the identification of few common proteins like glutathione S-transferase, ATP synthase, formate dehydrogenase, alpha-xylosidase, indole-3-acetonitrile nitrilase, ferredoxin-NADPH(+)-oxidoreductase involved in metabolism and redox homeostasis. In addition, three reports involving arabidopsis-*Fusarium moniliforme* elicitor [27], arabidopsis-fungal elicitor [34] and arabidopsis-*Fusarium oxysporum* elicitor [31] were focused

Table 10.1 A comprehensive list of proteomic analyses in plant-fungal interaction

Clade	Plant family	Organism	Pathogen	Fungal class	Reference[a]
Dicot	Brassicaceae	Arabidopsis	*Alternaria brassicicola*	Ascomycota	[23]
			Aspergillus ochraceus	Ascomycota	[26]
			Botrytis cineria	Ascomycota	[24]
			Fungal Elicitor	–	[34]
			Fungal Elicitor	–	[36]
			Fusarium moniliforme (elicitor)	Ascomycota	[27]
			Fusarium oxysporum (elicitor)	Ascomycota	[31]
			Fusarium sporotrichioides	Ascomycota	[30]
			Verticillium longisporum	Ascomycota	[19]
		Brassica	*Sclerotinia sclerotiorum*	Ascomycota	[21]
		Mustard	*Leptosphaeria maculans,*	Ascomycota	[28]
		Mustard	*Leptosphaeria maculans,*	Ascomycota	[29]
		Brassica	*Albugo candida*	Oomycetes	[32]
		Rapeseed	*Alternaria brassicae*	Ascomycota	[18]
			Sclerotinia sclerotiorum	Ascomycota	[20]
			Sclerotinia sclerotiorum	Ascomycota	[22]
			Sclerotinia sclerotiorum	Ascomycota	[25]
			Verticillium longisporum	Ascomycota	[35]
		Cabbage	*Hyaloperonospora parasitica*	Oomycetes	[33]
	Fabaceae	Groundnut	*Phaeoisariopsis personata*	Ascomycota	[50]
			Aspergillus flavus	Ascomycota	[45]
			Aspergillus flavus	Ascomycota	[43]
		Bean	*Uromyces appendiculatus*	Basidiomycota	[55]
			Trichoderma	Ascomycota	[40]
		Chickpea	*Fusarium oxysporum*	Ascomycota	[46]
			Fusarium oxysporum	Ascomycota	[44]
		Cowpea	*Colletotrichum gloeosporioides*	Ascomycota	[47]
		Medicago	*Aphanomyces euteiches*	Oomycetes	[56]
			Glomus intraradices	Glomeromycota	[38]
			Glomus intraradices	Glomeromycota	[37]
			Aphanomyces euteiches	Oomycetes	[53]
			Yeast (elicitor)	Ascomycota	[42]
		Pea	*Nectria haematococca*	Ascomycota	[25]
			Peronospora viciae	Oomycetes	[54]
			Erysiphe pisi	Ascomycota	[39]
			Sclerotinia sclerotiorum	Ascomycota	[45]
			Mycosphaerella pinodes	Ascomycota	[41]
		Common bean	*Trichoderma harzianum (ALL-42)*	Ascomycota	[48]
			Phakopsora pachyrhizi	Basidiomycota	[57]
			Phytophthora sojae elicitor	Oomycetes	[51]

(continued)

Table 10.1 (continued)

Clade	Plant family	Organism	Pathogen	Fungal class	Reference[a]
			Phytophthora sojae strain P6497	Oomycetes	[52]
	Solanaceae	Nicotiana	*Moniliophthora perniciosa*	Basidiomycota	[67]
		Potato	*Aspergillus terreus*	Ascomycota	[66]
			Phytophthora infestans	Oomycetes	[63]
			Phytophthora infestans	Oomycetes	[65]
			Phytophthora infestans	Oomycetes	[64]
		Tomato	*Botrytis cineria*	Ascomycota	[61]
			Fusarium oxysporum	Ascomycota	[60]
			Fusarium oxysporum	Ascomycota	[59]
			Fusarium oxysporum	Ascomycota	[58]
			Rhizopus nigricans	Zygomycota	[62]
	Vitaceae	Vitis	*Botrytis cineria*	Ascomycota	[73]
			Erysiphe necator	Ascomycota	[69]
			Plasmopara viticola	Oomycetes	[71]
			Diplodia seriata	Ascomycota	[68]
			Plasmopara viticola	Ascomycota	[70]
			Neofusicoccum parvum and Diplodia seriata	Ascomycota	[72]
	Rosaceae	Apple	*Alternaria alternata*	Ascomycota	[77]
			Botrytis cineria and yeast	Ascomycota	[76]
		Cherry	*Penicillium expansum*	Ascomycota	[75]
		Peach	*Penicillium expansum*	Ascomycota	[74]
		Strawberry	*Colletotrichum fragariae*	Ascomycota	[79]
			Fusarium oxysporum	Ascomycota	[78]
			Fusarium oxysporum	Ascomycota	[80]
	Malvaceae	Cotton	*Thielaviopsis basicola*	Ascomycota	[81]
			Verticillium dahliae	Ascomycota	[82]
	Rubiaceae	Coffee	*Hemileia vastatrix*	Basidiomycota	[83]
	Lamiaceae	Mint	*Alternaria alternata*	Ascomycota	[84]
			Alternaria alternata	Ascomycota	[85]
	Amaranthceae	Sugar beet	*Fusarium oxysporum*	Ascomycota	[86]
	Apiaceae	Carrot	*Mycocentrospora acerina*	Ascomycota	[87]
	Cucurbitaceae	Cucumber	*Trichoderma asperellum strain T34*	Ascomycota	[88]
	Lauraceae	Avocado	*Phytophthora cinnamomi*	Oomycetes	[89]
	Musaceae	Banana	*Fusarium oxysporum*	Ascomycota	[90]
	Anacardiaceae	Cashew	*Lasiodiplodia theobromae*	Ascomycota	[91]
	Myrtaceae	Eucalyptus	*Calonectria pseudoreteaudii*	Ascomycota	[93]
	Salicaceae	Poplar	*Melampsora medusae*	Basidiomycota	[92]
	Cannabaceae	Lupulus	*Verticillium dahliae*	Ascomycota	[95]
	Caryophyllaceae	Carnation	*Fusarium oxysporum*	Ascomycota	[96]
	Fagaceae	Fagus	*Phytophthora*	Oomycetes	[94]

(continued)

Table 10.1 (continued)

Clade	Plant family	Organism	Pathogen	Fungal class	Reference[a]
Monocot	Poaceae	Barley	*Fusarium graminearum*	Ascomycota	[124]
			Fusarium graminearum	Ascomycota	[123]
			Fusarium graminearum and F. culmorum	Ascomycota	[103]
			Puccinia hordei	Basidiomycota	[126]
			Fusarium graminearum	Ascomycota	[125]
		Barley and wheat	*Fusarium graminearum*	Ascomycota	[106]
			Fusarium verticillioides elicitor	Ascomycota	[34]
		Pearl millet	*Sclerospora graminicola*	Oomycetes	[127]
		Rice	*Cochliobolus miyabeanus*	Ascomycota	[116]
			Magnaporthe grisae	Ascomycota	[121]
			Magnaporthe grisae	Ascomycota	[120]
			Magnaporthe grisea	Ascomycota	[118]
			Magnaporthe oryzae	Ascomycota	[117]
			Puccinia striiformis	Basidiomycota	[122]
			Puccinia triticina	Basidiomycota	[120]
			Rhizoctonia solani	Basidiomycota	[119]
		Triticum	*Fusarium graminearum and Fusarium culmorum*	Ascomycota	[103]
		Wheat	*Blumeria graminis*	Ascomycota	[108]
			Fusarium graminearum	Ascomycota	[105]
			Fusarium graminearum	Ascomycota	[107]
			Fusarium graminearum	Ascomycota	[101]
			Fusarium graminearum	Ascomycota	[97]
			Fusarium graminearum	Ascomycota	[99]
			Puccinia triticina	Basidiomycota	[114]
			Puccinia triticina	Basidiomycota	[115]
			Pyrenophora tritici	Ascomycota	[112]
			Pyrenophora tritici	Ascomycota	[102]
			Pyrenophora tritici	Ascomycota	[100]
			Septoria tritici	Ascomycota	[106]
			Zymoseptoria tritici	Ascomycota	[113]
			Zymoseptoria tritici	Ascomycota	[110]
			Beauveria bassiana	Ascomycota	[122]
			Fusarium graminearum	Ascomycota	[111]
			Stagonospora nodorum effector protein	Ascomycota	[109]
			Fusarium graminearum	Ascomycota	[98]
	Arecaceae	Datepalm	*Beauveria bassiana*	Ascomycota	[128]
		Oil Palm	*Ganoderma boninense*	Basidiomycota	[129]
		Oil Palm	*Ganoderma boninense*	Basidiomycota	[130]

[a]Based on PubMed search dated November 5, 2015

on extracellular matrix protein profiling and identified matrix resident protein involved in cell wall mediated sensing of fungal pathogen. While, other two studies depicted apoplast proteome patterning during arabidopsis-*Verticillium longisporum* and *Brassica napus-Verticillium longisporum* interaction that provided clues for the cross-talks between various defense pathways and regulatory networks [19, 35]. Furthermore, one study on Brassicacease pathosystem (*Arabidopsis*-chitosan) encompasses post-translational modification analysis of phosphoproteome and led to the identification of 1186 phosphoproteins involved in pathostress-related signal transduction processes [36]. To our surprise, all the studies conducted on Brassicaceae involved gel-based proteomic approach and points toward the common mechanism of action affecting flavonoid pathway and ROS signaling which is further responsible for displaying morphological variation and diverse defense strategies.

10.5.2 Fabaceae-Fungal Pathosystem Dynamics: Social Class Versus Diversity

We have compared the experimentally determined Fabaceae pathosystem proteomes against fungal invasion, namely *Arachis diogoi-Phaeoisariopsis personata, Arachis hypogaea-Aspergillus flavus,* bean-*Uromyces appendiculatus,* bean-*Trichoderma,* chickpea-*Fusarium oxysporum,* cowpea-*Colletotrichum gloeosporioides,* medicago-*Aphanomyces euteiches,* medicago-*Glomus intraradices, Medicago truncatula-Yeast (elicitor),* pea-*Nectria haematococca,* pea-*Peronospora viciae,* pea-*Uromyces pisi,* pea-*Erysiphe pisi,* pea-*Sclerotinia sclerotiorum,* pea-*Mycosphaerella pinodes, Phaseolus vulgaris-Trichoderma harzianum,* soybean-*Phytophthora sojae* (Table 10.1). The *modus operandi* in investigating the proteomes of available Fabaceae species was the extensive literature and the availability of relevant databases. The proteins identified in these works were classified into functional categories. This classification is only tentative, since the biological role of many of the proteins identified has not been established experimentally. Furthermore, we applied a cross-species comparison on the available datasets. A logical strategy was used to maximise efficiency and the overall comparative results.

The findings from the already available proteomes suggest that up till now 9247 fungal stress responsive proteins from Fabaceae plant family have been identified including arachis, bean, chickpea, cowpea, medicago, pea, phaseolus and soyabean. Briefly, 87 proteins were identified from symbiotic interaction between medicago and *Glomus intraradices* [37, 38], of which 78 belongs to plasma membrane [37]. Remaining 9160 proteins were identified from pathogenic interaction involving ascomycetes (*Phaeoisariopsis personata, Aspergillus flavus, Fusarium oxysporum, Colletotrichum gloeosporioides, Erysiphe pisi, Sclerotinia sclerotiorum, Mycosphaerella pinodes, Trichoderma harzianum*) and basidiomycetes (*Uromyces appendiculatus, Uromyces pisi, Phakopsora pachyrhizi*) [25, 39–50] (Table 10.1).

Extended research showed that superoxide dismutase, cutinase, lipoxygenase and triose phosphate isomerase were common proteins among the pathosystem encompassing hemibiotrophic fungal pathogens [44, 46, 47, 51, 52]. To investigate and address key consequences of biotrophic invasion [50, 53–57], we compared the fungal stress responsive proteomic data available which revealed that cyclophilin, resistance proteins, ABC transport protein, malate dehydrogenase, and enolase were among the common proteins. Our comparative analysis of proteome remodelling during necrotrophic attack identified catalase, ascorbate peroxidase, actin,elongation factor tu, ribosomal proteins and chaperones [41, 49]. Exploring functional contexts of symbiotic association of fungal species with Facaceae family showed that in medicago-*Glomus intraradices* symbiotic interaction identified proteins were specific and structurally integrated. The rearranged protein pool contribute to multiple aspect of essential functions namely, nutrient supply (especially nitrogen, phosphorous and sulfur), resistance against patho-stress and abiotic stress factors, support of photosynthesis by providing of vitamin B12, N-signaling and detoxification of metabolites [37, 38].

Post-translational modification of proteins, such as phosphorylation, has been recognized to be vital in the signal transduction cascades that trigger plant defense responses. We observed that till date two reports depict the contribution of phosphorylation in fungal stress response in Fabaceae [56, 57]. Phosphoproteome were analyzed from medicago-*Aphanomyces euteiches* and soyabean-*Phakopsora pachyrhizi* exhibiting biotrophic mode of invasion in both the pathosystems. Therefore, micro-characterization of patho-stress phosphoproteome revealed that broad-spectrum basal defense response and the response controlled by *R*-genes were mainly influenced by phosphorylation and its mirror icon dephosphorylation.

10.5.3 Solanaceae-Fungal Pathosystem: Diaspora of Overlapping and Unique Proteins

Solanaceae family of flowering plants comprises about 100 genera and 2500 species, many of which are agriculturally important species including tomato (*Solanum lycopersicum*), potato (*Solanum tuberosum*), tobacco (*Nicotiana tabacum*), and pepper (*Capsicum annuum*) (Table 10.1). A descriptive proteomics approach on potato and tomato has dominated the investigation of Solanaceae-fungal interaction and provided insight into the mechanisms of fungal invasion in context of the host and fungus. Fungal stress-related responses are distinct in tomato and have led new perspectives on plant-microbe interactions, mode of invasion in particular. Data sets on tomato pathosystem proteomics have unwrapped the fact that studies were concentrated on fusarium species using gel-based approach and identified 66 proteins from total cell extract and xylem sap of leaves and root [58–60] (Table 10.1). Necrotrophic fungi, *Botrytis cineria* and tomato interaction proteome study pointing towards cell death and necrosis at later stages of invasion [61]. Altogether, reports showed that 187 proteins were identified from tomato and 47 fungal proteins were detected shedding light on the mechanism that regulates pathogenic interactions [61] (Table 10.1). The study on tomato-*Rhizopus nigricans* (Mucorales) proteome exhibited the pathogen specific responses in both intra- and extracellular spaces [62]. The differentially expressed proteins were integrated into several pathways operating patho-stress in tomato fruit. The composition of protein species and putative functions of the identified proteins specify their roles in plant-pathogen interactions. Collectively, results provide evidences that several regulatory pathways encompassing 1-aminocyclopropane-1-carboxylate oxidase homolog, methionine sulfoxide reductase, and family of heat shock protein contribute to the pathogen induced resistance in tomato fruit. Categorization of another agriculturally important crop of same family, potato revealed an imperative corollary shows overall studies uptill now have been conducted on ascomycetes (*Aspergillus terreus*) and oomycetes (*Phytophthora infestans*), which identified in total 150 potato proteins [63–66]. To comprehend further, potato interaction with *Phytophthora infestans* were studied for total cell extract, apoplast and cytoplasm using 1-DE, LC-MS/MS, and iTRAQ labeling and showed that defense response is a network scaffold and regulatory hub of interacting proteins requires rapid turnover [63–65]. The identified protein revealed the predominance of peptidase, proteinase, ROS regulated and protein degradation related stress protein. Nicotiana-*Moniliophthora perniciosa* proteome research spotlight total repertoire of stress related protein in response to basidiomycetes invasion [67]. It was apparent that primary metabolism, cell wall associated metabolic processes and multiple defense pathways are rapidly induced by fungal invasion in Solanaceae family members that causes enhanced resistance.

10.5.4 Vitaceae-Fungal Pathosystem: A Diverse Regime

Studies of the Vitaceae family member proteome in response to fungal stress are mainly centered on grapevine. The milestones among these studies are represented by the fact that diverse fungal genera attack grapevine yard through varied infection strategies. Till date, six reports have been published on grapevine invaded with *Botrytis cineria, Erysiphe necator, Plasmopara viticola, Diplodia seriata, Neofusicoccum parvum*, wherein all were ascomycetes except *Plasmopara viticola*, an oomycetes (Table 10.1). These reports identified 1361 fungal stress-responsive proteins from fruit, flower and veraison stage [68–73]. An usual rearrangement of host cellular machinery was observed in response to *Botrytis cineria* promoting secondary metabolites reallocation, whereas protein mediated oxidative homeostasis was reported in grapevine infected with *Diplodia seriata* [68, 73]. Major changes in biological processes were reported due to *Neofusicoccum parvum* invasion affecting both primary and secondary metabolism, photosynthesis, fruit quality, embryogenesis and development [72].

10.5.5 Rosaceae - Fungal Pathosystem: Shared and Unique Proteins

Assorted comparison of Rosaceae family member proteome in response to fungal invasion revealed that chimeric evolution was the main cause of proteome diversity in patho-stress response. For example, in apple invaded with *Alternaria alternate* and *Botrytis cineria*, transcription and translational machinery proteins were affected (Table 10.1). In case of cherry and peach invaded with *Penicillium expansum* differentially expressed proteins were found to be conserved [74–77]. Likewise, when strawberry was attacked by *Fusarium oxysporum* and *Colletotrichum fragariae,* 133 and 49 proteins were identified (Table 10.1). The study revealed while proteins involved in glycolysis, TCA cycle, pentose phosphate pathway and lipid metabolism showed differential expression in response to *Fusarium* infection; chaperones, photosynthesis associated protein were some of the diverse class of proteins exhibiting differential expression in *Colletotrichum fragariae* invaded tissue [78–80]. Overall, 572 proteins were identified from Rosaceae during fungal invasion (Table 10.1). It was found that strawberry and apple were better explored in comparison to cherry, peach and coffee. Therefore, comparison of their proteome, may not yield the postulated results as defined by genome analysis.

10.5.6 Less Studied Plant Family Proteome in Response to Fungus: Functional and Modular Protein Patterns

Case-by-case proteomics studies of less explored plant families revealed an evolutionary divergence as well as specificity with few conserved proteins. While investigating the Malavaceae family for fungal pathosystem interaction proteomics, it was found that 23 and 68 proteins were identified as patho-stress responsive proteins from cotton plants invaded with hemibiotroph *Thielaviopsis basicola* and necrotroph *Verticillium dahliae*, respectively [81, 82] (Table 10.1). These patho-stress proteins were seemed to be involved in defense responses, such as, induction of PR proteins and formation of isoprenoids. Cell wall biosynthesis, degradation, metabolism represented the affected protein classes in *Coffea arabica* (Rubiaceae) apopalast in response to necrotrophic basidiomycetes, *Hemileia vastatrix* [83]. Studies on Lamiaceae family member, wild-type mint and transgenic mint invaded with necrotrophic pathogen *Alternaria alternate* suggested a cross-talk between various defense pathways, regulatory networks and physiological conditions [84, 85]. Few social class member proteins like RuBisCo, ATP synthase β subunit, oxygen evolving complex I were down-regulated in transgenic mint but up-regulated in wild-type mint. Down-accumulation of putative tyrosine phosphatase, a possible target for H_2O_2 and known to be associated with signal transduction, was observed only in transgenic mint. Whereas, when vegetables like sugar beet (Amaranthceae) and carrot (Apiaceae) invaded with ascomycetes fungi *Fusarium oxysporum* and *Mycocentrospora acerina* respectively were analyzed, it was found that isoprene pathway enzymes were affected displaying common mode of action by two different fungi [86, 87]. Survey of non-climacteric fruit, cucumber (Cucurbitaceae) invaded with *Trichoderma asperellum* strain T34 identified a total of 28 patho-stress responsive proteins of heterogeneous nature pointing towards the involvement of isoprenoid and ethylene biosynthesis during invasion of the pathogen switching metabolic pathway to a non-assimilatory state [88]. Commercially important fruit-fungal interaction proteomics engross the agriculturally important aspects for fine modulation of protein content and yield characteristics. 63, 38, 13 proteins were identified during avocado (Lauraceae)-*Phytophthora cinnamomi* interaction [89], banana (Musaceae)-*Fusarium oxysporum* [90] and cashew (Anacardiaceae)-*Lasiodiplodia theobromae* interaction proteomics [91], respectively. These studies reflect that trafficking and signaling components, involving small GTPases, SNARE receptors proteins and dynamin were differentially affected during patho-stress. Furthermore, proteins implicated in the reactive oxygen species (ROS) and energy metabolism were also increased significantly which include glutathione peroxidase, glutathione S-transferase, catalase, thioredoxins, superoxide dismutase, and ascorbate peroxidase. Tree pathosystem interaction proteomics study in eucalyptus invaded with *Calonectria pseudoreteaudii* and poplar attacked by *Melampsora medusae* identified more than 1000 proteins either from total cell extract or apoplast [92, 93]. While TCA cycle was found to be largely affected in poplar, glycolysis pathway was the major target by the fungus in eucalyptus. Interaction proteomes in *Humulus lupulus* (Cannabaceae)-*Verticillium dahliae*, *Dianthus caryophyllus* (Caryophyllaceae)-*Fusarium oxysporum* and *Fagus* (Fagaceae)-*Phytophthora* showed varied response by the presence of heat shock protein 17.6 in the former two and heat shock protein 60 in the later, illustrating that nature has vastly invented different solutions to a common problem, viz. fungal stress [94–96].

10.6 Taxonomical Evaluation of Monocot Proteomes in Response to Fungal Attack

10.6.1 Poaceae-Fungal Pathosystem: Co-operative and Regulatory Theme

Poaceae family of monocots represents the most extensively investigated member for assessing plant-fungal interaction using proteomic analysis that includes forty four studies till date. Differential protein dataset is being generated with a remarkably fast pace in Poaceae with technical and fiscal challenges presented by the typically large, repetitive genomes. Indeed, interaction of nineteen fungi belonging to ascomycetes, basidiomycetes and oomycetes representing either biotropic, hemibiotropic or necrotropic mode of invasion in wheat, rice, maize and barley have been reported (Table 10.1). With respect to agricultural value of genera and species within the Poaceae, proteome-scale datasets of wheat-fungal pathosystem represent a more extensive base with nineteen interaction studies (*Aspergillus nidulans, Beauveria bassiana, Blumeria graminis, Cochliobolus miyabeanus, Curvularia lunata, Fusarium graminearum, Fusarium culmorum, Fusarium verticillioides, Magnaporthe grisae, Puccinia hordei, Puccinia striiformis, Puccinia triticina, Pyrenophora tritici, Rhizoctonia solani, Sclerospora graminicola, Septoria tritici, Stagonospora nodorum effector protein, Trichoderma harzianum T22, Zymoseptoria tritici*) (Table 10.1). Of these, fifteen studies focused on total cellular patterning, while three investigations emphasized on patho-stress responsive secretome and apoplast proteins [97–113]. Glimpse of proteomics analysis in wheat invaded by ascomycetes yielded a total of 950 host proteins from 14 studies and 30 fungal proteins from *Fusarium graminearum* and *Puccinia triticina* in two separate studies. Proteome studies of wheat infected with basidiomycetes *Puccinia triticina* have identified 40 defense related proteins [114, 115]. Unlike wheat, rice-fungal pathogen interaction is less studied with ten proteomics reports, of which six in response to ascomycetes (*Cochliobolus miyabeanus, Magnaporthe grisae, Magnaporthe oryzae*), three in response to basidiomycetes (*Puccinia triticina, Puccinia striiformis, Rhizoctonia solani*) and one in response to fungal elicitor (Table 10.1). Extracellular and intracellular proteome analysis of rice revealed that proteins related to metabolic processes, antioxidant processes and pathogenesis-related (PR) proteins were differentially expressed in response to *Cochliobolus miyabeanus* [116]. When rice suspension culture was induced by fungal elicitor, 10 proteins including pathogen-related protein class 10 (OsPR-10), isoflavone reductase like protein, b-glucosidase, and putative receptor-like protein kinase were found to be differentially expressed [117]. Furthermore, 169 proteins associated with photosynthesis, antioxidant systems, amino acid metabolism, defense response, molecular chaperones, protein synthesis, proteolysis, carbohydrate metabolism, secondary metabolism, signal transduction, antifungal activity have been identified through total proteome analysis [117–122]. Additionally, 732 fungal responsive proteins were identified during apoplast study of rice pathosystem [117]. Molecular basis of disease response of barley during Fusarium head blight (FHB), seedling blight and leaf rust disease caused by *Fusarium graminearum* and *Puccinia hordei,* respectively was elucidated using gel-based proteomics approach. Pathogenesis-related (PR) proteins, proteins involved in energy metabolism, secondary metabolism, protein synthesis and FHB responsive proteins associated with oxidative burst and oxidative stress response, such as malate dehydrogenase and peroxidase were identified in different barley genotypes (CI4196, Svansota, Harbin, CDC Bold, Scarlett and Stander), indicating that host cells might have prepared the terrain for fungal invasion during FHB disease [123, 124]. During seedling blight disease proteins involved in primary metabolism and detoxification, heat shock proteins and antioxidant enzymes were found to be up-regulated, whereas the plant protease inhibitors were majorly down-regulated illustrating the link between increased energy metabolism and oxidative stress in the germinating barley seeds in response to the necrotrophic fungal pathogen

F. graminearum [125]. Proteomics study of barley response during leaf rust disease, caused by the biotrophic fungal pathogen *Puccinia hordei* has identified 18 proteins associated with photosynthesis and energy metabolism, carbohydrate metabolism, protein degradation and defence [126]. In parallel, six studies involving maize pathosystem (*Curvularia lunata, Fusarium graminearum, Fusarium verticillioides, Trichoderma harzianum, Trichoderma virens*) have led to the identification of 13 and 208 proteins from extracellular matrix (ECM) and total proteome, respectively (Table 10.1). These fungal responsive proteins belong to diverse functional categories such as photosynthesis, carbohydrate metabolism, amino acid metabolism, cell wall metabolism, stress and defense responses, genetic information processing, chitinases, xylanase inhibitors, proteinase inhibitors, peroxidise, protein synthesis, or in protein folding and stabilization, oxidative stress tolerance, heat shock proteins. 63 Disease-responsive proteins have also been identified through gel-based proteomics analysis of powdery mildew disease in pearl millet caused by oomycetes fungal pathogen, *Sclerospora graminicola* [127].

10.6.2 Arecaceae - Fungal Pathosystem: Integrated Protein Profiles

Arecaceae family is the least studied family in monocots with only three very recent proteomic reports available on oilpalm and datepalm pathosystems (*Beauveria bassiana, Ganoderma boninense*) (Table 10.1). Total proteome analysis of datepalm during endophytic colonization of ascomycota fungi *Beauveria bassiana* has led to the identification of 16 proteins [128]. While 61 proteins associated with photosynthesis, metabolism and defence response have been identified from proteome analysis of oilpam infected with basidiomycetes fungal pathogen *Ganoderma boninense* in two separate studies [129, 130].

10.7 Conclusion and Future Perspectives

Throughout the evolutionary time, plants have confronted and mitigated the impacts of various environmental cues, including patho-stress. Plant cells have evolved mechanisms to perceive and integrate different signals during such stresses and to respond by modulating the appropriate gene and protein expression profiles. Proteomics is a powerful tool to study plant-pathogen interactions since patho-stress responsive proteins can be detected by comparing protein patterns of infected and control samples. Family based proteomics analysis provides additional information of the evolutionary conserved and divergent stress-responsive pathways in different crop species with varying phylogenetic depths. The basic reason of this divergence could be protein length divergence, protein copy number variation, different isoforms, epigenetic modification and post-translational modifications. A precise idea of fungus associated pathosystem proteome illustrates that some proteins are unique while some are common among different pathosystems. To explore contextual information of protein variants involved in plant-fungal interaction during hemibiotrophic, biotrophic and necrotrophic mode of invasion, fungal stress associated responses and pathways were observed at translational level. Till date, proteomics studies comprising plant-fungal interaction are majorly based on gel-based approaches. Very few reports are available showing fungal induced responses in plants through non gel-based approach. In last decade, proteomics field has been revolutionized with the incorporation of new technologies which enables the analysis of more complex biological systems. Further, it is useful in studying the molecular changes in both host plant and fungal counterparts offering opportunities to investigate the molecular basis of development and stress-related system evolution. Detailed investigation of these processes will deepen the understanding of induced resistance mechanism in plant kingdom.

Acknowledgments This research work was supported by grants from the Department of Biotechnology (DBT), Ministry of Science and Technology, Govt. of India (Grant Nos. BT/HRD/35/01/05/2013 and BT/PR10796/BRB/10/621/2008) and the National Institute of Plant Genome Research, India. K. N. and P. A. are the recipient of pre-doctoral fellowship from Council of Scientific and Industrial Research (CSIR), Govt. of India.

References

1. Arora DK (2003) Fungal biotechnology in agricultural, food, and environmental applications. CRC Press, Boca Raton
2. Fisher MC, Henk DA, Briggs CJ, Brownstein JS, Madoff LC, Mccraw SL et al (2012) Emerging fungal threats to animal, plant and ecosystem health. Nature 484:186–194
3. Coumans J, Harvey J, Backhouse D, Poljak A, Raftery M, Nehl D et al (2011) Proteomic assessment of host-associated microevolution in the fungus Thielaviopsis basicola. Environ Microbiol 13:576–588
4. Devos S, Laukens K, Deckers P, Van Der Straeten D, Beeckman T, Inzé D et al (2006) A hormone and proteome approach to picturing the initial metabolic events during Plasmodiophora brassicae infection on Arabidopsis. Mol Plant Microbe Interact 19: 1431–1443
5. Choudhary MK, Basu D, Datta A, Chakraborty N, Chakraborty S (2009) Dehydration-responsive nuclear proteome of rice (*Oryza sativa* L.) illustrates protein network, novel regulators of cellular adaptation, and evolutionary perspective. Mol Cell Proteomics 8:1579–1598
6. Ballini E, Morel J-B, Droc G, Price A, Courtois B, Notteghem J-L et al (2008) A genome-wide meta-analysis of rice blast resistance genes and quantitative trait loci provides new insights into partial and complete resistance. Mol Plant Microbe Interact 21:859–868
7. Varshney RK, Mir RR, Bhatia S, Thudi M, Hu Y, Azam S et al (2014) Integrated physical, genetic and genome map of chickpea (*Cicer arietinum* L.). Funct Integr Genomics 14:59–73

8. Lambel S, Lanini B, Vivoda E, Fauve J, Wechter WP, Harris-Shultz KR et al (2014) A major QTL associated with Fusarium oxysporum race 1 resistance identified in genetic populations derived from closely related watermelon lines using selective genotyping and genotyping-by-sequencing for SNP discovery. Theor Appl Genet 127:2105–2115
9. Buerstmayr H, Ban T, Anderson JA (2009) QTL mapping and marker-assisted selection for Fusarium head blight resistance in wheat: a review. Plant Breeding 128:1–26
10. Massman J, Cooper B, Horsley R, Neate S, Dill-Macky R, Chao S et al (2011) Genome-wide association mapping of Fusarium head blight resistance in contemporary barley breeding germplasm. Mol Breeding 27:439–454
11. Aghnoum R, Marcel TC, Johrde A, Pecchioni N, Schweizer P, Niks RE (2010) Basal host resistance of barley to powdery mildew: connecting quantitative trait Loci and candidate genes. Mol Plant Microbe Interact 23:91–102
12. Coram TE, Pang EC (2006) Expression profiling of chickpea genes differentially regulated during a resistance response to Ascochyta rabiei. Plant Biotechnol J 4:647–666
13. Ashraf N, Ghai D, Barman P, Basu S, Gangisetty N, Mandal MK et al (2009) Comparative analyses of genotype dependent expressed sequence tags and stress-responsive transcriptome of chickpea wilt illustrate predicted and unexpected genes and novel regulators of plant immunity. BMC Genom 10:415
14. Bedre R, Rajasekaran K, Mangu VR, Timm LES, Bhatnagar D, Baisakh N (2015) Genome-wide transcriptome analysis of cotton (*Gossypium hirsutum* L.) identifies candidate gene signatures in response to aflatoxin producing fungus Aspergillus flavus. PLoS ONE 10: e0138025
15. Winnenburg R, Urban M, Beacham A, Baldwin TK, Holland S, Lindeberg M et al (2008) PHI-base update: additions to the pathogen–host interaction database. Nucleic Acids Res 36: D572–D576
16. Chan SK, Hsing M, Hormozdiari F, Cherkasov A (2007) Relationship between insertion/deletion (indel) frequency of proteins and essentiality. BMC Bioinformatics 8:1
17. Ajawatanawong P, Baldauf SL (2013) Evolution of protein indels in plants, animals and fungi. BMC Evol Biol 13:140
18. Sharma N, Rahman MH, Strelkov S, Thiagarajah M, Bansal VK, Kav NN (2007) Proteome-level changes in two Brassica napus lines exhibiting differential responses to the fungal pathogen Alternaria brassicae. Plant Sci 172:95–110
19. Floerl S, Majcherczyk A, Possienke M, Feussner K, Tappe H, Gatz C et al (2012) Verticillium longisporum infection affects the leaf apoplastic proteome, metabolome, and cell wall properties in Arabidopsis thaliana. PLoS ONE 7:e31435
20. Garg H, Li H, Sivasithamparam K, Barbetti MJ (2013) Differentially expressed proteins and associated histological and disease progression changes in cotyledon tissue of a resistant and susceptible genotype of Brassica napus infected with Sclerotinia sclerotiorum. PLoS ONE 8: e65205
21. Liang Y, Strelkov SE, Kav NN (2009) Oxalic acid-mediated stress responses in Brassica napus L. Proteomics 9:3156–3173
22. Liang Y, Srivastava S, Rahman MH, Strelkov SE, Kav NN (2008) Proteome changes in leaves of Brassica napus L. as a result of Sclerotinia sclerotiorum challenge. J Agric Food Chem 56:1963–1976
23. Mukherjee AK, Carp M-J, Zuchman R, Ziv T, Horwitz BA, Gepstein S (2010) Proteomics of the response of Arabidopsis thaliana to infection with Alternaria brassicicola. J Proteomics 73:709–720
24. Mulema JM, Okori P, Denby KJ (2013) Proteomic analysis of the Arabidopsis thaliana-Botrytis cinerea interaction using two-dimensional liquid chromatography. Afr J Biotechnol 10:17551–17563
25. Wen F, Vanetten HD, Tsaprailis G, Hawes MC (2007) Extracellular proteins in pea root tip and border cell exudates. Plant Physiol 143:773–783

26. Hao J, Wu W, Wang Y, Yang Z, Liu Y, Lv Y et al (2015) Arabidopsis thaliana defense response to the ochratoxin A-producing strain (Aspergillus ochraceus 3.4412). Plant Cell Rep 34:705–719
27. Ndimba BK, Chivasa S, Hamilton JM, Simon WJ, Slabas AR (2003) Proteomic analysis of changes in the extracellular matrix of Arabidopsis cell suspension cultures induced by fungal elicitors. Proteomics 3:1047–1059
28. Subramanian B, Bansal VK, Kav NN (2005) Proteome-level investigation of Brassica carinata-derived resistance to Leptosphaeria maculans. J Agric Food Chem 53:313–324
29. Sharma N, Hotte N, Rahman MH, Mohammadi M, Deyholos MK, Kav NN (2008) Towards identifying Brassica proteins involved in mediating resistance to Leptosphaeria maculans: a proteomics-based approach. Proteomics 8:3516–3535
30. Asano T, Kimura M, Nishiuchi T (2012) The defense response in Arabidopsis thaliana against Fusarium sporotrichioides. Proteome Sci 10:1
31. O'brien JA, Daudi A, Finch P, Butt VS, Whitelegge JP, Souda P et al (2012) A peroxidase-dependent apoplastic oxidative burst in cultured Arabidopsis cells functions in MAMP-elicited defense. Plant Physiol 158:2013–2027
32. Kaur P, Jost R, Sivasithamparam K, Barbetti MJ (2011) Proteome analysis of the Albugo candida–Brassica juncea pathosystem reveals that the timing of the expression of defence-related genes is a crucial determinant of pathogenesis. J Exp Botany 62(3): 1285–1298
33. Sun C, Wang L, Hu D, Riquicho ARM, Liu T, Hou X et al (2014) Proteomic analysis of non-heading Chinese cabbage infected with Hyaloperonospora parasitica. J Proteomics 98:15–30
34. Chivasa S, Hamilton JM, Pringle RS, Ndimba BK, Simon WJ, Lindsey K et al (2006) Proteomic analysis of differentially expressed proteins in fungal elicitor-treated Arabidopsis cell cultures. J Exp Bot 57:1553–1562
35. Floerl S, Druebert C, Majcherczyk A, Karlovsky P, Kües U, Polle A (2008) Defence reactions in the apoplastic proteome of oilseed rape (*Brassica napus* var. *napus*) attenuate Verticillium longisporum growth but not disease symptoms. BMC Plant Biol 8:1
36. Benschop JJ, Mohammed S, O'flaherty M, Heck Aj, Slijper M, Menke FL (2007) Quantitative phosphoproteomics of early elicitor signaling in Arabidopsis. Mol Cell Proteomics 6:1198–1214
37. Valot B, Negroni L, Zivy M, Gianinazzi S, Dumas-Gaudot E (2006) A mass spectrometric approach to identify arbuscular mycorrhiza-related proteins in root plasma membrane fractions. Proteomics 6 Suppl 1:S145–S155. doi:10.1002/pmic.200500403
38. Aloui A, Recorbet G, Gollotte A, Robert F, Valot B, Gianinazzi-Pearson V et al (2009) On the mechanisms of cadmium stress alleviation in Medicago truncatula by arbuscular mycorrhizal symbiosis: a root proteomic study. Proteomics 9:420–433
39. Curto M, Camafeita E, Lopez JA, Maldonado AM, Rubiales D, Jorrín JV (2006) A proteomic approach to study pea (Pisum sativum) responses to powdery mildew (Erysiphe pisi). Proteomics 6 Suppl 1:S163–S174. doi:10.1002/pmic.200500396
40. Marra R, Ambrosino P, Carbone V, Vinale F, Woo SL, Ruocco M et al (2006) Study of the three-way interaction between Trichoderma atroviride, plant and fungal pathogens by using a proteomic approach. Curr Genet 50:307–321
41. Castillejo MÁ, Curto M, Fondevilla S, Rubiales D, Jorrín JV (2010) Two-dimensional electrophoresis based proteomic analysis of the pea (*Pisum sativum*) in response to Mycosphaerella pinodes. J Agric Food Chem 58:12822–12832
42. Lei Z, Chen F, Watson BS, Nagaraj S, Elmer AM, Dixon RA et al (2010) Comparative proteomics of yeast-elicited Medicago truncatula cell suspensions reveals induction of isoflavonoid biosynthesis and cell wall modifications. J Proteome Res 9:6220–6231
43. Wang T, Zhang E, Chen X, Li L, Liang X (2010) Identification of seed proteins associated with resistance to pre-harvested aflatoxin contamination in peanut (*Arachis hypogaea* L). BMC Plant Biol 10:1

44. Palomares-Rius JE, Castillo P, Navas-Cortés JA, Jiménez-Díaz RM, Tena M (2011) A proteomic study of in-root interactions between chickpea pathogens: The root-knot nematode Meloidogyne artiellia and the soil-borne fungus *Fusarium oxysporum f. sp. ciceris* race 5. Journal of proteomics 74:2034–2051
45. Wang Z, Yan S, Liu C, Chen F, Wang T (2012) Proteomic analysis reveals an aflatoxin-triggered immune response in cotyledons of Arachis hypogaea infected with Aspergillus flavus. J Proteome Res 11:2739–2753
46. Chatterjee M, Gupta S, Bhar A, Chakraborti D, Basu D, Das S (2014) Analysis of root proteome unravels differential molecular responses during compatible and incompatible interaction between chickpea (*Cicer arietinum* L.) and Fusarium *oxysporum f. sp. ciceri* Race1 (Foc1). BMC Genom 15:949
47. Moura HFN, Vasconcelos IM, Souza CEA, Silva FD, Moreno FB, Lobo MD et al (2014) Proteomics changes during the incompatible interaction between cowpea and Colletotrichum gloeosporioides (Penz.) Penz and Sacc. Plant Sci 217:158–175
48. Pereira JL, Queiroz RM, Charneau SO, Felix CR, Ricart CA, Da Silva FL et al (2014) Analysis of Phaseolus vulgaris response to its association with Trichoderma harzianum (ALL-42) in the presence or absence of the phytopathogenic fungi Rhizoctonia solani and Fusarium solani. PLoS ONE 9:e98234
49. Jain A, Singh A, Singh S, Singh V, Singh HB (2015) Comparative proteomic analysis in pea treated with microbial consortia of beneficial microbes reveals changes in the protein network to enhance resistance against Sclerotinia sclerotiorum. J Plant Physiol 182:79–94
50. Kumar D, Kirti PB (2015) Transcriptomic and proteomic analyses of resistant host responses in Arachis diogoi challenged with late leaf spot pathogen, *Phaeoisariopsis personata.* PloS One 10:e0117559
51. Subramanian S, Cho U-H, Keyes C, Yu O (2009) Distinct changes in soybean xylem sap proteome in response to pathogenic and symbiotic microbe interactions. BMC Plant Biol 9:1
52. Zhao J, Zhang Y, Bian X, Lei J, Sun J, Guo N et al (2013) A comparative proteomics analysis of soybean leaves under biotic and abiotic treatments. Mol Biol Rep 40:1553–1562
53. Colditz F, Nyamsuren O, Niehaus K, Eubel H, Braun H-P, Krajinski F (2004) Proteomic approach: identification of Medicago truncatula proteins induced in roots after infection with the pathogenic oomycete Aphanomyces euteiches. Plant Mol Biol 55:109–120
54. Amey RC, Schleicher T, Slinn J, Lewis M, Macdonald H, Neill SJ et al (2008) Proteomic analysis of a compatible interaction between *Pisum sativum* (pea) and the downy mildew pathogen Peronospora viciae. Eur J Plant Pathol 122:41–55
55. Lee J, Feng J, Campbell KB, Scheffler BE, Garrett WM, Thibivilliers S et al (2009) Quantitative proteomic analysis of bean plants infected by a virulent and avirulent obligate rust fungus. Mol Cell Proteomics 8:19–31
56. Trapphoff T, Beutner C, Niehaus K, Colditz F (2009) Induction of distinct defense-associated protein patterns in Aphanomyces euteiches (Oomycota)-elicited and-inoculated Medicago truncatula cell-suspension cultures: a proteome and phosphoproteome approach. Mol Plant Microbe Interact 22:421–436
57. Cooper B, Campbell KB, Feng J, Garrett WM, Frederick R (2011) Nuclear proteomic changes linked to soybean rust resistance. Mol BioSyst 7:773–783
58. Rep M, Dekker HL, Vossen JH, De Boer AD, Houterman PM, Speijer D et al (2002) Mass spectrometric identification of isoforms of PR proteins in xylem sap of fungus-infected tomato. Plant Physiol 130:904–917
59. Houterman PM, Speijer D, Dekker HL, Cg DEK, Cornelissen BJ, Rep M (2007) The mixed xylem sap proteome of Fusarium oxysporum-infected tomato plants. Mol Plant Pathol 8:215–221
60. Mazzeo MF, Cacace G, Ferriello F, Puopolo G, Zoina A, Ercolano MR et al (2014) Proteomic investigation of response to FORL infection in tomato roots. Plant Physiol Biochem 74:42–49

61. Shah P, Powell AL, Orlando R, Bergmann C, Gutierrez-Sanchez G (2012) Proteomic analysis of ripening tomato fruit infected by *Botrytis cinerea*. J Proteome Res 11:2178–2192
62. Pan X, Zhu B, Luo Y, Fu D (2013) Unraveling the protein network of tomato fruit in response to necrotrophic phytopathogenic Rhizopus nigricans. PLoS ONE 8:e73034
63. Fernández MB, Pagano MR, Daleo GR, Guevara MG (2012) Hydrophobic proteins secreted into the apoplast may contribute to resistance against Phytophthora infestans in potato. Plant Physiol Biochem 60:59–66
64. Lim S, Borza T, Peters RD, Coffin RH, Al-Mughrabi KI, Pinto DM et al (2013) Proteomics analysis suggests broad functional changes in potato leaves triggered by phosphites and a complex indirect mode of action against Phytophthora infestans. J Proteomics 93:207–223
65. Burra DD, Berkowitz O, Hedley PE, Morris J, Resjö S, Levander F et al (2014) Phosphite-induced changes of the transcriptome and secretome in Solanum tuberosum leading to resistance against Phytophthora infestans. BMC Plant Biol 14:254
66. Louis B, Waikhom SD, Roy P, Bhardwaj PK, Singh MW, Chandradev SK et al (2014) Invasion of *Solanum tuberosum* L. by Aspergillus terreus: a microscopic and proteomics insight on pathogenicity. BMC Res Notes 7:350
67. Villela-Dias C, Camillo LR, De Oliveira GA, Sena JA, Santiago AS, De Sousa ST et al (2014) Nep1-like protein from *Moniliophthora perniciosa* induces a rapid proteome and metabolome reprogramming in cells of Nicotiana benthamiana. Physiol Plant 150:1–17
68. Cobos R, Barreiro C, Mateos RM, Coque J-JR (2010) Cytoplasmic-and extracellular-proteome analysis of *Diplodia seriata*: a phytopathogenic fungus involved in grapevine decline. Proteome science 8:1
69. Marsh E, Alvarez S, Hicks LM, Barbazuk WB, Qiu W, Kovacs L et al (2010) Changes in protein abundance during powdery mildew infection of leaf tissues of Cabernet Sauvignon grapevine (*Vitis vinifera* L.). Proteomics 10:2057–2064
70. Milli A, Cecconi D, Bortesi L, Persi A, Rinalducci S, Zamboni A et al (2012) Proteomic analysis of the compatible interaction between Vitis vinifera and Plasmopara viticola. J Proteomics 75:1284–1302
71. Palmieri MC, Perazzolli M, Matafora V, Moretto M, Bachi A, Pertot I (2012) Proteomic analysis of grapevine resistance induced by Trichoderma harzianum T39 reveals specific defence pathways activated against downy mildew. J Exp Bot 63:6237–6251
72. Spagnolo A, Larignon P, Magnin-Robert M, Hovasse A, Cilindre C, Van Dorsselaer A et al (2014) Flowering as the most highly sensitive period of grapevine (*Vitis vinifera* L. cv Mourvèdre) to the Botryosphaeria dieback agents Neofusicoccum parvum and Diplodia seriata infection. Int J Mol Sci 15:9644–9669
73. Dadakova K, Havelkova M, Kurkova B, Tlolkova I, Kasparovsky T, Zdrahal Z et al (2015) Proteome and transcript analysis of Vitis vinifera cell cultures subjected to Botrytis cinerea infection. J Proteomics 119:143–153
74. Chan Z, Qin G, Xu X, Li B, Tian S (2007) Proteome approach to characterize proteins induced by antagonist yeast and salicylic acid in peach fruit. J Proteome Res 6:1677–1688
75. Chan Z, Wang Q, Xu X, Meng X, Qin G, Li B et al (2008) Functions of defense-related proteins and dehydrogenases in resistance response induced by salicylic acid in sweet cherry fruits at different maturity stages. Proteomics 8:4791–4807
76. Kwasiborski A, Bajji M, Renaut J, Delaplace P, Jijakli MH (2014) Identification of metabolic pathways expressed by Pichia anomala KH6 in the presence of the pathogen Botrytis cinerea on apple: new possible targets for biocontrol improvement. PLoS ONE 9:e91434
77. Zhang C-X, Tian Y, Cong P-H (2015) Proteome analysis of pathogen-responsive proteins from apple leaves induced by the alternaria blotch *Alternaria alternata*. PLoS ONE 10: e0122233
78. Fang X, Jost R, Finnegan PM, Barbetti MJ (2013) Comparative proteome analysis of the strawberry-fusarium oxysporum f. sp. fragariae pathosystem reveals early activation of defense responses as a crucial determinant of host resistance. J Proteome Res 12:1772–1788
79. Fang X, Chen W, Xin Y, Zhang H, Yan C, Yu H et al (2012) Proteomic analysis of strawberry leaves infected with *Colletotrichum fragariae*. J Proteomics 75:4074–4090

80. Fang X, Barbetti MJ (2014) Differential protein accumulations in isolates of the strawberry wilt pathogen *Fusarium oxysporum* f. sp. *fragariae* differing in virulence. J Proteomics 108:223–237
81. Coumans JV, Poljak A, Raftery MJ, Backhouse D, Pereg-Gerk L (2009) Analysis of cotton (Gossypium hirsutum) root proteomes during a compatible interaction with the black root rot fungus *Thielaviopsis basicola*. Proteomics 9:335–349
82. Wang FX, Ma YP, Yang CL, Zhao PM, Yao Y, Jian GL et al (2011) Proteomic analysis of the sea-island cotton roots infected by wilt pathogen *Verticillium dahliae*. Proteomics 11:4296–4309
83. Guerra-Guimarães L, Tenente R, Pinheiro C, Chaves I, Do Céu Silva M, Cardoso FM et al (2015) Proteomic analysis of apoplastic fluid of Coffea arabica leaves highlights novel biomarkers for resistance against *Hemileia vastatrix*. Front Plant Sci 6:478. doi:10.3389/fpls.2015.00478 eCollection 2015
84. Sinha R, Chattopadhyay S (2011) Changes in the leaf proteome profile of *Mentha arvensis* in response to *Alternaria alternata* infection. J Proteomics 74:327–336
85. Sinha R, Bhattacharyya D, Majumdar AB, Datta R, Hazra S, Chattopadhyay S (2013) Leaf proteome profiling of transgenic mint infected with *Alternaria alternata*. J Proteomics 93:117–132
86. Larson RL, Hill AL, Nuñez A (2007) Characterization of protein changes associated with sugar beet (*Beta vulgaris*) resistance and susceptibility to Fusarium oxysporum. J Agric Food Chem 55:7905–7915
87. Louarn S, Nawrocki A, Edelenbos M, Jensen DF, Jensen ON, Collinge DB et al (2012) The influence of the fungal pathogen *Mycocentrospora acerina* on the proteome and polyacetylenes and 6-methoxymellein in organic and conventionally cultivated carrots (*Daucus carota*) during post harvest storage. J Proteomics 75:962–977
88. Segarra G, Casanova E, Bellido D, Odena MA, Oliveira E, Trillas I (2007) Proteome, salicylic acid, and jasmonic acid changes in cucumber plants inoculated with *Trichoderma asperellum* strain T34. Proteomics 7:3943–3952
89. Acosta-Muniz CH, Escobar-Tovar L, Valdes-Rodriguez S, Fernandez-Pavia S, Arias-Saucedo LJ, La Cruz De, Espindola Barquera M et al (2012) Identification of avocado (*Persea americana*) root proteins induced by infection with the oomycete *Phytophthora cinnamomi* using a proteomic approach. Physiol Plant 144:59–72
90. Li X, Bai T, Li Y, Ruan X, Li H (2013) Proteomic analysis of *Fusarium oxysporum f. sp. cubense* tropical race 4-inoculated response to Fusarium wilts in the banana root cells. Proteome science 11:1
91. Cipriano AK, Gondim DM, Vasconcelos IM, Martins JA, Moura AA, Moreno FB et al (2015) Proteomic analysis of responsive stem proteins of resistant and susceptible cashew plants after *Lasiodiplodia theobromae* infection. J Proteomics 113:90–109
92. Pechanova O, Hsu C-Y, Adams JP, Pechan T, Vandervelde L, Drnevich J et al (2010) Apoplast proteome reveals that extracellular matrix contributes to multistress response in poplar. BMC Genom 11:674
93. Chen Q, Guo W, Feng L, Ye X, Xie W, Huang X et al (2015) Transcriptome and proteome analysis of Eucalyptus infected with Calonectria pseudoreteaudii. J Proteomics 115:117–131
94. Valcu C-M, Junqueira M, Shevchenko A, Schlink K (2009) Comparative proteomic analysis of responses to pathogen infection and wounding in Fagus sylvatica. J Proteome Res 8:4077–4091
95. Mandelc S, Timperman I, Radisek S, Devreese B, Samyn B, Javornik B (2013) Comparative proteomic profiling in compatible and incompatible interactions between hop roots and *Verticillium albo-atrum*. Plant Physiol Biochem 68:23–31
96. Ardila HD, Fernandez RG, Higuera BL, Redondo I, Martinez ST (2014) Protein extraction and gel-based separation methods to analyze responses to pathogens in carnation (*Dianthus caryophyllus* L). Methods Mol Biol 1072:573–591
97. Wang Y, Yang L, Xu H, Li Q, Ma Z, Chu C (2005) Differential proteomic analysis of proteins in wheat spikes induced by *Fusarium graminearum*. Proteomics 5:4496–4503

98. Zhou W, Kolb FL, Riechers DE (2005) Identification of proteins induced or upregulated by Fusarium head blight infection in the spikes of hexaploid wheat (*Triticum aestivum*). Genome 48:770–780
99. Zhou W, Eudes F, Laroche A (2006) Identification of differentially regulated proteins in response to a compatible interaction between the pathogen *Fusarium graminearum* and its host, *Triticum aestivum*. Proteomics 6:4599–4609
100. Cao T, Kim YM, Kav NN, Strelkov SE (2009) A proteomic evaluation of *Pyrenophora tritici-repentis*, causal agent of tan spot of wheat, reveals major differences between virulent and avirulent isolates. Proteomics 9:1177–1196
101. Dornez E, Croes E, Gebruers K, Carpentier S, Swennen R, Laukens K et al (2010) 2-D DIGE reveals changes in wheat xylanase inhibitor protein families due to *Fusarium graminearum* ΔTri5 infection and grain development. Proteomics 10:2303–2319
102. Kim YM, Bouras N, Kav NN, Strelkov SE (2010) Inhibition of photosynthesis and modification of the wheat leaf proteome by Ptr ToxB: A host-specific toxin from the fungal pathogen *Pyrenophora tritici-repentis*. Proteomics 10:2911–2926
103. Eggert K, Pawelzik E (2011) Proteome analysis of *Fusarium* head blight in grains of naked barley (*Hordeum vulgare subsp. nudum*). Proteomics 11:972–985
104. Eggert K, Zörb C, Mühling K, Pawelzik E (2011) Proteome analysis of *Fusarium* infection in emmer grains (*Triticum dicoccum*). Plant Pathol 60:918–928
105. Gunnaiah R, Kushalappa AC, Duggavathi R, Fox S, Somers DJ (2012) Integrated Metabolo-Proteomic approach to decipher the mechanisms by which wheat QTL (*Fhb1*) contributes to resistance against *Fusarium graminearum*. PLoS ONE 7:e40695
106. Yang F, Melo-Braga MN, Larsen MR, Jørgensen HJ, Palmisano G (2013) Battle through signaling between wheat and the fungal pathogen Septoria tritici revealed by proteomics and phosphoproteomics. Mol Cell Proteomics 12:2497–2508
107. Magliano TMA, Ortega LM, Astoreca AL, Pritsch C (2013) Proteomic approaches to analyze wheat-*Fusarium graminearum* interaction. In: Chulze NS, Alconada Magliano MT (eds) Fusarium head blight in Latin America. Springer, Netherlands, Dordrecht, pp 123–140
108. Mandal MS, Fu Y, Zhang S, Ji W (2014) Proteomic analysis of the defense response of wheat to the powdery mildew fungus, *Blumeria graminis f. sp. tritici*. Protein J 33:513–524
109. Winterberg B, Du Fall LA, Song X, Pascovici D, Molloy M, Ohms S et al (2014) The necrotrophic effector protein SnTox3 re-programs metabolism and elicits a strong defence response in susceptible wheat leaves. BMC Plant Biol 14:1
110. M'barek SB, Cordewener JH, Van Der Lee TA, America AH, Gohari AM, Mehrabi R et al (2015) Proteome catalog of Zymoseptoria tritici captured during pathogenesis in wheat. Fungal Genet Biol 79:42–53
111. Chetouhi C, Bonhomme L, Lecomte P, Cambon F, Merlino M, Biron DG et al (2015) A proteomics survey on wheat susceptibility to Fusarium head blight during grain development. Eur J Plant Pathol 141:407–418
112. Day J, Gietz RD, Rampitsch C (2015) Proteome changes induced by *Pyrenophora tritici-repentis* ToxA in both insensitive and sensitive wheat indicate senescence-like signaling. Proteome Sci 13:1
113. Yang F, Li W, Derbyshire M, Larsen MR, Rudd JJ, Palmisano G (2015) Unraveling incompatibility between wheat and the fungal pathogen Zymoseptoria tritici through apoplastic proteomics. BMC Genom 16:1
114. Rampitsch C, Bykova NV, Mccallum B, Beimcik E, Ens W (2006) Analysis of the wheat and *Puccinia triticina* (leaf rust) proteomes during a susceptible host-pathogen interaction. Proteomics 6:1897–1907
115. Maytalman D, Mert Z, Baykal AT, Inan C, Günel A, Hasançebi S (2013) Proteomic analysis of early responsive resistance proteins of wheat (*Triticum aestivum*) to yellow rust (*Puccinia striiformis f. sp. tritici*) using ProteomeLab PF2D. Plant Omics 6:24
116. Kim JY, Wu J, Kwon SJ, Oh H, Lee SE, Kim SG et al (2014) Proteomics of rice and *Cochliobolus miyabeanus* fungal interaction: insight into proteins at intracellular and extracellular spaces. Proteomics 14:2307–2318

117. Kim SG, Wang Y, Lee KH, Park Z-Y, Park J, Wu J et al (2013) In-depth insight into in vivo apoplastic secretome of rice-*Magnaporthe oryzae* interaction. J Proteomics 78:58–71
118. Kim ST, Kim SG, Hwang DH, Kang SY, Kim HJ, Lee BH et al (2004) Proteomic analysis of pathogen-responsive proteins from rice leaves induced by rice blast fungus, *Magnaporthe grisea*. Proteomics 4:3569–3578
119. Lee J, Bricker TM, Lefevre M, Pinson SR, Oard JH (2006) Proteomic and genetic approaches to identifying defence-related proteins in rice challenged with the fungal pathogen Rhizoctonia solani. Mol Plant Pathol 7:405–416
120. Li H, Goodwin PH, Han Q, Huang L, Kang Z (2012) Microscopy and proteomic analysis of the non-host resistance of *Oryza sativa* to the wheat leaf rust fungus, *Puccinia triticina f. sp. tritici*. Plant Cell Rep 31:637–650
121. Li Y, Nie Y, Zhang Z, Ye Z, Zou X, Zhang L et al (2014) Comparative proteomic analysis of methyl jasmonate-induced defense responses in different rice cultivars. Proteomics 14: 1088–1101
122. Zhao J, Yang Y, Kang Z (2014) Proteomic analysis of rice nonhost resistance to *Puccinia striiformis f. sp.* tritici using two-dimensional electrophoresis. Int J Mol Sci 15:21644–21659
123. Geddes J, Eudes F, Laroche A, Selinger LB (2008) Differential expression of proteins in response to the interaction between the pathogen *Fusarium graminearum* and its host, *Hordeum vulgare*. Proteomics 8:545–554
124. Yang F, Jensen JD, Svensson B, Jørgensen HJ, Collinge DB, Finnie C (2010) Analysis of early events in the interaction between *Fusarium graminearum* and the susceptible barley (*Hordeum vulgare*) cultivar Scarlett. Proteomics 10:3748–3755
125. Yang F, Svensson B, Finnie C (2011) Response of germinating barley seeds to *Fusarium graminearum*: the first molecular insight into *Fusarium* seedling blight. Plant Physiol Biochem 49:1362–1368
126. Bernardo L, Prinsi B, Negri AS, Cattivelli L, Espen L, Valè G (2012) Proteomic characterization of the Rph15 barley resistance gene-mediated defence responses to leaf rust. BMC Genom 13:1
127. Anup CP, Melvin P, Shilpa N, Gandhi MN, Jadhav M, Ali H et al (2015) Proteomic analysis of elicitation of downy mildew disease resistance in pearl millet by seed priming with beta-aminobutyric acid and Pseudomonas fluorescens. J Proteomics 120:58–74
128. Gómez-Vidal S, Salinas J, Tena M, Lopez-Llorca LV (2009) Proteomic analysis of date palm (*Phoenix dactylifera L.*) responses to endophytic colonization by entomopathogenic fungi. Electrophoresis 30:2996–3005
129. Al-Obaidi JR, Mohd-Yusuf Y, Razali N, Jayapalan JJ, Tey CC, Md-Noh N et al (2014) Identification of proteins of altered abundance in oil palm infected with *Ganoderma boninense*. Int J Mol Sci 15:5175–5192
130. Jeffery Daim LD, Ooi TE, Ithnin N, Mohd Yusof H, Kulaveerasingam H, Abdul Majid N et al (2015) Comparative proteomic analysis of oil palm leaves infected with *Ganoderma boninense* revealed changes in proteins involved in photosynthesis, carbohydrate metabolism, and immunity and defense. Electrophoresis 36:1699–1710

Chapter 11
Insight into Physiological, Molecular, and Proteomic Changes Associated with Phytoplasma Infection in Crop Plants

Javad Gharechahi, Mohsen Mardi and Ghasem Hosseini Salekdeh

Abstract Phytoplasmas are insect transmitted bacterial pathogens that bring about devastating diseases in a wide range of plants including crops, ornamental plants, fruit trees, and vegetables. Phytoplasma diseased plants often display symptoms that thought to be resulted from altered plant development. Knowledge about the molecular basis of plant-phytoplasma interaction is limited because they are unable to be cultured under in vitro conditions and largely inaccessible in their host plants. The detailed response of several plant species to phytoplasma infection has been explored at transcriptome level using conventional and recently developed next-generation sequencing (NGS) based approaches. Considering the broad changes that can occur in proteome in terms of abundance, post translational modification, sub-cellular localization, and protein-protein interactions in a plant in response to pathogen infection, proteomics provides valuable information that are essential for in-depth understanding of plant-pathogen interaction. Till now, little progress has been made in understanding plant-phytoplasma interaction from proteomics view point. Here, we first briefly outline physiological and transcriptional changes associated with phytoplasma infection in some plant species and then review changes in proteome of plants in response to phytoplasma infection with a particular emphasis on comparative studies that dealt with changes in abundance of proteins.

Keywords Phytoplasma · Plant-pathogen interaction · Proteome · Transcriptome · Gene expression

---

J. Gharechahi
Chemical Injuries Research Center, Baqiyatallah University of Medical Sciences, Tehran, Iran

M. Mardi · G.H. Salekdeh (✉)
Department of Systems Biology, Agricultural Biotechnology Research Institute of Iran, Agricultural Research, Education, and Extension Organization, Karaj, Iran
e-mail: hsalekdeh@yahoo.com

G.H. Salekdeh (ed.), *Agricultural Proteomics Volume 2*,
DOI 10.1007/978-3-319-43278-6_11

11.1 The Biology and Life Cycle of Phytoplasmas

Phytoplasmas are mycoplasma-like bacterial pathogens that cause devastating diseases with sever yield losses in a diverse ranges of economically important species including crops, ornamental plants, fruit trees, and vegetables [1, 2]. Same as mycoplasmas, phytoplasmas belong to the class Mollicutes and the phylum Tenericutes [3]. Members of the class Mollicutes are thought to be diverged from the gram positive bacteria of the phylum Firmicutes through loss of the cell wall and a reduction of genome size [1, 4]. Within the class mollicutes, phytoplasmas show the high similarity in terms of codon usage and metabolic pathways to acholeplasmas than to the mycoplasmas and spiroplasmas [1]. In contrast to mycoplasmas that are human and animal pathogens, phytoplasma and some species of spiroplasmas are plant pathogens that are also able to infect sap-feeding insects of the order Hemiptera including leafhoppers, planthoppers, and psyllids that serve as vectors for these pathogens [5]. In addition, several species of Mycoplasma and Spiroplasma have successfully been cultured under in vitro conditions but phytoplasmas are yet to be cultured and therefore, they are assigned to the provisional genus name *Candidatus* Phytoplasma [6].

Phytoplasmas are non-helical and pleiomorphic in shape with sizes similar to those of mycoplasmas, 200–800 nm in diameter, surrounded by a single membrane. Compared with other bacteria, they have small genomes (between 0.5 and 1.5 Mbp) with relatively low G+C content [7]. The complete genome sequence of 4 phytoplasmas has recently been determined indicating that they have been subjected to an evolutionary genome condensation [8–11]. Interestingly, genome shrinkage in these microorganisms has resulted in the loss of many essential genes including those responsible for the ATP synthesis, the biosynthesis of amino acids, fatty acids, phosphotransferase system, de novo synthesis of nucleotides, as well as some enzymes of tricarboxylic acid cycle (TCA cycle) [7, 8]. Therefore, phytoplasmas are considered as obligate intracellular biotrophic parasites that are largely depend on host cells for their nutritional and energy requirements. Interestingly, to compensate for the loss of metabolic genes, phytoplasmas have acquired many genes encoding for transporters including those for malate, metal-ions and amino-acids [12]. This may lead to aggressive import of essential metabolites from host cells resulting in disturbed metabolic balance and the development of disease symptoms. In addition, loss of essential metabolic genes has also hindered in vitro culture of phytoplasmas in cell-free medium and therefore, greatly slowed down their molecular characterization.

Phytoplasmas are unique in their ability to infect and replicate in organisms of two different kingdoms, namely plants (Plantae) and insects (Animalia). In addition, they are also capable of invading and multiplying in cells of different tissues and organs of the insect vector [1]. This remarkable capacity makes them of utmost importance in the study of molecular mechanisms of host adaptation [13]. In plants, phytoplasmas are largely restricted to the phloem sieve elements where they can easily spread systematically throughout the plant by passive movement along with the phloem sap [14]. Phloem sap is rich in nutrients such as carbohydrates, minerals, proteins, amino acids, and ATP, which are required for the growth and colonization of phytoplasmas [15]. Phytoplasmas are acquired by insect vectors feeding from the phloem sap of infected plants, during a process known as acquisition feeding. Once within insects, they must traverse from the gut epithelium and enter into the insect haemolymph where they can replicate and invade other insect organs. To be transferred to a new plant, they have to colonize the salivary glands of the insect vector from which they are introduced back into the phloem of a new healthy plant during feeding and salivation [7].

Phytoplasmas have a broad host rage among monocots and dicots. They are known as causative agents of more than 1000 diseases and in several cases they bring about severe damaging effects on plant growth, development, and yield production. They are also accused of being responsible for several disease outbreaks in plant species with high economic importance. For example, in 2001, an outbreak of *Candidatus* phytoplasma mali in apple trees resulted in an economical loss of ca. 100 million € in Italy and 25 million € in Germany [16]. As another example, witch's broom disease of acid lime (WBDL), which is caused by *Candidatus* phytoplasma aurantifolia has also severely affected traditional lime cultivation and greatly reduced Mexican lime production in the southern regions of Iran. With gradual increase in world mean temperature and concomitant changes in climate, diseases associated with phytoplasmas are expected to increase because the spread of these bacterial pathogens largely depends on insect vectors that are best adapted to warm climatic conditions [16]. This highlights the importance of and the urgent need for the development of new approaches for control and management of diseases associated with these bacterial pathogens.

11.2 Symptom Development in Phytoplasma Infected Plants

Phytoplasma infection usually associates with broad changes in physiological, biochemical, molecular and developmental processes (Fig. 11.1). In diseased plants, phytoplasmas mainly localize in sink tissues including roots and developing leaves and flower, whereas source leaves remain largely uninfected [7, 17]. Accordingly, disease symptoms mainly appear in the young and developing tissues of infected plants. Phytoplsamas elicit symptoms that resemble abnormal plant development. Typical phytoplasma disease symptoms include development of many secondary shoots with small internodes and dense leaves (also known as witches' broom), the growth of shoots from floral organs (proliferation), conversion of flowers into leaf-like structures (phyllody), the greening of non-green flower organs (virescence), growth of elongated stalks (bolting), reddening of leaves and stems (purple top), overall yellowing, sterility of flowers, formation of bunchy fibrous secondary roots, decline and dwarfism of plants, and phloem necrosis

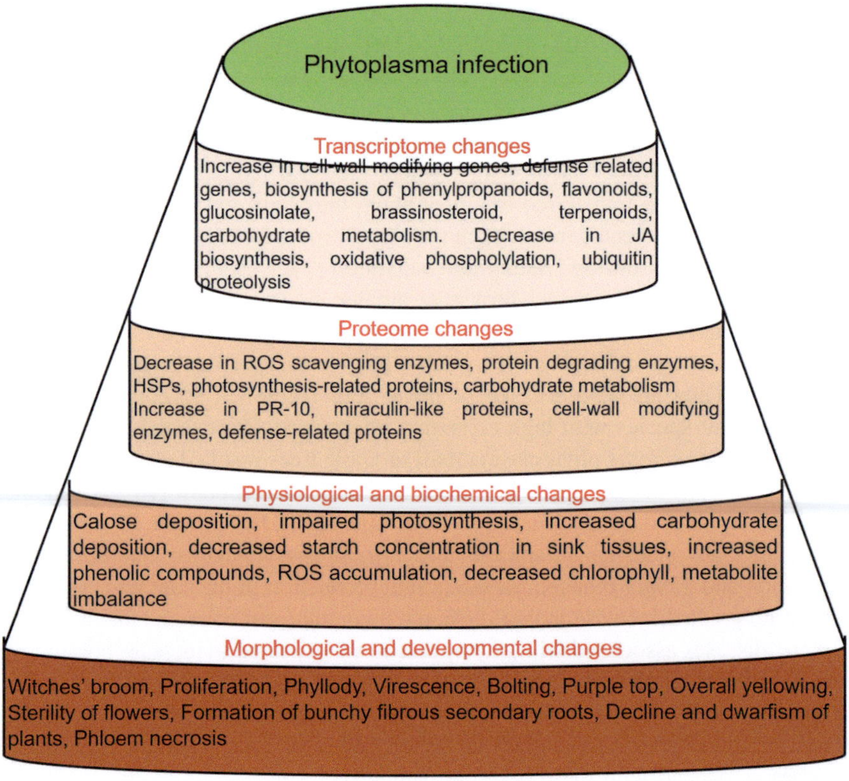

Fig. 11.1 Molecular, physiological, morphological, and developmental changes associated with phytoplasma infection

[7, 14]. The molecular mechanisms governing these morphological changes and their consequences for the infecting bacteria are largely unknown. Indeed, phytoplasma disease symptoms were first believed to be evoked by side effects of phytoplasma infection such as aggressive consumption of plant metabolites, which may lead to an imbalance in metabolic homeostasis in host cells [18]. However, the morphological and developmental changes associated with phytoplasmas infection are thought to be resulted from altered normal balance of plant growth regulators and are often powered by changes in plant gene expression [19–22]. As an example, peanut Witches' Broom phytoplasma (PnWB)-infected *Catharanthus roseus* plants show various floral abnormalities, including discoloration, virescence, and phyllody. Gene expression analysis of PnWB-infected C. *roseus* plants showed that genes related to the flowering pathways were repressed while those responsible for leaf development were induced in phytoplasma infected plants indicating leafy flower transition [23].

Recent evidence suggested that phytoplasmas may actively induce specific symptoms by secretion of effector proteins and modification of plant gene expression [19–21]. For example, it has been shown that *Ca*. Phytoplasma asteris (OY-M strain) encodes a small secretory protein called TENGU, which when ectopically expressed in tobacco and Arabidopsis plants, induces typical phytoplasma disease symptoms including witches' broom and dwarfism [19]. Gene expression analysis of TENGU-expressing plants showed that the expression of many auxin-related genes was significantly down-regulated suggesting that TENGU inhibits auxin-related pathways, thereby modulating plant development. Recently, MacLean et al. [24] showed that upon infection, Aster Yellows Phytoplasma strain Witches' Broom (AY-WB) releases an effector protein (SAP54) into host cells causing the development of leaf-like flowers which are more attractive for leafhopper, which serves as vector for AY-WB phytoplasma and thereby aiding propagation of the bacteria. This led to a speculation that phytoplasma-infected plants are converted into sterile zombies that mainly serve to propagate phytoplasmas [24–26].

11.3 Physiological Changes Associated with Phytoplasma Infection

In spite of their economic importance and their interesting biological features, phytoplasmas have largely remained uncharacterized largely owing to their inability to grow under in vitro cell-free conditions and their inaccessibility in host plants [7, 12, 18]. Therefore, there is limited information about phytoplasma pathogenicity and their unique mode of interaction with host plants. Since phytoplasmas mainly inhabit the sieve elements of phloem tissues, they exert several physiological and physical modifications to phloem tissues of host plants. Interestingly, phytoplasma-infected plants show aberrant deposition of callose in the sieve-plates of phloem tubes which is followed by a collapse of sieve elements and companion cells and blockage of phloem transport [27]. Callose deposition at the phloem sieve plates is thought to be a non-specific defense mechanism that plants utilize to hinder the spread and invasion of the phytoplasma [22, 27]. In addition, increased phytoplasma concentration in the lumen of sieve cells may also lead to the occlusion of sieve elements and further decrease phloem sap flow [12]. However, it should be noted that mechanisms involved in phloem impairment might differ in plants with different levels of susceptibility to phytoplasma infection.

Phytoplasma-infected plants show impaired photosynthesis, increased carbohydrate deposition in mature leaves, and decreased starch concentration in sink tissues such as roots, which might be due to the inhibition of phloem transport in these plants [12, 28–30]. This indicates that the source-sink relationship is changed in phytoplasma-infected plants and in these plants, infected leaves mainly serve as sink but not source organs [29]. On the other hand, accumulation of carbohydrate assimilates in photosynthetic source tissues induces feedback inhibition of photosynthesis causing chlorosis [31]. In addition, phytoplasma-infected plants display increased concentration of phenolic compounds, elevated level of hydrogen peroxide and superoxide, and decreased chlorophyll content [30, 32, 33].

Phytoplasma diseased plants show unique metabolite profile compared with healthy plants [30, 34]. Interestingly, in fully symptomatic phytoplasma-infected leaves of lime (180 days after phytoplasma inoculation), a significant reduction in concentration of essential metabolites including arginine, glutamine, alanine, fructose-6P, ribose-5P, citrate, 2-oxoglutarate, succinate, and salicylate was observed [34]. This indicates that at the final stages of disease development, phytoplasmas aggressively consume plant metabolites leading to a significant metabolic burden on host plants. In addition, metabolome analysis showed that phytoplasma infection is associated with significant changes in the cellular level of plant hormones abscisic acid (ABA) and cytokinin. It is also associated with increased and decreased content of γ-aminobutric acid (GABA) in leaves and phloem sap, respectively [30]. The increased cellular level of GABA is correlated with pathogen infection and it is thought to be involved in protection of plant cells against oxidative damage caused by invading pathogens [30, 35].

11.4 Molecular Changes in Phytoplasma-Infected Plants

Understanding how plants and their invading pathogens recognize and respond to each other is of great importance for deciphering their mutual interactions. In recent years, significant progress has been made in understanding pathogenesis and mode of interaction of many bacterial pathogens with their corresponding host plants. However, little information is available concerning the molecular basis of plant-phytoplasma interaction. Generally, similar pathogens share common pathogenesis mechanisms even thought that their host plants are different. Plants have evolved general defense mechanisms that invading pathogens need to overcome in order to successfully colonize and multiply [36]. Conversely, to survive and to remain healthy, plants also have to overcome the virulence of pathogens. In face of pathogen infection, plants activate an array of molecular and biochemical responses, which almost exclusively depend on changes in gene expression. As an example, upon pathogen recognition, plants trigger a cascade of signaling pathways that lead to the overproduction of reactive oxygen species (ROS) such as H_2O_2 and O_2^{-1}, a reaction known as hypersensitivy response (HR) which is followed by programmed cell death and tissue necrosis [37]. The function of the HR is to limit the spread and invasion of pathogens. However, to overcome the HR reaction, pathogens induce the expression of genes encoding for ROS scavenging enzymes such as catalase (CAT) and superoxide dismutase (SOD) [36]. Pathogens also stimulate the expression and secretion of extracellular enzymes, such as pectin esterases, polygalacturonases, xylanases, pectato lyases, and cellulases, which promote degradation of the cell walls of host plant and therefore, facilitate pathogen penetration [22, 36]. In addition, modification of host cell wall polysaccharides might facilitate pathogen infection via the conversion of cell wall polymers into nutritional substrates required for growth and colonization of the pathogen. Cell wall degradation also paves the way for the invading pathogens to effectively release virulence factors into host cells to target host molecules and to enable pathogen to hijack the host plant for its own multiplication.

Advances in gene expression analysis including the emergence of novel approaches based on next-generation sequencing (NGS) technologies for relative and absolute quantification of transcriptome have revolutionized genomics research. In particular, NGS has allowed to systematically analyze transcriptome of nearly all plant species even those with limited genome sequence or expressed sequence tag (EST), those with complex genomes such as polyploids, and those in which genome sequencing is not cost-efficient [22]. In combination with traditional differential display and microarray, these approaches have provided a wealth of information regarding the molecular basis of plant-pathogen interaction and have significantly improved our understanding about defense mechanisms employed by plants to defect pathogens with different mode of pathogenesis [22, 38–41]. In an attempt to characterize changes in gene expression caused by phytoplasma infection, our research group compared the transcriptome of Mexican lime trees infected by *Ca.* P. aurantifolia with that of corresponding healthy plants [22]. Interestingly, we found an increased abundance of transcripts related to cell wall biogenesis and degradation including cellulase, chitinase, glucosidase, xylanase, glucanase, pectate lyase, polygalacturonase, pectinesterase, and expansins in phytoplasma diseased plants. These results suggested that phytoplasmas exploit similar strategy, as discussed for other pathogens, to overcome host cell-wall barriers and to seize plant resources for their own multiplication.

Plants also trigger innate immunity response to phytoplasma infection. This include the induction of genes encoding for pathogenesis-related proteins (PR proteins) [22, 23, 40, 42, 43], LRR receptor-like serine/threonine-protein kinase (FLS2), disease resistance proteins, somatic embryogenesis receptor kinases (SERKs), Leucine-rich repeat (LRR) receptor kinase brassinosteroid insensitive 1 (BAK1), chitin elicitor receptor kinase 1 (CERK1), and wall-associated kinases (WAKs) [22, 43]. Plant innate immunity is activated in response to the recognition of pathogen-associated molecular patterns (PAMPs, such as lipopolysaccharides, flagellin, EF-Tu, and chitin) by transmembrane or cytosolic pattern recognition receptors (PRRs) including FLS2, EFR, and CERK1 resulting in PAMP-triggered immunity (PTI) that can suppress pathogen colonization [44]. PTI is associated with an array of responses including the induction of the HR reaction, profound change in gene expression, and the production of antimicrobial proteins [45]. The HR reaction is an extremely effective defense mechanism to combat with biotrophic pathogens such as phytoplasma [46].

At transcriptome level, phytoplasma infection is associated with increased abundance of genes related to the biosynthesis of certain secondary metabolites including phenylpropanoids, flavonoids, and terpenoids [22, 40, 42, 43]. Phenylpropanoids constitute a diverse class of secondary metabolites that are known to be involved in mounting defense responses to numerous plant pathogens

[47]. Certain derivatives of phenylpropanoids and flavonoids are converted into phytoalexins such as stilbenes, coumarins, and isoflavonoids, and phenolic compounds such as acetosyringone and salicylic acid (SA), which are produced by diseased plants and are involved in plant-pathogen interactions.

Terpenoids are also involved in plant-pathogen interaction. Within this category of secondary metabolites, there are several bioactive naturally-occurring diterpenoids that are known as plant growth regulators, namely gibberellins (GAs) [48]. GAs are plant hormones that regulate diverse developmental processes including seed germination, stem elongation, leaf expansion, fruit development, pollen maturation, and flowering induction [49]. The exact role of GAs in plant-pathogen interaction has largely remained uncharacterized. The expression of genes encoding for GA biosynthesis pathway showed conflicting results in different phytoplasma-infected plants. In tomato plants infected with purple top phytoplasma, the expression of genes related to the GA biosynthesis pathway decreased resulting in disruption of GA homeostasis and the development of disease symptoms [50]. However, in phytoplasma diseased lime trees, an extensive up-regulation of the genes encoding for the GA biosynthesis pathway was detected [22]. The components of GA signaling pathway was also changed in phytoplasma diseased plants including a decrease in transcripts of DELLA protein and gibberellin receptor GID1 and an increase in phytochrome-interacting factor 3 (PIF3) [22]. These results suggest that GAs are implicated in plant-phytoplasma interaction and they might also be involved in the development of disease symptoms associated with phytoplasma infection.

Phytoplasma infection is also associated with reduced level of jasmonate (JA) biosynthesis [20]. Interestingly, lipoxygenase and allene oxide synthase which catalyze critical steps in the JA biosynthesis pathway displayed decreased abundances in phytoplasma-diseased plants [20, 22]. The role of JA in defense against necrotrophic pathogens has been well established [51], however, its implication in plants interaction with biotrophic pathogens such as phytoplasmas has remained elusive.

Carbohydrate metabolism is also subjected to change in phytoplasma-infected plants. Indeed, accumulation of soluble sugars such as glucose, fructose, sucrose, and starch is a common phenomenon observed in source leaves of phytoplasma-infected plants. Sucrose is the major form of photosynthesis carbohydrate assimilate loaded into the phloem of photosynthetic source leaves. As a consequence of genome shrinkage, phytoplasmas have lost the enzymes of the phosphotransferase system, which is utilized by most bacteria as an energy-efficient way for simultaneously importing and phosphorylating sugars such as sucrose, glucose, and fructose [15]. Therefore, sucrose has to be broken down into glucose and fructose and the phosphorylated form of glucose and fructose are used as carbon source by the phytoplasma. In plants, sucrose is cleaved by invertase and sucrose synthase, which both have been shown to increase in phytoplasma diseased plants [22, 40, 43]. In addition, alpha amylase, which breaks down starch into glucose, has also shown an increased abundance in phytoplasma diseased plants [22, 43]. In phloem tissues, sucrose synthase supplies UDP-glucose for the biosynthesis of callose, which is deposited at the sieve plates in phytoplasma infected plants and hinders the systemic spread of phytoplasma.

11.5 Proteomic Changes Associated with Phytoplasma Infection

Transcriptome analysis has enabled detailed exploration of the changes in gene expression, which are evoked by pathogens to host plants. However, transcriptome does not reflect complete cellular regulatory mechanisms since there exist multiple post-transcriptional regulatory check points that fine tune the abundance and the function of proteins as end-products of genes. Therefore, exploring dynamic changes in cellular proteome is thought to be much more informative than transcriptome. Proteomics provides an opportunity to explore cellular proteome in terms of abundance, post-translational modifications (PTMs), subcellular localization, and protein-protein interactions. Indeed, proteomics offers complementary information to transcriptomics which are essential for in-depth understanding of complex biological processes at molecular level [52]. The success of proteomics largely depends on the availability of genome sequence for the species of interest or its relatives for accurate protein identification using mass spectrometry (MS)-based approaches.

Although transcriptome changes of plant-phytoplasma interaction has been well explored, only a few studies have investigated on related proteome changes [53–57]. In fact, these studies are comparative proteome analyses that have utilized the power of two-dimensional gel electrophoresis (2-DE) or label free shotgun proteomics to explore the changes in host plant proteome in terms of abundance in response to phytoplasma infection (Table 11.1). In an effort to characterize changes in the proteome of Mexican lime trees affected with WBDL, which is brought about by *Ca.* P. aurantifolia, Taheri et al. [57] applied a 2-DE coupled with MS for protein identification. WBDL is a devastating disease that has significantly affected lime cultivation in the southern regions of Iran as well as the adjoining countries including Oman, United Arab Emirates, and Pakistan [22, 56]. In Iran, in particular this disease has eradicated more than 500,000 Mexican lime trees since 2000 [58]. Newly infected trees locally develop secondary shoots with shorted internodes covered by many yellow or pale-green leaves (so called witches' broom). At the later stages of the disease development, all shoots of the tree develop witches' broom and leaves become dry and the tree eventually declines within 4 to 5 years [57]. By comparing the leaf proteome of phytoplasma-infected and healthy plants, Taheri and collogues identified 39 proteins as being responsive to phytoplasma infection. Interestingly, ROS scavenging enzymes including APX, Cu/Zn-SOD, and glutathione-S-transferase (GST) were decreased in phytoplasma diseased plants. Decreased abundance of these proteins might be associated with increased ROS production, which is thought to be essential for efficiently combating with phytoplasma infection through the HR reaction. PR-10 and some miraculin-like proteins displayed an increased abundance in phytoplasma diseased lime trees. PR-10 is an intracellular pathogenesis-related protein that shows broad antimicrobial and in vitro ribonuclease activities [58]. Miraculins are protease inhibitors whose exact role in plant-pathogen interaction has remained elusive. However,

Table 11.1 Proteomic analyses conducted to explore the interaction of plant with Phytoplasma

Plant species	Infecting phytoplasma	Type of tissue	Proteomic approach	Significant findings	References
Mulberry	Mulberry dwarf phytoplasma	Leaf	2-DE; MS	Photosynthetic proteins showed enhanced degradation in diseased plants	[53]
Grapevine	Flavescence dorée phytoplasma	Leaf	2-DE; MS	Anti-oxidative response proteins were decreased in diseased plants; proteins related to photosynthesis, response to stress and the antioxidant system differentially phosphorylated in diseased plants	[55]
Grapevine	Flavescence dorée phytoplasma	Leaf	2-DE; MS	Proteins related to metabolism, energy processes, protein synthesis and degradation, cell rescue, defense and virulence were increased, while photosynthesis related proteins decreased in abundance in diseased plants	[54]
Mexican lime	*Ca.* Phytoplasma aurantifolia	Leaf	2-DE; MS	ROS scavenging enzymes and photosynthesis related proteins were mainly decreased in infected plants. PR-10 and some miraculin-like proteins were increased in diseased plants	[57]
Mexican lime	*Ca.* Phytoplasma aurantifolia	Leaf	Shotgun proteomics	Phytoplasma responsive proteins were related to stress response, metabolism, growth and development, signal transduction, photosynthesis, cell cycle, and cell wall organization	[56]

These studies comparatively analyzed the leaf proteome of phytoplasma-infected and that of healthy plants

recent evidence suggests that they might be involved in defense against some pathogens and insects [59–61].

In a complementary study, Monavarfeshani et al. [56] applied label-free quantitative shotgun proteomics to further explore proteomic changes associated with WBDL in Mexican lime. Out of 990 proteins quantified in this study 448 showed reproducible changes in abundance in diseased compared to healthy plants. In agreement with Taheri's findings, ROS scavenging enzymes including APX (both cytosolic and chloroplastic isoforms), dehydroascorbate reductase 1, glutathione peroxidase (GPX), Mn-SOD, and catalase (CAT) were decreased in phytoplasma challenged plants. Similarly, in mulberry plants infected by phytoplasma, the activity of ROS scavenging enzymes including SOD, CAT, and peroxidase decreased [53]. However, in grapevines infected by flavescence dore'e phytoplasma, the abundance of the same set of ROS detoxifying enzymes increased [55], suggesting that the response of oxidative stress related proteins largely depends on plant species and degree of susceptibility to phytoplasma infection.

Enzymes associated with cell wall modification showed an increased abundance in phytoplasma diseased plants including β-1,4-endoglucanase, cellulase, pectin methylesterases, and expansins [54, 56], which was also in accordance with the transcriptome data [22, 40, 43]. As discussed earlier in this chapter, host cell wall modification is thought to facilitate phytoplasma infection. Interestingly, changes in abundance and activities of pectin methylesterases and expansins have been shown to be associated with increased susceptibility to pathogen infection [62, 63].

An interesting finding of Monavarfeshani's study, which also agreed with transcriptome data was extensive down regulation of proteins involved in degradation and turnover of proteins. These include 16 proteases and 5 subunits of proteasome degradation machinery. Down-regulation of components of proteasome has also been reported in other phytoplasma infected plants [55]. The plant proteolytic system plays an important role in removal of damaged, misfolded, or those proteins whose function is no longer required. In addition, it might be involved in degradation of pathogen specific proteins as part of plant defense mechanisms. How down-regulation of proteolysis associated proteins contributes to plant-phytoplasma interaction is unknown. However, evidence suggests that during evolution, some pathogens have acquired the ability to disrupt host plant proteasomal system to facilitate their own multiplication [64]. This suggests that phytoplasmas might have acquired the same strategy to overcome their host plants.

Proteins related to signaling pathways were also subjected to modification in phytoplasma-infected plants. Interestingly, the abundance of three leucine-rich repeat (LRR) receptors that are known to play a critical role in plant innate immunity response was increased in phytoplasma diseased plants [56]. Likewise, some defense related proteins including thaumatin-like and osmotin-like proteins as well as PR17 were increased in phytoplasma-infected plants [53, 54].

Phytoplasma infection was also associated with decreased abundance of heat shock proteins (HSPs) and chaperone-like AAA^+ family proteins [53, 56, 57]. In plants, HSPs play an important role in signal transduction pathways leading to the establishment of defense against pathogens infection. A typical example is the suppression of the expression of HSP70 and HSP90 in *Nicotiana benthamiana* that compromises both the HR and non-host resistance [65, 66]. Thus, down-regulation of HSPs in diseased plants might be associated with increased susceptibility to phytoplasma infection. Within the AAA^+ family proteins, RuBisCo activase is a critical chaperone-like protein that plays an important role in maintaining RuBisCo in an active conformation. In mulberry plants infected with phytoplasma, RuBisCo activase showed an extensive degradation, a phenomenon observed under other stresses as well [53].

The abundance of photosynthesis-related proteins was markedly decreased in phytoplasma-infected plants suggesting that photosynthesis is among the processes to be significantly affected by phytoplasma infection [53, 54, 56, 57]. This is also in accordance with physiological measurements indicating suppressed photosynthetic capacity of phytoplasma-infected plants [30, 53]. This reduction might be largely due to the feedback inhibition triggered by the accumulation of carbohydrate assimilates in photosynthetic leaves that itself is the consequence of the blockage of phloem seive elements by phytoplasmas. In addition, declined photosynthesis in phytoplasma diseased plants is also associated with considerable reduction in photosynthetic pigments, including chlorophyll a, chlorophyll b, and carotenoids as well as decreased activity of PSII [29, 53]. Decreased content of photosynthetic pigments results in overall yellowing and chlorosis of leaves, which are common symptoms observed in phytoplasma diseased plants. Likewise, phytoplasma infection was associated with the degradation of photosynthesis-related proteins including RuBisCo activase, sedoheptulose-1,7-bisphosphatase, and RuBisCo large subunit [53]. This degradation was likely resulted from the accumulation of ROS which is a common phenomenon associated with phytoplasma infection.

Proteomics has also proven change in carbohydrate metabolism due to phytoplasma infection. Particularly, proteins involving in starch metabolism including starch synthase, starch branching enzyme II, isoamylase-type starch-debranching enzyme 2, starch phosphorylase, and starch ADP/ATP carrier proteins decreased in response to phytoplasma infection [56]. However, the abundance of enzymes such as sucrose synthase and fructokinase increased in phytoplasma-infected plants. As discussed earlier in this chapter, sucrose synthase and fructokinase provide phosphorylated forms of carbohydrate monomers as preferred carbon sources for phytoplasmas.

References

1. Sugio A, Maclean AM, Kingdom HN, Grieve VM, Manimekalai R, Hogenhout SA (2011) Diverse targets of phytoplasma effectors: from plant development to defense against insects. Annu Rev Phytopathol 49:175–195
2. Bertaccini A (2007) Phytoplasmas: diversity, taxonomy, and epidemiology. Front Biosci 12:673–689
3. Ludwig W, Schleifer K-H, Whitman WB (2009) Revised road map to the phylum Firmicutes. In: Bergey's manual® of systematic bacteriology. Springer, Berlin, pp 1–13
4. Weisburg WG, Tully JG, Rose DL, Petzel JP, Oyaizu H, Yang D et al (1989) A phylogenetic analysis of the mycoplasmas: basis for their classification. J Bacteriol 171:6455–6467
5. Weintraub PG, Beanland L (2006) Insect vectors of phytoplasmas. Annu Rev Entomol 51:91–111
6. Group IPSWT–PT (2004) '*Candidatus* Phytoplasma', a taxon for the wall-less, non-helical prokaryotes that colonize plant phloem and insects. Int J Syst Evol Microbiol 54:1243–1255

7. Christensen NM, Axelsen KB, Nicolaisen M, Schulz A (2005) Phytoplasmas and their interactions with hosts. Trends Plant Sci 10:526–535
8. Oshima K, Kakizawa S, Nishigawa H, Jung HY, Wei W, Suzuki S et al (2004) Reductive evolution suggested from the complete genome sequence of a plant-pathogenic phytoplasma. Nat Genet 36:27–29
9. Bai X, Zhang J, Ewing A, Miller SA, Jancso Radek A, Shevchenko DV et al (2006) Living with genome instability: the adaptation of phytoplasmas to diverse environments of their insect and plant hosts. J Bacteriol 188:3682–3696
10. Tran-Nguyen LT, Kube M, Schneider B, Reinhardt R, Gibb KS (2008) Comparative genome analysis of "*Candidatus* Phytoplasma australiense" (subgroup tuf-Australia I; rp-A) and "*Ca.* Phytoplasma asteris" Strains OY-M and AY-WB. J Bacteriol 190:3979–3991
11. Kube M, Schneider B, Kuhl H, Dandekar T, Heitmann K, Migdoll AM et al (2008) The linear chromosome of the plant-pathogenic mycoplasma '*Candidatus* Phytoplasma mali'. BMC Genomics 9:306
12. Bertaccini A, Duduk B (2010) Phytoplasma and phytoplasma diseases: a review of recent research. Phytopathol Mediterr 48:355–378
13. Makarova O, Maclean AM, Nicolaisen M (2015) Phytoplasma adapt to the diverse environments of their plant and insect hosts by altering gene expression. Physiol Mol Plant Pathol 91:81–87
14. Hogenhout SA, Oshima K, El Ammar D, Kakizawa S, Kingdom HN, Namba S (2008) Phytoplasmas: bacteria that manipulate plants and insects. Mol Plant Pathol 9:403–423
15. Kube M, Mitrovic J, Duduk B, Rabus R, Seemuller E (2012) Current view on phytoplasma genomes and encoded metabolism. Sci World J 2012:185942
16. Strauss E (2009) Microbiology. Phytoplasma research begins to bloom. Science 325:388–390
17. Sugio A, Hogenhout SA (2012) The genome biology of phytoplasma: modulators of plants and insects. Curr Opin Microbiol 15:247–254
18. Oshima K, Maejima K, Namba S (2013) Genomic and evolutionary aspects of phytoplasmas. Front Microbiol 4:230
19. Hoshi A, Oshima K, Kakizawa S, Ishii Y, Ozeki J, Hashimoto M et al (2009) A unique virulence factor for proliferation and dwarfism in plants identified from a phytopathogenic bacterium. Proc Natl Acad Sci USA 106:6416–6421
20. Sugio A, Kingdom HN, Maclean AM, Grieve VM, Hogenhout SA (2011) Phytoplasma protein effector SAP11 enhances insect vector reproduction by manipulating plant development and defense hormone biosynthesis. Proc Natl Acad Sci USA 108:E1254–E1263
21. Himeno M, Neriya Y, Minato N, Miura C, Sugawara K, Ishii Y et al (2011) Unique morphological changes in plant pathogenic phytoplasma-infected petunia flowers are related to transcriptional regulation of floral homeotic genes in an organ-specific manner. Plant J 67:971–979
22. Mardi M, Karimi Farsad L, Gharechahi J, Salekdeh GH (2015) In-depth transcriptome sequencing of mexican lime trees infected with *Candidatus* Phytoplasma aurantifolia. PLoS ONE 10:e0130425
23. Liu LY, Tseng HI, Lin CP, Lin YY, Huang YH, Huang CK et al (2014) High-throughput transcriptome analysis of the leafy flower transition of *Catharanthus roseus* induced by peanut witches'-broom phytoplasma infection. Plant Cell Physiol 55:942–957
24. Maclean AM, Orlovskis Z, Kowitwanich K, Zdziarska AM, Angenent GC, Immink RG et al (2014) Phytoplasma effector SAP54 hijacks plant reproduction by degrading MADS-box proteins and promotes insect colonization in a RAD23-dependent manner. PLoS Biol 12: e1001835
25. Rumpler F, Gramzow L, Theissen G, Melzer R (2015) Did convergent protein evolution enable phytoplasmas to generate 'Zombie plants'? Trends Plant Sci 20:798–806
26. Du Toit A (2014) Bacterial pathogenicity: phytoplasma converts plants into zombies. Nat Rev Microbiol 12:393

27. Musetti R, Buxa SV, De Marco F, Loschi A, Polizzotto R, Kogel KH et al (2013) Phytoplasma-triggered Ca(2+) influx is involved in sieve-tube blockage. Mol Plant Microbe Interact 26:379–386
28. Maust BE, Espadas F, Talavera C, Aguilar M, Santamaria JM, Oropeza C (2003) Changes in carbohydrate metabolism in coconut palms infected with the lethal yellowing phytoplasma. Phytopathology 93:976–981
29. Tan Y, Wei HR, Wang JW, Zong XJ, Zhu DZ, Liu QZ (2015) Phytoplasmas change the source-sink relationship of field-grown sweet cherry by disturbing leaf function. Physiol Mol Plant Pathol 92:22–27
30. Gai YP, Han XJ, Li YQ, Yuan CZ, Mo YY, Guo FY et al (2014) Metabolomic analysis reveals the potential metabolites and pathogenesis involved in mulberry yellow dwarf disease. Plant Cell Environ 37:1474–1490
31. Paul MJ, Foyer CH (2001) Sink regulation of photosynthesis. J Exp Bot 52:1383–1400
32. Junqueira A, Bedendo I, Pascholati S (2004) Biochemical changes in corn plants infected by the maize bushy stunt phytoplasma. Physiol Mol Plant Pathol 65:181–185
33. Musetti R, Di Toppi LS, Ermacora P, Favali MA (2004) Recovery in apple trees infected with the apple proliferation phytoplasma: an ultrastructural and biochemical study. Phytopathology 94:203–208
34. Mollayi S, Zadali R, Farzaneh M, Ghassempour A (2015) Metabolite profiling of Mexican lime (*Citrus aurantifolia*) leaves during the progression of witches' broom disease. Phytochem Lett 13:290–296
35. Bouche N, Fromm H (2004) GABA in plants: just a metabolite? Trends Plant Sci 9:110–115
36. Mehta A, Brasileiro AC, Souza DS, Romano E, Campos MA, Grossi-De-Sa MF et al (2008) Plant-pathogen interactions: what is proteomics telling us? FEBS J 275:3731–3746
37. O'brien JA, Daudi A, Butt VS, Bolwell GP (2012) Reactive oxygen species and their role in plant defence and cell wall metabolism. Planta 236:765–779
38. Mark GL, Dow JM, Kiely PD, Higgins H, Haynes J, Baysse C et al (2005) Transcriptome profiling of bacterial responses to root exudates identifies genes involved in microbe-plant interactions. Proc Natl Acad Sci USA 102:17454–17459
39. Moy P, Qutob D, Chapman BP, Atkinson I, Gijzen M (2004) Patterns of gene expression upon infection of soybean plants by *Phytophthora sojae*. Mol Plant Microbe Interact 17:1051–1062
40. Hren M, Nikolic P, Rotter A, Blejec A, Terrier N, Ravnikar M et al (2009) 'Bois noir' phytoplasma induces significant reprogramming of the leaf transcriptome in the field grown grapevine. BMC Genomics 10:460
41. Eulgem T (2005) Regulation of the Arabidopsis defense transcriptome. Trends Plant Sci 10:71–78
42. Fan G, Cao X, Niu S, Deng M, Zhao Z, Dong Y (2015) Transcriptome, microRNA, and degradome analyses of the gene expression of Paulownia with phytoplamsa. BMC Genomics 16:896
43. Albertazzi G, Milc J, Caffagni A, Francia E, Roncaglia E, Ferrari F et al (2009) Gene expression in grapevine cultivars in response to Bois Noir phytoplasma infection. Plant Sci 176:792–804
44. Jones JD, Dangl JL (2006) The plant immune system. Nature 444:323–329
45. Jones DA, Takemoto D (2004) Plant innate immunity – direct and indirect recognition of general and specific pathogen-associated molecules. Curr Opin Immunol 16:48–62
46. Coll NS, Epple P, Dangl JL (2011) Programmed cell death in the plant immune system. Cell Death Differ 18:1247–1256
47. Dixon RA, Achnine L, Kota P, Liu CJ, Reddy MS, Wang L (2002) The phenylpropanoid pathway and plant defence—a genomics perspective. Mol Plant Pathol 3:371–390
48. Cheng AX, Lou YG, Mao YB, Lu S, Wang LJ, Chen XY (2007) Plant terpenoids: biosynthesis and ecological functions. J Integr Plant Biol 49:179–186
49. Daviere JM, Achard P (2013) Gibberellin signaling in plants. Development 140:1147–1151

50. Ding Y, Wu W, Wei W, Davis RE, Lee IM, Hammond RW et al (2013) Potato purple top phytoplasma-induced disruption of gibberellin homeostasis in tomato plants. Ann Appl Biol 162:131–139
51. Robert-Seilaniantz A, Grant M, Jones JD (2011) Hormone crosstalk in plant disease and defense: more than just jasmonate-salicylate antagonism. Annu Rev Phytopathol 49:317–343
52. Zhang Z, Wu S, Stenoien DL, Pasa-Tolic L (2014) High-throughput proteomics. Annu Rev Anal Chem (Palo Alto Calif) 7:427–454
53. Ji X, Gai Y, Zheng C, Mu Z (2009) Comparative proteomic analysis provides new insights into mulberry dwarf responses in mulberry (*Morus alba* L.). Proteomics 9:5328–5339
54. Margaria P, Abba S, Palmano S (2013) Novel aspects of grapevine response to phytoplasma infection investigated by a proteomic and phospho-proteomic approach with data integration into functional networks. BMC Genomics 14:38
55. Margaria P, Palmano S (2011) Response of the *Vitis vinifera* L. cv. 'Nebbiolo' proteome to Flavescence doree phytoplasma infection. Proteomics 11:212–224
56. Monavarfeshani A, Mirzaei M, Sarhadi E, Amirkhani A, Khayam Nekouei M, Haynes PA et al (2013) Shotgun proteomic analysis of the Mexican lime tree infected with "*Candidatus* Phytoplasma aurantifolia". J Proteome Res 12:785–795
57. Taheri F, Nematzadeh G, Zamharir MG, Nekouei MK, Naghavi M, Mardi M et al (2011) Proteomic analysis of the Mexican lime tree response to "*Candidatus* Phytoplasma aurantifolia" infection. Mol BioSyst 7:3028–3035
58. Liu X, Huang B, Lin J, Fei J, Chen Z, Pang Y et al (2006) A novel pathogenesis-related protein (SsPR10) from Solanum surattense with ribonucleolytic and antimicrobial activity is stress- and pathogen-inducible. J Plant Physiol 163:546–556
59. Tsukuda S, Gomi K, Yamamoto H, Akimitsu K (2006) Characterization of cDNAs encoding two distinct miraculin-like proteins and stress-related modulation of the corresponding mRNAs in Citrus jambhiri lush. Plant Mol Biol 60:125–136
60. Mozoruk J, Hunnicutt LE, Cave RD, Hunter WB, Bausher MG (2006) Profiling transcriptional changes in *Citrus sinensis* (L.) Osbeck challenged by herbivory from the xylem-feeding leafhopper Homalodisca coagulata (Say) by cDNA macroarray analysis. Plant Sci 170:1068–1080
61. Maserti BE, Del Carratore R, Croce CM, Podda A, Migheli Q, Froelicher Y et al (2011) Comparative analysis of proteome changes induced by the two spotted spider mite *Tetranychus urticae* and methyl jasmonate in citrus leaves. J Plant Physiol 168:392–402
62. Pelloux J, Rusterucci C, Mellerowicz EJ (2007) New insights into pectin methylesterase structure and function. Trends Plant Sci 12:267–277
63. Fu J, Liu H, Li Y, Yu H, Li X, Xiao J et al (2011) Manipulating broad-spectrum disease resistance by suppressing pathogen-induced auxin accumulation in rice. Plant Physiol 155:589–602
64. Citovsky V, Zaltsman A, Kozlovsky SV, Gafni Y, Krichevsky A (2009) Proteasomal degradation in plant-pathogen interactions. Semin Cell Dev Biol 20:1048–1054
65. Kanzaki H, Saitoh H, Ito A, Fujisawa S, Kamoun S, Katou S et al (2003) Cytosolic HSP90 and HSP70 are essential components of INF1-mediated hypersensitive response and non-host resistance to *Pseudomonas cichorii* in *Nicotiana benthamiana*. Mol Plant Pathol 4:383–391
66. Lu R, Malcuit I, Moffett P, Ruiz MT, Peart J, Wu AJ et al (2003) High throughput virus-induced gene silencing implicates heat shock protein 90 in plant disease resistance. EMBO J 22:5690–5699

Chapter 12
Insect Pest Proteomics and Its Potential Application in Pest Control Management

Mehdi Dastranj, Javad Gharechahi and Ghasem Hosseini Salekdeh

Abstract Proteomics has been considered as a high-throughput approach for large-scale analysis of proteins particularly their structures and functions. Thanks to the recent advances in mass spectrometry for protein identification, entomologists have greatly benefited from proteomics to unravel the molecular mechanisms behind insect feeding, diapause, metamorphosis, vitellogenesis, embryogenesis, etc. This capability is particularly exploited for the development of novel bio-pesticides in which identification of new target molecules is considered critical. Genome annotation and phylogenetic studies are the other entomological applications of proteomics. Current mass spectrometry-based proteomics strategies have enabled researchers to reproducibly, accurately, quantitatively, and comprehensively survey on proteome content from cells and tissues to whole body of an organism. Here, proteomics development in diverse insect pests and its potential application in insect pest management have been reviewed.

Keywords Proteomics · Mass spectrometry · Insect pest management · Gene discovery · Bio-rational pesticides

---

M. Dastranj · G.H. Salekdeh (✉)
Department of Systems Biology, Agricultural Biotechnology Research Institute of Iran, Agricultural Research, Education, and Extension Organization, Karaj, Iran
e-mail: hsalekdeh@yahoo.com

M. Dastranj
Department of Plant Protection, College of Agriculture and Natural Resources, University of Tehran, Karaj, Iran

J. Gharechahi
Chemical Injuries Research Center, Baqiyatallah University of Medical Sciences, Tehran, Iran

G.H. Salekdeh (ed.), *Agricultural Proteomics Volume 2*,
DOI 10.1007/978-3-319-43278-6_12

12.1 Introduction

Insects are considered as the largest and remarkably-diverse group of organisms in the world accounting for 75 % of all known species [1]. Moreover, they play numerous roles in different ecosystems around the globe because of their enormous diversity in form, function, and life style, as well as their interactions with other organisms [2]. Some insect species particularly those involved in pollination [3], honey production [4], silk production [5], predation, and parasitism [6], are brought about great benefits to human beings. On the other hand, some insect species can negatively affect human activities by transmission of diseases and phytophagy which cause considerable damages to agricultural production [7, 8].

The application of chemical insecticides is the main approach used for insect pest control. However, pesticide application has raised concerns regarding the risks posed to the ecosystem by perturbing natural and/or environmental balance and due to the high persistence of these compounds as well as their toxicity to non-target organisms [9]. Public and academic concerns over the use of pesticides with respect to food safety, especially about their residual toxicity in contaminated crops on one hand and the induction of resistance in insects on the other hand have motivated research on the development of novel alternative pest control options [10].

Studying insect omics including transcriptomics, proteomics, peptidomics, and metabolomics has been shown to hold great promises for discovery of novel and species-specific target molecules. Once a specific target is identified, a disrupting agent can be designed and used as bio-rational insecticide. These disrupting agents are known to interfere with physiological functions specific to a group of insects and are characterized by their improved selectivity and environmentally friendly nature marking them ideal biopesticides [1].

The development of high-throughput genome sequencing technologies and advances in mass spectrometry-based proteomics have now enabled investigation of complex organisms, from tissue to whole body. Since the genome of *Drosophila* was first sequenced, many great advances in the field of insect physiology and endocrinology have been achieved [11]. The genome sequences of several other insect pests including red flour beetle (*Tribolium castaneum*), pea aphid (*Acyrthosiphon pisum*), and the mountain pine beetle (*Dendroctonus ponderosae*) are now available and significance of such data is explained by the i5k project (http://www.arthropodgenomes.org/wiki/i5K) whose goal is to sequence the genome of 5000 insects and arthropods. Since the 5ki project was launched in 2011 to address genome sequencing of most insect pest in the world, proteomics, a prominent tool for genome annotation, have also been used for targeted and genome-wide analysis of proteins complement of the genomes. In fact, insect proteomics is quickly developing, wherein attempts are being made to analyze organellar or tissue proteome of insects. For instance, insect proteomics has been used to explore hemolymph proteins following feeding, immune-challenging, and parasitism, insect-plant interaction, molting fluid content, components of the cuticle, and to monitor toxicity of pesticides (Table 12.1) [12, 13]. Insect proteomics

Table 12.1 insect pest proteomics

Insect	Tissue	Developmental stage	Method	Detected proteins	Biological question	References
Spodoptera littoralis	Hemolymph	Larva	2-DE, Edman degradation	16	Changes in the larval haemolymph proteome induced by parasitoid	[30]
Dendroctonus ponderosae	Whole body	Larva	iTRAQ	1507	Physiological and biochemical determinants of overwintering capacity	[53]
Tribolium casteneum	Midgut	Larva	2-DE, MALDI-MS	10	Response/symptoms to a benzoylphenyl urea insecticide	[13]
Tribolium casteneum	Female spermatheca	Adult	1-D, MALDI TOF/TOF MS/MS	13	Enzyme determinants of sperm protection in seminal fluid	[70]
	Male seminal vesicles			61		
Tribolium casteneum	Gut	Larva	2-DE, MALDI-TOF	37	Characterization of midgut protein	[46]
			2-DE, Nano-HPLC and ESI-MS/MS	98		
Phaedon cochleariae	Midgut lumen	Larva	2-DE, LC-MS/MS	11	Plant cell wall-degrading enzymes	[47]
Heliconius melpomene	Saliva	Adult	1-D, ion trap LC-MS/MS	30	Identification of saliva proteins	[22]
Plutella xylostella	Midgut	Larva	2D-DIGE, MALDI-TOF, TOF MS	31	Midgut proteins related with Cry1Ac resistance	[39]
Helicoverpa armeigera	Midgut	Larva	2-DE, MS/MS	40	Identification of major gut lumen proteins	[23]

(continued)

Table 12.1 (continued)

Insect	Tissue	Developmental stage	Method	Detected proteins	Biological question	References
Macrosiphum euphorbiae	Whole body	Adult (wingless and winged)	2-D, PMF or LC-MS/MS	22	Effects of fluctuating heat and UV stress regimes of the potato aphids	[59]
Sitodiplosis mosellana	Whole body	Larva	2-DE, MALDI-TOF-MS	7	Molecular mechanisms of wheat blossom midge diapause	[74]
Macrosiphum euphorbiae	Whole body	Adult	2-DE, SELDI-TOF/MS or LC-MS/MS	24	Molecular responses of aphid to parasitoids	[60]
Helicoverpa armeigera	Peritrophic matrix	Larva	LC-MS/MS	41	Identification of peritrophic matrix proteins	[25]
Plutella xylostella	Hemolymph	Larva	2-DE, Q-TOF MS/MS	26	Identification of plasma proteome after parasitism	[29]
Manduca sexta	Hemolymph	Larva	1-DE and 2-DE, LC-MS	123	Identification of hemolymph proteome	[32]
Manduca sexta	Sperm	Adult	1-D, LC-MS/MS	896	Identification of heteromorphic sperm proteome	[35]
Aphis gossypii	Whole body	Adult	2-DE, MALDI-TOF/MS	26	protein involved in spirotetramat resistance	[58]
Aphis gossypii	Honeydew	Adult	2D-DIGE, MALDI/MS	98	The proteome content of honeydew	[62]
Nilavaparta lugens	Whole body	Adult (high-fecundity populations)	2-DE, MALDI-TOF/TOF MS	39	Mechanism of the insect fecundity	[15]
		Adult (low-fecundity populations)		54		

(continued)

Table 12.1 (continued)

Insect	Tissue	Developmental stage	Method	Detected proteins	Biological question	References
Frankliniella occidentalis	Whole body	Adult	2-DE, MALDI-TOF/TOF MS	37	Mechanism of viral infection	[64]
Locusta migratoria	–	Egg	iTRAQ, LC-ESI-MS/MS	1078	Molecular mechanism of diapause	[71]
Helicoverpa armigera	Brain	Larva	2-DE, MALDI-TOF	37	Molecular mechanism of diapause	[14]
Sesamia nonagrioides	Hemolymph	Larva	2-DE, MALDI-TOF/TOF MS/MS	11	Molecular mechanism of diapause	[75]

analyses have now greatly improved our understanding about the molecular mechanism of insect diapause, metamorphosis, vitellogenesis, embryogenesis, and their regulation (Fig. 12.1) [14–16]. Combining proteomics findings with those of the other omics as well as physiological data would provide an integrated systems biology view of signaling, regulatory, and metabolic networks governing insect-specific processes [17].

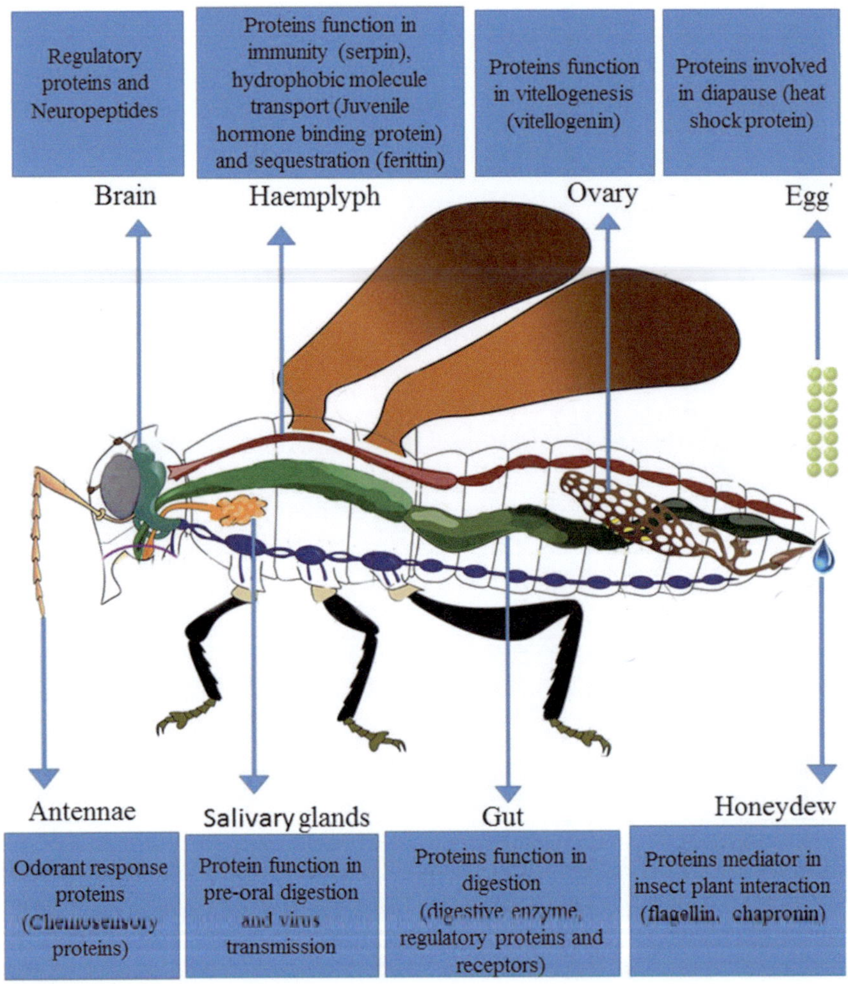

Fig. 12.1 Insect pest proteomics. Proteomics has been extensively utilized for studying insect proteins of different tissue origins. Insect brain proteins have been explored for the identification of neuropeptides and also diapause-related proteins. Hemolymph proteome has been extensively explored for the identification of metamorphosis, diapause, and viral transmission related proteins. Insect gut proteins have been studied for the identification of digestive enzymes, as well as regulatory and receptor proteins. Insect egg proteins have been studied for the characterization of diapause specific proteins

The conservativity of the protein sequences can be exploited for functional annotation of proteins in insect species of limited genome sequence information. In addition, homology-driven proteomics approaches have also been used for phylogenetic analysis and taxonomic classification of certain taxa [18]. For this purpose, highly-conserved proteins/peptides are employed for reconstruction of deep phylogenetic tree of different species and for achieving a closer insight into their evolutionary relationships. For instance, proteomics analysis showed that insect neuropeptides are evolutionary-conserved proteins that can be used for phylogenetic analysis, in particular, for complementing molecular, morphological, and other taxonomic data [18].

In this chapter, we briefly review proteomics findings in diverse insect pest systems, highlighting the achievements obtained in multiple insect processes with an especial emphasis on the identification of target proteins for the development of novel environmentally-friendly biopesticides.

12.1.1 Lepidopteran Pest Proteomics

The order lepidoptera comprises more than 160,000 species of butterflies and moths, characterized by their complete metamorphosis, during which they go through four distinct stages namely, egg, larva (caterpillar), pupa, and adult. During the larva stage of growth, they largely feed on plant materials by chewing or biting green or non-green plant tissues. On the other hand, adult insects of this order are largely anthophilous, meaning that they develop a specific mouthpart, proboscis, for imbibing nectar and other liquid substances from flowers. The role of lepidopteran as pollinators in the ecosystem has been well established and there are several cases in which insect proboscis has been adapted to match the morphology of the flower [19].

Unlike other butterflies that feed solely on floral nectar, those belonging to the neotropical genus *Heliconius* have been evolved to feed on both nectar and pollen [20]. They collect and store pollens on the surface of their proboscis during nectar feeding. Pollens are then moisturized with the salvia exuded from the proboscis and are subsequently disrupted and their released amino acids are consumed by the insect [21]. The molecular mechanism of pollen feeding behavior in these insects has largely remained unknown. To elucidate the role of insect-secreted proteins in the degradation and consumption of pollen-derived nutrients, Harpel et al. [22] employed a shotgun proteomics approach to identify saliva proteins in *H. melpomene*. Using a combined SDS-PAGE protein fractionation and liquid chromatography-tandem mass spectrometry (LC-MS/MS), 31 proteins were identified, most of which were secreted proteins containing signal peptides. It should be noted that salvia proteins are mainly involved in proteolysis and carbohydrate hydrolysis to facilitate the absorption of the proteins and carbohydrate contained in nectar and pollen by the insect.

Despite the positive role of adult lepidopteran in pollination, their larvae or caterpillars cause significant yield losses in various crop species and are considered as problematic pests in agriculture. Most caterpillars are herbivorous animals whose digestive tracts have been evolved to aquire high amounts of plant materials in order to meet their high energy requirements caused by their fast growth rate [23]. The digestive system of these insects is a simple tube that has been structurally adapted based on the type of the food they consume. Proteins secreted into the digestive tube of herbivores insects are mainly involved in digestion, detoxification, sequestration, and defense against plant anti-herbivory mechanisms [23]. Proteomics has been applied to explore secreted midgut lumen proteins in caterpillars of *Helicoverpa armigera* with an aim to identify stable gut proteins involved in insect nutrition [23]. Two-dimensional electrophoresis (2-DE) coupled with MS/MS analysis enabled the identification of major insect secreted gut proteins involved in the digestion of proteins, carbohydrates, and lipids. In addition, proteomic analysis showed the presence of stable non-digestive proteins such as lipocalin, an arginine kinase, as well as proteins of unknown function in the lumen, where is considered to has a harsh environment.

Similar to many other arthropods, the digestive tube of lepidopteran contains a peritrophic matrix (PM) composed of chitinous fibers cross-linked by proteins [24]. The major function of this semi-permeable membrane is to protect the gut cells from abrasion by food particles and pathogens. It also functions as support for immobilizing digestive enzymes and toxins and to compartmentalize the gut into smaller endo- and ecto-peritrophic spaces. MS-based proteomics analysis allowed to identify PM proteins of the midgut in the caterpillar, *H. armigera* following chitin depolimerization using anhydrous trifluoromethanesulfonic acid treatment [25]. Chitin deacetylase-like and mucin-like proteins were identified as major protein constituents of the PM, indicating their critical role in PM structure and function. In addition, several polycalin and aminopeptidase proteins were also detected in the PM fraction. These proteins are known to bind to *Bacillus thurengiensis* Cry1A crystal toxin proteins and contribute to the insect immunity against pathogens [26].

Insects use innate immunity for effective defense against foreign pathogenic microorganisms, viruses, and parasites. To overcome insects immunity, some endoparasites suppress the host defense system and elicite profound changes in metabolism, food consumption, pigmentation, and even development and behavior of the host insect [27]. For example, when *Plutella xylostella*, a pest of cruciferous plants, is parasitized by *Cotesia plutellae*, its innate immunity is significantly suppressed, the development of the larvae is delayed, and the insect will not undergo metamorphosis, suggesting that the parasite reprograms host developmental processes [28]. Therefore, the study of host-parasite interaction could provide a deeper insight into the mechanisms which could be used to develop novel biopesticides targeting immunity and developmental processes in insect pests. To achieve this Song et al. [29] applied 2-DE and MS/MS analysis to explore plasma proteome of *P. xylostella* larvae following parasitism by *C. plutellae*. Proteomics analysis revealed that the majority of parasitism responsive proteins were involved in the modulation of immune response, including immune-related proteins, serpin 2, protease inhibitor, translationally controlled tumor protein, histidine kinase, and apolipophorin-III.

The egg–larval parasitoid *Chelonus inanitus* employs polydnavirus and venom injection to establish parasitism and to manipulate the host physiology. The infection of *Spodoptera littoralis* with the parasite *C. inanitus* is followed by developmental changes including early onset metamorphosis and failure to complete prepupal stage. Proteomic analysis of the haemolymph of *S. littoralis* upon parasitism with *C. inanitus* showed minor differences in the haemolymph proteome and only several minor abundant viral or viral-induced proteins were detected [30].

Among lepidopteran, tobacco hornworm, *Manduca sexta*, is an excellent insect model organism for studying physiological, developmental, and metamorphosis-associated processes in insects [31]. To understand the biology of this damaging insect pest, the hemolymph proteome of fifth instar larvae was analysed using one or 2-DE based proteomics [32]. LC-MS analysis allowed to identify 58 non-redundant proteins in the hemolymph of *M. sexta*, most of which were related to defense, transport and metabolism, storage, and metamorphosis processes. Some lepidopteran species undergo an especial type of spermatogenesis in which male insects produce two sperm morphs (heteromorphism), i.e., eupyrene; a nucleated and less abundant fertilization competent form, and apyrene; an anucleated high abundant fertilization incompetent type (accounting for up to 96–99 % of the sperm produced) [33, 34]. To understand the sperm proteome composition and to identify sperm proteins contributing to the formation of sperms with varying fecundity potential, quantitative label-free proteomics was conducted using *M. sexta* as a model of heteromorphic spermatogenesis [35]. A significant proportion of *M. sexta* sperm proteins were found evolutionary conserved displaying sequence homology with proteins of non-lepidopteran origin. However, lepidopteran-specific sperm proteins were among the highly abundant proteins, suggesting their critical role in heteromorphic spermatogenesis, in particular, in apyrene sperm functionality.

As an alternative to the application of chemical insecticides in pest management programs, microbial-based strategies, mainly dominated by *B. thuringiensis* (*Bt*), have gained popularity largely due to their environmentally friendly nature [36]. Applications of *Bt* against lepidopteran pests are largely based on either the formulations of spore-crystal or the generation of genetically modified (GM) plants expressing *Bt* toxins [37]. Although, *Bt* protects agricultural crops against lepidopteran and coleopteran pests, insect resistance to *Bt* toxins in the field have been reported for several lepidopteran species [38]. Understanding the molecular mechanisms of resistance to *Bt* toxins could provide opportunities to postpone the development of *Bt* resistance in insects and to extend the sustainability of the engineered GM crops. To explore the proteome changes in Cry1Ac susceptible and resistant *P. xylostella* larvae, midgut proteins were analyzed using two-dimensional difference in-gel electrophoresis (2D-DIGE) and ligand blotting [39]. Among the identified proteins, ATP-binding cassette (ABC) transporter subfamily G member 4 (ABCG4), trypsin, heat shock protein 70 (HSP70), vacuolar H+-ATPase, glycosylphosphatidylinositol anchor attachment 1 protein (GAA1), and solute carrier family 30 member 1 (SLC30A1) were reported as being involved in *Bt* resistance.

Additionally, proteins such as ABC transporter subfamily C member 1 (ABCC1), SLC36A1, NADH dehydrogenase iron-sulfur protein 3, prohibitin, and Rap1 GTPase-activating protein 1 were identified as Cry1Ac-binding proteins in Cry1Ac-susceptible *P. xylostella* larvae, indicating their critical role in susceptibility of the insect to CryA1c toxin.

12.1.2 Coleopteran Pest Proteomics

Coleoptera consisting of beetles and weevils are considered to be the largest insect order representing over 360,000 species or 40 % of the known insect species in the world [40]. Members of this order play different roles in various ecosystems, from agricultural pest such as Colorado potato beetle (*Leptinotarsa decemlineata*) and the boll weevil (*Anthonomus grandis*), to natural enemies of the insect pests such as ladybirds (Coccinellidae) [41–43].

The red flour beetle, *T. castaneum*, is a pest of stored products that has been used as an excellent insect model for functional genomics analysis using RNA interfierence (RNAi) in coleopteran insects. Indeed, *T. castaneum* is the first coleopteran whose complete genome sequence became available [44]. The unique feature of *T. castaneum* is related to its systemic and widespread response to RNAi, allowing genome-wide and large scale RNAi screening [45]. Therefore, most omics-based studies for the identification, characterization, and functional analysis of genes in this order are performed in *T. castaneum.*

In a study, comprehensive proteomics and transcriptomics analyses of the red flour beetle elytra and hidwings, representatives of rigid and soft cuticles, respectively, were performed to understand gene expression signatures of structural proteins which may influence the mechanical properties of cuticles [12]. Results demonstrated distinct differences between two tissues in terms of proteome and gene expression profiles. Matrix-assisted laser desorption ionization-time of flight (MALDI-TOF) MS analysis led to the identification of 18 known cuticular proteins (CP) and a single novel low complexity protein enriched by charged residues. Microarray analysis revealed greater differences in gene expression between elytra and hidwings. More specifically, 372 genes showed up to 10-fold variations in abundance. Interestingly, transcripts encoding for cuticular proteins belonging to the Rebers and Riddiford family (CPR) type 2 and low complexity (CPLC) proteins enriched by glycine or proline, were expressed in higher abundance in elytra. While transcripts with greater expression in hidwings mostly belonged to the CPR type 1 and the Tweedle families. Proteomics has also been used for the identification of gut lumen and epithelium-specific proteins in *T. castaneum* [46].

Unlike other herbivorus insects that rarely encode for plant cell-wall degrading enzymes, phytophagous beetles, species belonging to the superfamilies Chrysomeloidea and Curculionoide, have multigene families in their genome that encode for a diverse class of cell-wall degrading enzymes [47–49]. In order to demonstrate whether this cell-wall degrading enzymes are expressed and contribute to the degradation of plant lignocellulosic materials, a proteomics analysis was conducted to identify secreted cell-wall degrading enzymes in larval gut contents of mustard leaf beetle, *Phaedon cochleariae* [47]. Using a combination of gel electrophoresis and activity assay for the detection of cellulose-, pectin- and xylan-degrading enzymes, 13 proteins belonging to three glycoside hydrolase families; i.e., xylanases, polygalacturonases or pectinases, and β-1,4-glucanases or cellulases were identified. In addition, MS analysis also allowed to identify proteins of plant origin including various proteolysis-stable pathogenesis related (PR) and polygalacturonase-inhibiting proteins in the gut content of *P. cochleariae*.

Whole Mountain Pine Beetle (MPB), *Dendroctonus ponderosae*, undergoes physiological stress during host colonization as it has to be able to cooperatively invade and kill its host tree in order to successfully reproduce [50, 51]. To identify changes taking place during host colonization, the proteomes of starved and host-colonized female and male MPBs were monitored using isobaric tags for relative and absolute quantification (iTRAQ)-based proteomics approach [50]. Chaperone proteins and enzymes required for vitellogenesis were changed with feeding in female insects. In male insects, chaperones and motor proteins were responsive to feeding. These results indicated that the reproductive physiology of MPB females were rapidly affected by shifting in biological phase during colonization. Several detoxifying enzymes were also identified in female's proteome including cytochrome P450 suggesting their defensive role in plant-herbivore interaction. The differentially abundant proteins in males during phase change was mainly related to muscle cellular structure which might be due to the fact that they were no longer needed for flight. Altogether, MPB females conserve energy, store it, and then rapidly use the stored energy in their reproductive attempts, unlike males which consume much of their stored energy during the dispersal [50].

MBP overwinters as larvae under the bark of host trees and therefore, must be able to tolerate cold and freezing stresses during its life cycle. The freezing stress tolerance is thought to be mediated by redirecting energy metabolism toward the biosynthesis of cryoprotectants such as glycerol in order to avoid internal ice formation [52]. To uncover the cold hardening mechanism of MBP, the proteome profile of overwintering and developing larvae was compared using iTRAQ-based proteomics [53]. The comparison of the proteome of MBP larvae during autumn cooling and spring warming revealed differential abundance of 33 and 473 proteins, respectively. A closer look into the function of the identified proteins proved the implication of trehalose, 2-deoxyglucose, and antioxidative enzymes in overwintering physiology of MBP, further demonstrating the role of glycerol in cold hardening capacity of this insect.

12.1.3 Hemipteran Pest Proteomics

Hemipteran, which are also referred to as the true bugs, are an order of insects including cicadas, aphids, plant hoppers, leafhoppers, and shield bugs. Most of the hemipteran pests feed on plants through piercing and sucking plant phloem sap. Some insects of this order are considered as important agricultural pests, damaging crops through direct feeding or indirectly by being vectors of serious viral and phytoplasma diseases.

Among the insects of the order hemiptera, aphids constitute more than 4700 species, out of which 250 are considered as agricultural pests both through direct phloem feeding or indirect virus transmission [54, 55]. Like other types of plant-insect interactions, phloem feeding is also associated with plant defense responses that are thought to be mediated by particular molecules known as elicitors. In phloem-feeding insects, the injected saliva proteins may play a critical role in preventing the plant's wound response and may contribute to plant-insect interaction. To identify saliva proteins in *Myzus persicae*, Harmel et al. [55] applied 2-DE coupled with MS analysis. The identification of *M. persicae* proteins was made possible by quering the MS spectral data against the expressed sequence tag (EST) databases. Several proteins such as glucose oxidase, glucose dehydrogenase, NADH dehydrogenase, α-glucosidase, and α-amylase were identified as salvia proteins of *M. persicae*, indicating their possible role in plant-aphid interactions. Some of these proteins are already known as elicitors in various plant-insect interactions and have been detected in the salvia of herbivorous insects *Helicoverpa zea* and *Pieris brassicae* [56, 57].

Intensive application of insecticide against insects could ultimately lead to the development of insecticide resistance. Although genomics has enabled the discovery of genes responsible for many types of insecticide resistance, proteomics can provide additional information regarding the role of proteins, post translational modicications (PTMs), and protein-protein interactions during the development of insecticide resistance. Proteome analysis using a combination of 2-DE and MALDI-TOF MS allowed to identify proteins involved in spirotetramat resistance in cotton aphid, *Aphis gossypii* [58]. To achieve this, the proteome of a spirotetramat-resistance strain and that of a susceptible strain were compared, resulting in the identification of 26 proteins with a two-fold higher or lower adundances in the resistant strain compared with the sensitive one. Acetyl-CoA carboxylase (ACC), HSP70, ubiquitin-conjugating enzyme, fatty acid synthase, and UDP-glucose dehydrogenase were among the candidate proteins involved in the development of resistance to spirotetramat.

Proteomic analysis was also employed to elucidate physiological changes in winged and wingless forms of *Macrosiphum euphorbiae* in response to fluctuating heat and UV [59]. The negative effects of environmental stress observed as reduction in growth and fecundity, were more marked in the wingless than in the winged aphids. The proteome profile showed more variations in winged aphids indicating more metabolic resources and adaptive means for dealing with environmental stresses in this morph. The reduction in aphid performance under heat stress was associated with lower abundances of the enzymes involved in energy metabolism. The higher abundance of exoskeleton proteins in response to heat stress suggested that cuticle barrier enhancement at molting was an aphid adaptation to stressful thermal conditions. The UV had no significant effects on aphids movement, development, and fecundity [59].

Comparative proteomic analysis was also conducted to explore *M. euphorbiae* proteome in response to parasitism by *Aphidius nigripes*, a well-adapted parasitoid, and *Aphidius ervi*, a less-adapted parasitoid [60]. Both parasitoids successfully parasited the host, but *A. nigripes* established a full parasitism and finally killed the host, while the development of *A. ervi* was stopped at the primary egg stage indicating its sensitivity to the host defense mechanisms. Proteomic analysis showed a greater abundance of pro-phenol oxidase and some cuticule proteins in *A. ervi* parasited host compared with *A. nigripes* parasited one, suggesting their critical role in the establishment of host defense against parasitism. In addition, aphid immunity to *A. ervi* parasitism was associated with up-regulation of antioxidative, energy-related, cytoskeleton, and heat shock proteins. On the other hand, susceptibility to *A. nigripes* parasitism was largely correlated with changes in host nutrition and energy metabolism processes.

Aphids and some other hemipteran insects secrete honeydew during phloem feeding. Aphid honeydew is an important food source for many insects including honeybees, wasps, and predatory insects [61, 62]. Aphid honeydew was primarily considered to contain only sugars and amino acids. However, proteomic analysis of *A. pisum* honeydew revealed that it also contained a diverse set of proteins of different origins (i.e., the host aphid and its microbiota) [62]. Among aphid proteins, several enzymes including phosphoglycerate mutase, α-glucosidase, hydroxypyruvate reductase, cathepsin B, inorganic pyrophosphatase, and oxidoreductase were identified. Furthermore, in aphid honeydew monophosphatase, dihydrofolate reductase peroxidase were also detected. Almost half of the identified proteins were homologous with bacterial endosymbiont sequences. Moreover, most of the bacterial protein were associated with amino acid synthesis including acetyl-CoA synthetase, ATP phosphoribosyl-transferase, phosphoserine aminotransferase, and 2-isopropylmalate synthases for lysine, histidine, serine, and leucine production. In addition, aphid honeydew was found to contain proteins like chaperonin, GroEL and Dnak chaperones, elongation factor Tu (EF-Tu), and flagellin which might be involved in plant-aphid interactions.

Fecundity is an important characteristic contributing to the outbreak of insect pests. In natural populations, insects of the same species usually differ in fecundity. To analyze the molecular mechanisms governing the high fecunfity in the brown planthopper (BPH), *Nilavaparta lugens* (Hemipteran: Delphacidae), Zhai et al. [15] applied 2-DE proteomics and transcriptome sequencing to identify fecundity-related proteins and genes. BPH is a damaging pest of rice in both temperate and tropical regions. Proteome analysis showed higher energy metabolisms in the high-fecundity group compared with the low-fecundity group. Among many proteins and transcripts differentially expressed between the low and high-fecundity BPH, vitellogenin, pyruvate dehydrogenase, glutamine synthetase (GS), HSP90, and HSP70 consistently changed in abundance. Of these candidate fecundity-related genes, the role of GS was further analyzed using RNAi. Interestingly, GS-knocked down BPH displayed a significant reduction (64 %) in fecundity, failed in ovary development, and showed limited vitellogenin expression, suggesting a critical role for GS in fecundity of the insect.

12.1.4 Insect Vector Proteomics

Insects, mites, and ticks are the main arthropods involved in the transmission of many plant and animal-disease caused by viruses. Successful disease transmission by insect vectors is achieved by coordinated gene expression, post translational modification (PTMs) of proteins, and a complex network of protein-protein interactions involving insect vector, disease host, and the pathogen. Therefore, to understand the molecular mechanisms underling this complex phenomenon, proteomics can provide valuable information about proteins in each of these interacting organisms involved in the process. Identification of these proteins and their mode of interaction could provide opportunities for targeting disease transmition by insect vectors and for designing novel biorationale vector-based control strategies [63].

The genus *Tospovirus* contains plant-infecting viruses that are exclusively transmitted by insects in the order Tysonoptera. The interaction between tomato spotted wilt virus (TSWV), and its vector, the western flower thrips (*Frankliniella occidentalis*), were investigated using combined transcriptomic and proteomics approaches [64]. Using 2-DE proteomics, the proteome patterns of the infected first-instar larvae (L1) and that of non-infected larvae were compared resulting in the identification of 37 proteins with possible implication in vector-virus interaction. Among the identified proteins integrin, glutaredoxin 5, vitellogenin, and stress-induced phosphoprotein 1 might be involved in virus entry into the insect as well as in vector-virus interaction.

Among the insect vectors, aphids are the most important vectors in terms of the number of viruses they can transmit. Plant virus transmission by insect vectors is accomplished by an extremely species-specific mechanism, so that among many ingested viruses during phloem feeding by aphids, only some viruses are transmitted by the insect [65]. Some viruses can replicate in their vectors (propagative mode) while others are transmitted as virus particles (non-propagative mode). To be transferred to a new host, ingested viruses must be able to cross the hindgut layer and to enter into the hemolymph, from which they are transmitted and concentrated in the basal lamina of the salivary glands of the insect vector [65]. Transmission of virions across insect tissues is thought to be mediated by receptor-mediated endocytosis involving specific viral coat proteins and unknown proteins of insect vector origin [66]. There exists a high degree of variability in terms of the ability to transmit a virus in between the individuals of a population. Identification of proteins that contribute to these differences could lead to a detailed understanding of molecular basis of this complex phenomenon and consequently to the development of novel sustainable virus management practices [63].

Proteomics has been widely used for the identification of *Schizaphis graminarum* proteins contributing to the transmission of viruses of the family Luteoviridae [63, 65, 67, 68]. Cereal yellow dwarf virus-RPV (CYDV-RPV), belonging to the Luteoviridae, is transmitted as intact viral particles by *S. graminarum.* Comparative 2D-DIGE proteomics analysis of transmission-competent and refractive aphids coupled with co-immunopercipitation of virus binding proteins of *S. graminum* revealed the involvement of luciferase and cyclophilin in the transmission of the CYDV-RPV [65, 68]. Cilia et al. [67] applied a similar methodology to explore a F2 generation of *S. graminum* segregating for the capability to transmit CYDV-RPV. Their comparative proteomic analysis revealed the differential abundance of proteins originated from aphid or aphid's obligate bacterial endosymbiont, *Buchnera,* between transmission-competent and noncompetent aphids. Interestingly, it was found that the expression of aphid proteins contributing to the virus transmission phenotype was likely inherited from the competent male parent while changes in the abundance of endosymbiont proteins, which was correlated with the virus transmission phenotype, were likely inherited from the noncompetent females. In addition, differences in the ability to transmit viruses were associated with PTM of proteins and the heterogenity in the *Buchnera* population.

12.1.5 Diapause Specific Proteomics

Diapause is a seasonal adaptation to unfavorable environmental conditions in insects during which the physiological growth is hampered. Arrest in development can occur in diverse embryonic stages, different larval instars, pupae, pharate, and adults, but for any specified species, the potential for diapause is typically limited to a single stage. This process can be regulated through environmental and hormonal responses [69]. A significant reprogramming in gene expression occurs during diapaue and accordingly, many genes are silenced while many others are differentially expressed during the process. The main event before starting diapause is the synthesis of specific proteins which are released into the hemolymph throughout the process and disappear as diapause is terminated [70]. Proteomics has been proven to be a useful tool for understanding insect diapause processes [71].

The migratory locust has an embryonic diapause which is induced by various factors such as low temperature, short photoperiod, and high latitude [72]. To understand the molecular basis of embryonic diapause in migratory locust, *Locusta migratoria* L. (Orthoptera: Acridoidea), Tu et al. [71] applied a combined transcriptome and proteome analyses. Comparison between diapause and non-diapause eggs showed up-regulation of genes such as glutathiones-S-transferase (GST), UDP-glucuronosyl transferase, and transforming growth factor-β-receptor and down-regulation of hemocyanin subunit, hexamerin-like protein, and NADH dehydrogenase both at the transcriptional and translational levels. In addition, Cu/Zn superoxide dismutase (Cu/Zn-SOD), peroxiredoxin-1, juvenile hormone (JH) esterase, the enzyme hydrolyzing JH, were shown to be up-regulated at the protein level in diapause eggs compared with non-diapause eggs. The other proteins up-regulated following diapause included β-glucan synthase; an enzyme catalyzing the transfer of sugar moieties, eukaryotic translation initiation factor 4A; a protein associated with protein translation, and lipase which is essential for the digestion and transport of dietary lipids. Interestingly, transcriptome analysis showed significant up-regulation of genes related to carbohydrate metabolism in eggs following diapause, suggesting their critical role in circumventing energy requirement of diapause and post-diapause stages and also in the cryoprotection of eggs.

In cotton bollworm, *H. armigera*, the signal for diapause, i.e., a short daylength, is received by the larval brain while diapause is induced at the pupal stage [14]. Photoperiod information at the larval stage will be stored and processed in the brain where acts as a diapause program center. There are limited differences between nondiapause- and diapause-destined larvae indicating that a few genes are changed during diapause induction phase. Using a combined proteomics and metabolomics approaches, Zhang et al. [14] explored the molecular mechanisms behind *H. armigera* larval diapause. Metabolic analysis identified several amino acid neurotransmitters including glutamate, GABA, aspartate, glycine, and serine. Interestingly, glutamate content displayed a 1.5-fold increase in diapause-destined larvae suggesting a critical role for glutamate in the perception and the storage of diapause-induced memory signals. An increased content of glucose was also detected in the brain of diapause-destined larvae indicating its key role in memory formation during the diapause induction phase. Diapause-destined larvae also displayed increased contents of terhalose and grycerol, two metabolites involved in cryoprotection. 2-DE coupled to MALDI-TOF MS analysis enabled the identification of 37 diapause-related proteins in the brain of diapause-destined larvae. Interestingly, an increased abundance of GS was detected in diapause-destined larvae, an enzyme involved in removal of excess glutamate which is cytotoxic and causes neuronal damages [73]. Proteins related to metabolisms of carbohydrates, lipid, energy, and amino acids were also changed suggesting that adaptation of general metabolism was required for the successful establishment of diapause.

To uncover wheat blossom midge (*Sitodiplosis mosellana*) diapause proteome, proteins from larvae at different developmental stages including pre-diapause, over-summering diapause, over-wintering diapause, and post-diapause were analyzed using 2-DE coupled to MALDI-TOF MS for protein identification [74]. Among diapause-related proteins, proopiomelanocortin, a precursor polypeptide cleaved to give rise multiple peptide hormones, NADH dehydrogenase subunit 1, and F10F2.5 were significantly decreased during diapause. IKK interacting protein isoform 2 (IKIP2) was also displayed an increased abundance in diapause larvae.

The mechanism underlying larval diapause and development in the corn borer, *Sesamia nonagrioides*, a noctuid pest that feeds on maize, sugar cane, and sorghum, was explored using 2-DE-based proteomics [75]. Similar to *H. armigera*, the environmental signal for the induction of diapause in this insect is short day photoperiod. The hemolymph proteins were compared in diapausing and non-diapausing larvae of *S. nonagrioides* at two developmental stages (the beginning and the end of the sixth instar). Proteins such as fatty acid transport protein (FATP), soluble guanylate cyclase 89 Da, juvenile hormone binding protein, serine proteinase inhibitor, and arylphorin protein were shown increased abundances at the sixth instar stage of the diapausing larvae, suggesting that they might be involved in the diapause process. On the other hand, proteins such as diacylglycerol kinase, kinase C1, and heat shock protein 68 showed greater abundanes in non-diapausing larvae, indicating their possible role in the larval-pupal metamorphosis.

12.2 Proteomics and Bio-Pesticide Production

The most important application of insect pest proteomics is to identify target sites (receptor proteins, enzymes, signaling proteins, or transcription factors) for designing and producing bio-rational insecticides (Fig. 12.2). Bio-rational pesticides are chemicals/molecules that aim to interfere with insect physiological function [1]. These bio-pesticides can be designed to precisely target a particular molecule or process in an insect pest or a group of insects without affecting other non-target or beneficial insects. The insect specific mode of action of bio-rational insecticides make them more environmentally-friendly than conventional chemical pesticides [1]. The remarkable potentials of proteomics in the identification of

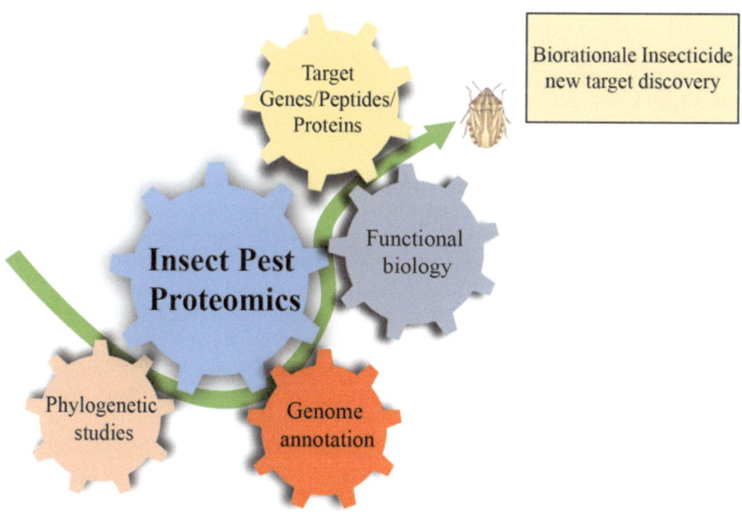

Fig. 12.2 Insect pest proteomics. Insect pest proteomics aims to link a biological function to insect proteins as well as to identify new insect-specific target proteins for the generation of bio-rational pesticides. The evolutionary conservativity of insect proteins can be exploited for the phylogenetic and taxonomic analysis of insect pests as well as for genome annotation

insect-specific proteins, peptides, PTMs, and protein-protein interactions could significantly pave the way towards bio-rational insecticide production. In particular, proteomics can be helpful in the identification of new insect species-specific targets. Once an insect-specific target protein is identified, suitable inhibitors, agonists, and/or antagonists can be designed to specifically target its function.

References

1. Cusson M (2008) The molecular biology toolbox and its use in basic and applied insect science. Bioscience 58:691–700
2. Jouanin L, Bonadé-Bottino M, Girard C, Morrot G, Giband M (1998) Transgenic plants for insect resistance. Plant Sci 131:1–11
3. Heard TA (1999) The role of stingless bees in crop pollination. Annu Rev Entomol 44:183–206
4. Thomson D (2004) Competitive interactions between the invasive European honey bee and native bumble bees. Ecology 85:458–470
5. Sutherland TD, Young JH, Weisman S, Hayashi CY, Merritt DJ (2010) Insect silk: one name, many materials. Annu Rev Entomol 55:171–188
6. Hawkins BA, Cornell HV, Hochberg ME (1997) Predators, parasitoids, and pathogens as mortality agents in phytophagous insect populations. Ecology 78:2145–2152
7. Perring TM (2001) The *Bemisia tabaci* species complex. Crop Prot 20:725–737
8. Mello MO, Silva-Filho MC (2002) Plant-insect interactions: an evolutionary arms race between two distinct defense mechanisms. Braz J Plant Physiol 14:71–81
9. Goulson D (2013) Review: an overview of the environmental risks posed by neonicotinoid insecticides. J Appl Ecol 50:977–987
10. Dastranj M, Bandani AR, Mehrabadi M (2013) Age-specific digestion of *Tenebrio molitor* (Coleoptera: Tenebrionidae) and inhibition of proteolytic and amylolytic activity by plant proteinaceous seed extracts. J Asia-Pac Entomol 16:309–315
11. Hewes RS, Taghert PH (2001) Neuropeptides and neuropeptide receptors in the *Drosophila melanogaster* genome. Genome Res 11:1126–1142
12. Dittmer NT, Hiromasa Y, Tomich JM, Lu N, Beeman RW, Kramer KJ et al (2011) Proteomic and transcriptomic analyses of rigid and membranous cuticles and epidermis from the elytra and hindwings of the red flour beetle, *Tribolium castaneum*. J Proteome Res 11:269–278
13. Merzendorfer H, Kim HS, Chaudhari SS, Kumari M, Specht CA, Butcher S et al (2012) Genomic and proteomic studies on the effects of the insect growth regulator diflubenzuron in the model beetle species *Tribolium castaneum*. Insect Biochem Mol Biol 42:264–276
14. Zhang Q, Lu YX, Xu WH (2012) Integrated proteomic and metabolomic analysis of larval brain associated with diapause induction and preparation in the cotton bollworm, *Helicoverpa armigera*. J Proteome Res 11:1042–1053
15. Zhai Y, Zhang L, Sun Z, Dong X, He Y, Kang K et al (2013) Proteomic and transcriptomic analyses of fecundity in the brown planthopper *Nilaparvata lugens* (Stal). J Proteome Res 12:5199–5212
16. Sehrawat N, Gakhar SK (2014) Mosquito proteomics: present and future prospective. Res Biotechnol 5:25–33
17. Boerjan B, Cardoen D, Verdonck R, Caers J, Schoofs L (2012) Insect omics research coming of age. Can J Zool 90:440–455
18. Roth S, Fromm B, Gäde G, Predel R (2009) A proteomic approach for studying insect phylogeny: CAPA peptides of ancient insect taxa (Dictyoptera, Blattoptera) as a test case. BMC Evol Biol 9:1

19. Krenn HW (2010) Feeding mechanisms of adult Lepidoptera: structure, function, and evolution of the mouthparts. Annu Rev Entomol 55:307
20. Gilbert LE (1972) Pollen feeding and reproductive biology of Heliconius butterflies. Proc Natl Acad Sci 69:1403–1407
21. Krenn HW, Eberhard MJB, Eberhard SH, Hikl A-L, Huber W, Gilbert LE (2009) Mechanical damage to pollen aids nutrient acquisition in Heliconius butterflies (Nymphalidae). Arthropod Plant Interact 3:203–208
22. Harpel D, Cullen DA, Ott SR, Jiggins CD, Walters JR (2015) Pollen feeding proteomics: salivary proteins of the passion flower butterfly, *Heliconius melpomene*. Insect Biochem Mol Biol 63:7–13
23. Pauchet Y, Muck A, Svatos A, Heckel DG, Preiss S (2008) Mapping the larval midgut lumen proteome of *Helicoverpa armigera*, a generalist herbivorous insect. J Proteome Res 7:1629–1639
24. Terra WR (2001) The origin and functions of the insect peritrophic membrane and peritrophic gel. Arch Insect Biochem Physiol 47:47–61
25. Campbell PM, Cao AT, Hines ER, East PD, Gordon KH (2008) Proteomic analysis of the peritrophic matrix from the gut of the caterpillar, *Helicoverpa armigera*. Insect Biochem Mol Biol 38:950–958
26. Angelucci C, Barrett-Wilt GA, Hunt DF, Akhurst RJ, East PD, Gordon KH et al (2008) Diversity of aminopeptidases, derived from four lepidopteran gene duplications, and polycalins expressed in the midgut of *Helicoverpa armigera*: identification of proteins binding the δ-endotoxin, Cry1Ac of *Bacillus thuringiensis*. Insect Biochem Mol Biol 38:685–696
27. Beckage N, Gelman D (2003) Wasp parasitoid disruption of host development: implications for new biologically based strategies for insect control. Annu Rev Entomol 49:299–330
28. Ibrahim AM, Kim Y (2006) Parasitism by Cotesia plutellae alters the hemocyte population and immunological function of the diamondback moth, *Plutella xylostella*. J Insect Physiol 52:943–950
29. Song KH, Jung MK, Eum JH, Hwang IC, Han SS (2008) Proteomic analysis of parasitized *Plutella xylostella* larvae plasma. J Insect Physiol 54:1270–1280
30. Kaeslin M, Pfister-Wilhelm R, Molina D, Lanzrein B (2005) Changes in the haemolymph proteome of *Spodoptera littoralis* induced by the parasitoid Chelonus inanitus or its polydnavirus and physiological implications. J Insect Physiol 51:975–988
31. Riddiford LM, Hiruma K, Zhou X, Nelson CA (2003) Insights into the molecular basis of the hormonal control of molting and metamorphosis from *Manduca sexta* and *Drosophila melanogaster*. Insect Biochem Mol Biol 33:1327–1338
32. Furusawa T, Rakwal R, Nam HW, Hirano M, Shibato J, Kim YS et al (2008) Systematic investigation of the hemolymph proteome of *Manduca sexta* at the fifth instar larvae stage using one- and two-dimensional proteomics platforms. J Proteome Res 7:938–959
33. Swallow JG, Wilkinson GS (2002) The long and short of sperm polymorphisms in insects. Biol Rev Camb Philos Soc 77:153–182
34. Friedlander M (1997) Control of the eupyrene-apyrene sperm dimorphism in Lepidoptera. J Insect Physiol 43:1085–1092
35. Whittington E, Zhao Q, Borziak K, Walters JR, Dorus S (2015) Characterisation of the *Manduca sexta* sperm proteome: genetic novelty underlying sperm composition in Lepidoptera. Insect Biochem Mol Biol 62:183–193
36. Ruiu L (2015) Insect pathogenic bacteria in integrated pest management. Insects 6:352–367
37. Alvarez-Alfageme F, Maharramov J, Carrillo L, Vandenabeele S, Vercammen D, Van Breusegem F et al (2011) Potential use of a serpin from Arabidopsis for pest control. PLoS ONE 6:e20278
38. Tabashnik BE, Gassmann AJ, Crowder DW, Carriere Y (2008) Insect resistance to Bt crops: evidence versus theory. Nat Biotechnol 26:199–202

39. Xia J, Guo Z, Yang Z, Zhu X, Kang S, Yang X et al (2016) Proteomics-based identification of midgut proteins correlated with Cry1Ac resistance in *Plutella xylostella* (L.). Pestic Biochem Physiol, in press
40. Sheffield NC, Song H, Cameron SL, Whiting MF (2008) A comparative analysis of mitochondrial genomes in Coleoptera (Arthropoda: Insecta) and genome descriptions of six new beetles. Mol Biol Evol 25:2499–2509
41. Hare JD (1990) Ecology and management of the Colorado potato beetle. Annu Rev Entomol 35:81–100
42. Fitt GP (1994) Cotton pest management: part 3. An Australian perspective. Annu Rev Entomol 39:543–562
43. Hodek I, Honêk A (2013) Ecology of coccinellidae. Springer, Berlin
44. Tribolium Genome Sequencing C, Richards S, Gibbs RA, Weinstock GM, Brown SJ, Denell R et al (2008) The genome of the model beetle and pest *Tribolium castaneum*. Nature 452:949–955
45. Ulrich J, Dao VA, Majumdar U, Schmitt-Engel C, Schwirz J, Schultheis D et al (2015) Large scale RNAi screen in Tribolium reveals novel target genes for pest control and the proteasome as prime target. BMC Genom 16:1–9
46. Morris K, Lorenzen MD, Hiromasa Y, Tomich JM, Oppert C, Elpidina EN et al (2009) *Tribolium castaneum* larval gut transcriptome and proteome: a resource for the study of the coleopteran gut. J Proteome Res 8:3889–3898
47. Kirsch R, Wielsch N, Vogel H, Svatoš A, Heckel DG, Pauchet Y (2012) Combining proteomics and transcriptome sequencing to identify active plant-cell-wall-degrading enzymes in a leaf beetle. BMC Genom 13:1–15
48. Consortium HG (2012) Butterfly genome reveals promiscuous exchange of mimicry adaptations among species. Nature 487:94–98
49. Pauchet Y, Wilkinson P, Chauhan R, Ffrench-Constant RH (2010) Diversity of beetle genes encoding novel plant cell wall degrading enzymes. PLoS ONE 5:e15635
50. Pitt C, Robert JA, Bonnett TR, Keeling CI, Bohlmann J, Huber DP (2014) Proteomics indicators of the rapidly shifting physiology from whole mountain pine beetle, *Dendroctonus ponderosae* (Coleoptera: Curculionidae), adults during early host colonization. PLoS ONE 9: e110673
51. Rudinsky J (1962) Ecology of scolytidae. Annu Rev Entomol 7:327–348
52. Bentz BJ, Mullins DE (1999) Ecology of Mountain Pine Beetle (Coleoptera: Scolytidae) cold hardening in the intermountain west. Environ Entomol 28:577–587
53. Bonnett TR, Robert JA, Pitt C, Fraser JD, Keeling CI, Bohlmann J et al (2012) Global and comparative proteomic profiling of overwintering and developing mountain pine beetle, *Dendroctonus ponderosae* (Coleoptera: Curculionidae), larvae. Insect Biochem Mol Biol 42:890–901
54. D'acier AC, Pérez-Hidalgo N, Petrović-Obradović O (2010) Aphids (Hemiptera, Aphididae) Chapter 9.2. BioRisk 4:435
55. Harmel N, Letocart E, Cherqui A, Giordanengo P, Mazzucchelli G, Guillonneau F et al (2008) Identification of aphid salivary proteins: a proteomic investigation of *Myzus persicae*. Insect Mol Biol 17:165–174
56. Eichenseer H, Mathews MC, Bi JL, Murphy JB, Felton GW (1999) Salivary glucose oxidase: multifunctional roles for *Helicoverpa zea*? Arch Insect Biochem Physiol 42:99–109
57. Mattiacci L, Dicke M, Posthumus MA (1995) beta-Glucosidase: an elicitor of herbivore-induced plant odor that attracts host-searching parasitic wasps. Proc Natl Acad Sci USA 92:2036–2040
58. Xi J, Pan Y, Wei Z, Yang C, Gao X, Peng T et al (2015) Proteomics-based identification and analysis proteins associated with spirotetramat tolerance in Aphis gossypii Glover. Pestic Biochem Physiol 119:74–80
59. Nguyen TT, Michaud D, Cloutier C (2009) A proteomic analysis of the aphid *Macrosiphum euphorbiae* under heat and radiation stress. Insect Biochem Mol Biol 39:20–30

60. Nguyen TT, Boudreault S, Michaud D, Cloutier C (2008) Proteomes of the aphid *Macrosiphum euphorbiae* in its resistance and susceptibility responses to differently compatible parasitoids. Insect Biochem Mol Biol 38:730–739
61. Moller H, Clapperton K, Gaze P, Sandlant G, Thomas B, Tilley J (1987) Honeydew: life blood of South Island beech forests. For Bird 18:14–16
62. Sabri A, Vandermoten S, Leroy PD, Haubruge E, Hance T, Thonart P et al (2013) Proteomic investigation of aphid honeydew reveals an unexpected diversity of proteins. PLoS ONE 8: e74656
63. Cilia M, Howe K, Fish T, Smith D, Mahoney J, Tamborindeguy C et al (2011) Biomarker discovery from the top down: protein biomarkers for efficient virus transmission by insects (Homoptera: Aphididae) discovered by coupling genetics and 2-D DIGE. Proteomics 11:2440–2458
64. Badillo-Vargas IE, Rotenberg D, Schneweis DJ, Hiromasa Y, Tomich JM, Whitfield AE (2012) Proteomic analysis of *Frankliniella occidentalis* and differentially expressed proteins in response to tomato spotted wilt virus infection. J Virol 86:8793–8809
65. Tamborindeguy C, Bereman MS, Deblasio S, Igwe D, Smith DM, White F et al (2013) Genomic and proteomic analysis of *Schizaphis graminum* reveals cyclophilin proteins are involved in the transmission of cereal yellow dwarf virus. PLoS ONE 8:e71620
66. Gildow F (1993) Evidence for receptor-mediated endocytosis regulating luteovirus acquisition by aphids. Phytopathol N Y Baltim Then St Paul 83:270
67. Cilia M, Tamborindeguy C, Fish T, Howe K, Thannhauser TW, Gray S (2011) Genetics coupled to quantitative intact proteomics links heritable aphid and endosymbiont protein expression to circulative polerovirus transmission. J Virol 85:2148–2166
68. Yang X, Thannhauser TW, Burrows M, Cox-Foster D, Gildow FE, Gray SM (2008) Coupling genetics and proteomics to identify aphid proteins associated with vector-specific transmission of polerovirus (luteoviridae). J Virol 82:291–299
69. Denlinger DL (2002) Regulation of diapause. Annu Rev Entomol 47:93–122
70. Xu WH, Lu YX, Denlinger DL (2012) Cross-talk between the fat body and brain regulates insect developmental arrest. Proc Natl Acad Sci USA 109:14687–14692
71. Tu X, Wang J, Hao K, Whitman DW, Fan Y, Cao G et al (2015) Transcriptomic and proteomic analysis of pre-diapause and non-diapause eggs of migratory locust, *Locusta migratoria* L. (Orthoptera: Acridoidea). Sci Rep 5:11402
72. Tanaka H (1994) Embryonic diapause and life cycle in the migratory locust, *Locusta migratoria* L. (Orthoptera: Acrididae), in Kyoto. Appl Entomol Zool 29:179–191
73. Neves SR, Ram PT, Iyengar R (2002) G protein pathways. Science 296:1636–1639
74. Cheng W-N, Li X-L, Yu F, Li Y-P, Li J-J, Wu J-X (2009) Proteomic analysis of pre-diapause, diapause and post-diapause larvae of the wheat blossom midge, *Sitodiplosis mosellana* (Diptera: Cecidomyiidae). Eur J Entomol 106:29
75. Perez-Hedo M, Sanchez-Lopez I, Eizaguirre M (2012) Comparative analysis of hemolymph proteome maps in diapausing and non-diapausing larvae of *Sesamia nonagrioides*. Proteome Sci 10:58

Chapter 13
Proteomic Research on Honeybee Diseases

Yue Hao and Jianke Li

Abstract Honeybee colonies are challenged with a wide range of diseases caused by various pathogens and abnormal environmental conditions in different ways. Honeybee diseases lead to disabled body functions and finally the dead infected bees will be removed. In most cases, if the new emergence of healthy bees cannot compensate the loss, the colony will rapidly collapse, which results in inefficient pollination and severe losses of agricultural economy. Proteomic tools have been used for decipher the mechanism of various honeybee diseases and aiming at finding useful biomarkers. This chapter briefly reviews the progress on proteomic studies on honeybee disease such as sacbrood disease, nosema, varroa destructor infection, and the effect of pesticides on honeybee brains. The valuable results gain new molecular insight into the pathological mechanism of honeybee diseases. Moreover, the identification of the biomarkers may have important implication for the diagnosis, prevention and treatment of the honeybee diseases.

Keywords Proteomics · Honeybee · Honeybee disease · Sacbrood disease · Nosema · Varroa destructor infection

13.1 Introduction

Similar to the other living organism, honeybee colonies are challenged with a wide range of diseases caused by various pathogens including parasites, bacteria, fungi and viruses. These pathogens threaten honeybee in different ways. Varroa mites feed on bodily fluid of honeybee, and pass viruses to honeybee at different stages. American foulbrood is caused by spore-forming paenibacillus in food, which hatch in the gut of bee larvae. Chalk brood is caused by fungus, which compete for food and also consume the larvae's body. Another kind of fungal disease, Nosema,

Y. Hao · J. Li (✉)
Institute of Apicultural Research, Chinese Academy of Agricultural Sciences, Beijing 100093, China
e-mail: apislijk@126.com

G.H. Salekdeh (ed.), *Agricultural Proteomics Volume 2*,
DOI 10.1007/978-3-319-43278-6_13

invades the intestinal tracts of honeybee, and is the most common and widespread adult honeybee disease. Honeybees infected with viruses will be inhibited in development thus their body functions are disabled. Infected bees will be dead and removed. If this loss can no longer be compensated by new emergence of healthy bees, the colony will rapidly collapse. The pathogenic mechanisms of honeybee diseases are not clear yet, thus the current treatment most often based on chemicals and mechanical control to maintain the infestation at a tolerate level [1–4].

In spite of diseases, abnormal environmental conditions may also result in collapses and death of the honeybee population. For example, cold weather is one of the risk factors for the population. Honeybees are incapable of making cleansing flight, due to the low temperature, will possibly result in dysentery. If the hives are not opened at the right temperature, or the house-keeping bees that keeping the temperature of the hive via clustering are removed, the brood might become chilled, deforming and dead. In addition to temperature causes, many of the chemicals used for controlling pests of agriculture are also the poisons of honeybee brood. This so-called pesticide loss is a major factor in pollinator decline.

In recent years, beekeepers from many countries and regions suffer from the sudden die-off of their colonies. This phenomenon, called Colony Collapse Disorder (CCD), was first discovered by David Hackenberg in Florida U.S. in the year 2006, when he set out 400 hives for the pollination of a pepper grower's farm and 90 % of the hives were empty after 3 weeks. Pollinator loss will cause pollination loss, which has global economic and ecological impact on human life. It is estimated that 12 % fruit and 6 % vegetables from current consumption will loss if without the pollinator [5]. The mechanisms of CCD are still unknown. Many causes have been considered and studies have been carried on for the aim of solving this pressing issue. Scientists now agree that it most likely to be multiple factors that cause CCD, including mites, viruses, insecticides, fungicides and environmental stresses [6]. A variety of potential microbes have been found in CCD colonies. Israeli acute paralysis virus (IAPV) is initially identified as a potential marker of CCD [7], yet the link between IAPV is still absent since subsequent research indicates that IAPV is not present in all CCD colonies [8]. Other pathogens have also been examined, but none of them are found consistently correlated with CCD or contribute to CCD significantly [8, 9]. Bromenshenk et al. used LC-MS to analyze bees from 3 CCD colonies in the U.S. (year 2006–2009) and revealed that the co-infection by invertebrate iridescent virus type 6, and Nosema caused the CCD. The combine of the two pathogens is more lethal to bees than either of them [6]. It was suggested that the disruption of the relation of the two pathogens might be an option to reduce honeybee mortality.

In addition, proteomic tools have also been used for decipher the mechanism of honeybee diseases including Sacbrood disease [10, 11], Nosema ceranae [12] and Varroa destructor infection [13]. These valuable results gain new molecular insight into the pathological mechanism of honeybee diseases, and will have important implication for the diagnosis, prevention and treatment of the honeybee diseases. In addition, significant findings regarding the impact of insecticide fipronil to honeybee brain [14]. All these results will serve as valuable resource for understanding the pathogenic mechanism of honeybee diseases and immune system. We will review these progresses in the following paragraphs.

13.1.1 Sacbrood Disease

The honeybee larvae infected by Sacbrood virus (SBV) will fail to pupate. And ecdysial fluid gathers around the integument of the larvae, forming the "sac" for which the disease is named. The infected larva is pale yellow in color, and form a dark brown gondola shape after death [15]. These obvious signs allow the infection to be easily diagnosed. The SBV infecting Chinese indigenous honeybee (*Apis cerana cerana*, Acc) is called CSBV. The comparative proteome of CSBV-infected ACC worker larvae and the healthy larvae using 2-DE based and MS-based proteomic analysis, 180 proteins and 19 phosphoproteins altered their expressions after infection with CSBV [10] (Fig. 13.1). The infected worker larvae are significantly down-regulated in many crucial biological pathways that supporting organ generation and tissue development, including carbohydrate and energy metabolism, development, protein metabolism, cytoskeleton, silk protein, anti-oxidation and protein folding. These account for over 3/4 of the differentially expressed proteins and nearly 2/3 of the differentially expressed phosphoproteins. It is indicated that these proteins are of great needs by healthy larvae for their normal growth and development, as the larvae go through series remarkable physiological changes to prepare for pupation and further metamorphosis. The terribly disruption of the above pathways due to the infection are lethal to cause the larval death. The more number of phosphoproteins are also up-regulated in the healthy larvae also explained this, as the phosphoproteins usually play essential roles in larval growth [16]. On the other hand, the up-regulated proteins in infectious larvae are mainly involved in small molecule catabolic process, cellular chemical homeostasis and protein degradation, meaning the sick larvae had stronger responses to against viral challenges. A number of heat shock proteins are up-regulated after the larvae challenged with CSBV, suggesting the stress-responding strategies has recruited to fight-against the viruses.

Compared with *Apis mellifera* (Am), *Apis cerana* (Ac) is more likely to be infected by CSBV. However, the resistant mechanism of Am to this disease is still unknown. The food of Ac larvae and Am larvae, RJc and RJm, respectively, are switched and then undergo the CSBV challenge. Interestingly, Ac bees feeding with RJm had significantly lower mortality rate compared with RJc-fed group. The RJm may protect Ac larvae from infection by activating the genes involved in energy metabolism pathways, antioxidation and ubiquitin-proteasome system. However, the food switch had no obvious effect on the virus-resistance of Am bees [11].

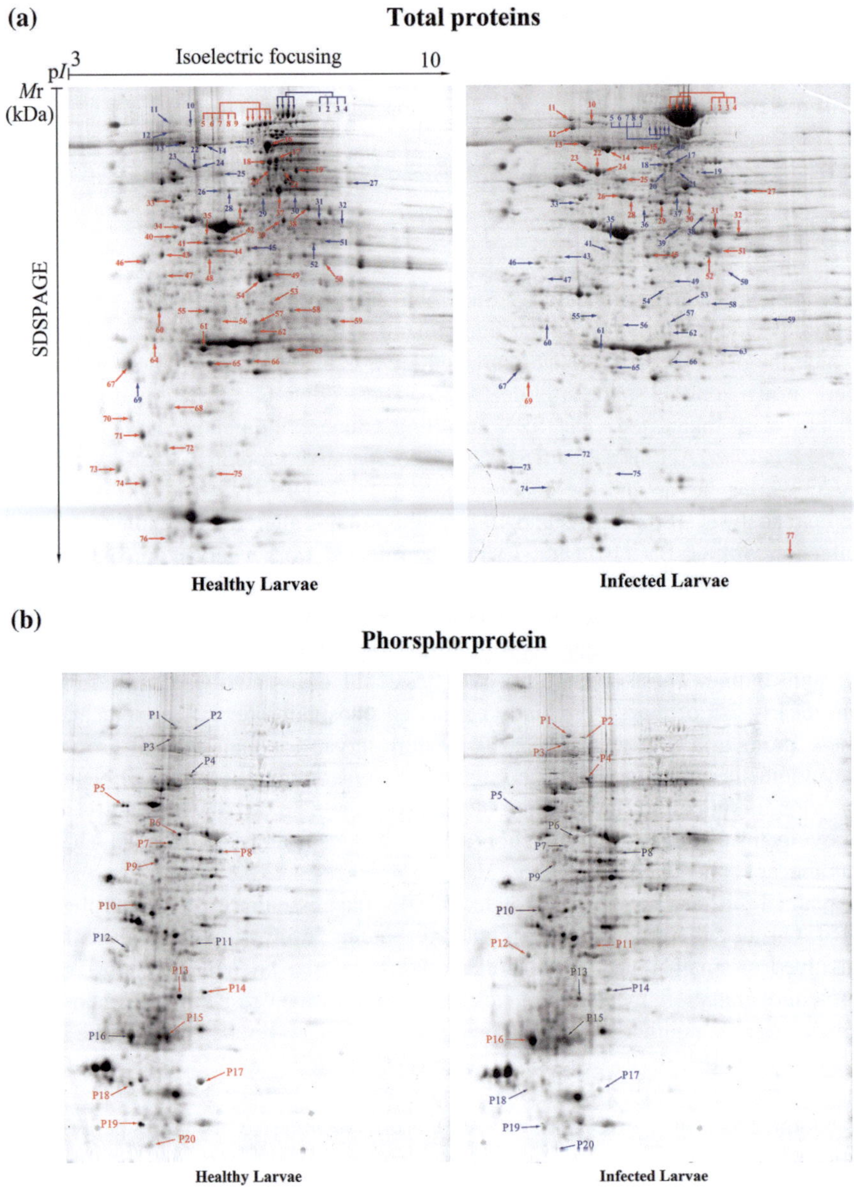

Fig. 13.1 2-DE images of healthy larvae and Chinese sacbrood virus infected worker larvae of ACC. **a** The 2-DE gels stained with the Comassiee Blue Brilliant (CBB, G-250). **b** The 2-DE gels stained with the phosphoprotein-specific fluorescent dye (Pro-Q Diamond). Proteins are separated on 17 cm IPG gel strips (pI 3–10 linear) with 450 μg of sample loading, followed by 12.5 % SDS-PAGE on a vertical slab gel. Differentially expression protein spots of known identity are labeled with color codes, where *red* indicates up-regulation and *blue* indicates down-regulation at each developmental stage [10]

13.1.2 Nosema Ceranae

Nosema ceranae is a major threaten to honeybee colonies [17]. The 2D-based protomics analysis on worker bees' midguts, the proliferation place of *Nosema ceranae*, indicates the fact that the *N. ceranae* can change the honeybee midgut proteome to be favorable for parasite development. Evidences including the up-regulation of carbohydrate catabolism and energy transfer and the down-regulation of energy supply proteins, which denote a higher food demand and lower energy consumption of infected bees. Accumulation of oxidative stress related proteins and lower immune-related proteins in infected bees suggest the host body sent out distress signal to breakdown the stress from infection [12].

13.1.3 Vorroa Destructor

Vorroa destructor (VD) is a parasitic mite that feed on the hemolymph of honeybee (Fig. 13.2). It is also the carrier of many viruses that threaten honeybee, such as acute bee paralysis virus, cloudy wing virus, deformed wing virus and Kakugo virus. A significant VD infestation will lead to the death of a honeybee colony. Studies suggested that VD might be a contributing factor to CCD [18]. *Apis cerana* colonies that resistant to VD are found to significantly up-regulated the proteins for metabolic, respiratory and other activities. This suggests a higher rate of metabolism in the resistant bees. The comparison between resistant bees and sensitive bees that challenged with VD imply that the biological pathways of defense response, mRNA variable shear, cell apoptosis and stress response are greatly enriched in the resistant bees, indicating these activities of honeybees are affected when challenged with VD [13].

Hemolymph plays great roles to distribute immune components throughout the honeybee, and is an powerful indicator of individual's physiological condition. The identification of virus proteins in honeybee hemolymph creates the possibilities of being used for biomarker research [19, 20]. To identify indicator proteins that

Fig. 13.2 *Vorroa destructor* parasite on the adult and pupal bees' body. *Source* These photos are provided by Ms Katrina Kellet

would allow the easy diagnosis of disease states, Chan et al. [19] searched hemolymph samples against all viral database, and first found strong matches between a drone sample and a polyprotein from Deformed wing virus (DWV). Then later this was found by Bogaerts et al. [20] from their worker bee's sample. Bogaerts et al. also found another two matches with *Varroa destructor* virus (VDV) and Kakugo virus (KV), respectively. Both DWV and VDV are transmitted by a parasitic mite *Varroa destructor*, who feeds on the hemolymph of honeybee thus spreads RNA viral agents to honeybee. KV is transmitted by oral infection in the colony and is believed to cause aggressive behavior in worker bees. In spite of different infection phenotypes, the polypeptides of these three viruses are extremely similar and all these viruses can cause serious infestation of the honeybee colony [21–24]. Whether these could be potential biomarkers for the diagnosis of honeybee diseases remains to be validated.

Honeybees performing hygienic behaviors are able to detect, uncap and quickly remove the brood infected with bacteria, fungus, and mites. *Varroa sensitive hygiene* (VSH) is less understood behavior, which will ultimately result in a high proportion of non-reproductive mites in the brood by uncapping and/or removing infected pupae [25, 26]. As a consequence, this behavior limits the spread of infection and reduces the rapidity of the reproduction of pathogens [27]. HB is heritable, thus the colonies that possess this behavior are economically important to beekeepers [27–29]. Proteome-wide correlation analyses ($\sim$1200 proteins total) in larval integument, which is the initial physical barrier to VD, and worker bees' antennae, which is the primary sensory organs of adult bees, identified that several proteins are related with VD infestation. Proteins related with larval chitin biosynthesis and/or structural regulation, immune responses and wounding responses are important for the disease resistance and infestation adaption. In addition, the speed of HB may be reflected by the changes in the antenna proteome. Chemosensory and neurological processes are also linked to the tolerance of VD infestation [30] (Fig. 13.3). A more recent correlation analysis was carried between HB and antennal protein profile of several hundred bees collected at two geographically distince sites over 3 years. It has been found that seven proteins involved in semiochemical sensing, nerve signal transduction and decay, are required to respond to an olfactory signal from freezed-killed brood. Thus these proteins might be promising biomarkers of HB [31]. These findings provide molecular clues of the complex honeybee behavioral adaptions, and will benefit the selective breeding in commercial apiculture.

13.1.4 Insecticide Fipronil

Insecticides are widely used for pest control in agriculture. However, honeybees can be exposed to the insecticides during their foraging activities, and the insecticides can be spread within the colony through contact with contaminated nectar and pollen. As a consequence, the brood with different stages can be exposed to

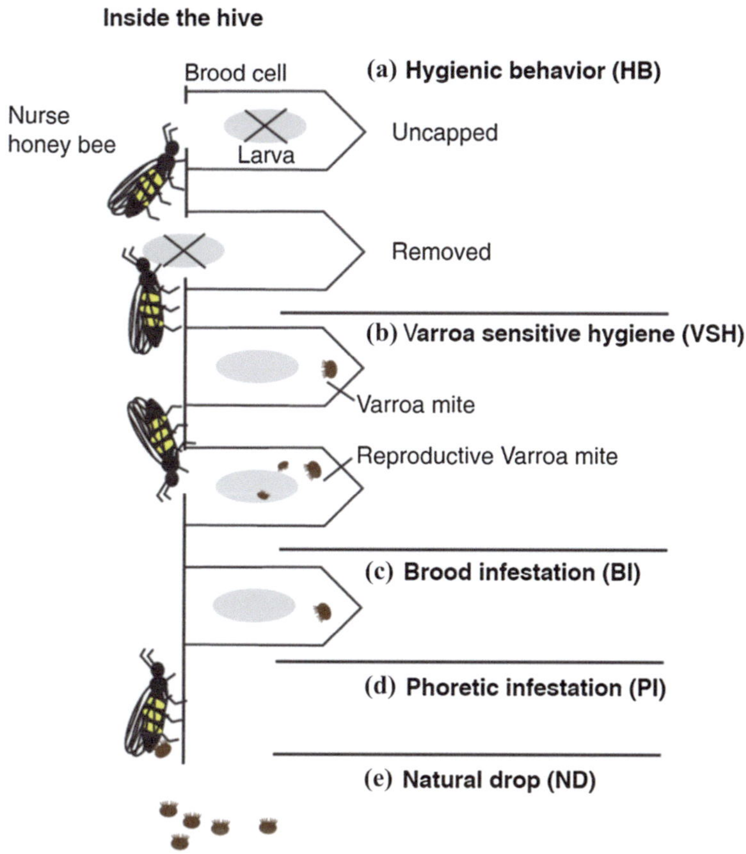

Fig. 13.3 Diagram depicting honey bee disease tolerant traits and infestation dynamics. **a** Hygienic behavior (HB) is composed of two component behaviors, 'uncapping' (uncapped, U) which involves the opening of the cell containing a dead pupa and 'removal' (removed, R) which involves the removal of the dead pupa from the cell after uncapping has occurred. These behaviors are not always performed by the same bee. HB was recorded over 24 h (rapid) and 48 h (slow) periods. **b** Varroa sensitive hygiene (VSH) was defined by determining the proportion of Varroa-infested cells in which no reproductively viable Varroa mite daughters were produced. Increases in this measure infer that greater proportions of mites remaining in the brood have had their reproduction suppressed because of infertility, death, the production of only males, or have had their reproduction delayed preventing sexual maturation of females. **c** Brood infestation (BI) is the percentage of brood cells infested by one or more mites regardless of the mite's reproductive status. **d** Phoretic infestation (PI) is an estimate of the density of mite phoresy on adult bees, and **e** natural drop (ND) is a normalized measure of the number of mites falling from the adult bees onto an adhesive board on the bottom board of colonies [30]

insecticides, which can be toxic or even cause deaths to honeybee [32–34]. Fipronil, an active compound that is present in many pesticides, acts on the nervous system of insects by blocking GABA receptors and inhibiting the ionotropic glutamate-gated chloride channels. GABA receptors are also found in neuropils of the honeybee brain [35]. Lethal dose of fipronil are verified to be harmful to newly emerged worker bees. Low-dose fipronil changes the metabolism of mushroom body but not the antennal lobes in aged worker bees [36]. Honeybee physiology and behavior can be affected by low-dose fipronil in many ways, such as sucrose sensitivity, memory and olfactory learning [37]. Proteomic analysis of the brain of worker bees signify that, the exposure of fipronil will possibly cause susceptibility to pathogens infection chemical stress, misfolding of neuronal proteins, higher occurrence of apoptosis, ischemia, visual problems, damage to synapses, brain degeneration, impairment of memory and learning. It is also pointed out that ATP production is encouraged due to the exposure of the insecticide because energy is called to remove the chemical stress and repair the damages [14].

References

1. Munawar MS, Raja S, Waghchoure ES, Barkat M (2010) Controlling American Foulbrood in honeybees by shook swarm method. Pak J Agric Res 23:53–58
2. Vandervalk LP, Nasr ME, Dosdall LM (2014) New miticides for integrated pest management of Varroa destructor (Acari: Varroidae) in honey bee colonies on the Canadian Prairies. J Econ Entomol 107:2030–2036
3. Grout RA (1946) The hive and the honeybee: a new book on beekeeping to succeed the book Langstroth on the hive and the honeybee. American Bee Journal, Hamilton
4. Higes M, Martin-Hernandez R, Garrido-Bailon E, Gonzalez-Porto AV, Garcia-Palencia P, Meana A et al (2009) Honeybee colony collapse due to *Nosema ceranae* in professional apiaries. Environ Microbiol Rep 1:110–113
5. Gallai N, Salles JM, Settele J, Vaissiere BE (2009) Economic valuation of the vulnerability of world agriculture confronted with pollinator decline. Ecol Econ 68:810–821
6. Bromenshenk JJ, Henderson CB, Wick CH, Stanford MF, Zulich AW, Jabbour RE et al (2010) Iridovirus and microsporidian linked to honey bee colony decline. PLoS One 5:e13181
7. Cox-Foster DL, Conlan S, Holmes EC, Palacios G, Evans JD, Moran NA et al (2007) A metagenomic survey of microbes in honey bee colony collapse disorder. Science 318:283–287
8. Vanengelsdorp D, Evans JD, Saegerman C, Mullin C, Haubruge E, Nguyen BK et al (2009) Colony collapse disorder: a descriptive study. PLoS One 4:e6481
9. Maori E, Lavi S, Mozes-Koch R, Gantman Y, Peretz Y, Edelbaum O et al (2007) Isolation and characterization of Israeli acute paralysis virus, a dicistrovirus affecting honeybees in Israel: evidence for diversity due to intra- and inter-species recombination. J Gen Virol 88: 3428–3438
10. Han B, Zhang L, Feng M, Fang Y, Li J (2013) An integrated proteomics reveals pathological mechanism of honeybee (*Apis cerena*) sacbrood disease. J Proteome Res 12:1881–1897
11. Zhang Y, Zhang G, Huang X, Han R (2014) Proteomic analysis of *Apis cerana* and *Apis mellifera* larvae fed with heterospecific royal jelly and by CSBV challenge. PLoS One 9: e102663
12. Vidau C, Panek J, Texier C, Biron DG, Belzunces LP, Le Gall M et al (2014) Differential proteomic analysis of midguts from *Nosema ceranae*-infected honeybees reveals manipulation of key host functions. J Invertebr Pathol 121:89–96
13. Ji T, Shen F, Liu Z, Yin L, Shen J, Liang Q et al (2015) Comparative proteomic analysis reveals mite (*Varroa destructor*) resistance-related proteins in Eastern honeybees (*Apis cerana*). Genet Mol Res 14:10103–10118

14. Roat TC, Dos Santos-Pinto JR, Dos Santos LD, Santos KS, Malaspina O, Palma MS (2014) Modification of the brain proteome of Africanized honeybees (*Apis mellifera*) exposed to a sub-lethal doses of the insecticide fipronil. Ecotoxicology 23:1659–1670
15. Bailey L (2015) Recent research on honeybee viruses. Bee World 56:55–64
16. Delom F, Chevet E (2006) Phosphoprotein analysis: from proteins to proteomes. Proteome Sci 4:15
17. Hernandez LG, Lu B, Da Cruz GC, Calabria LK, Martins NF, Togawa R et al (2012) Worker honeybee brain proteome. J Proteome Res 11:1485–1493
18. Guzman-Novoa E, Eccles L, Calvete Y, Mcgowan J, Kelly PG, Correa-Benitez A (2010) Varroa destructor is the main culprit for the death and reduced populations of overwintered honey bee (*Apis mellifera*) colonies in Ontario, Canada. Apidologie 41:443–450
19. Chan QW, Howes CG, Foster LJ (2006) Quantitative comparison of caste differences in honeybee hemolymph. Mol Cell Proteomics 5:2252–2262
20. Bogaerts A, Baggerman G, Vierstraete E, Schoofs L, Verleyen P (2009) The hemolymph proteome of the honeybee: gel-based or gel-free? Proteomics 9:3201–3208
21. De Miranda JR, Genersch E (2010) Deformed wing virus. J Invertebr Pathol 103(Suppl 1): S48–S61
22. Martin SJ, Highfield AC, Brettell L, Villalobos EM, Budge GE, Powell M et al (2012) Global honey bee viral landscape altered by a parasitic mite. Science 336:1304–1306
23. Moore J, Jironkin A, Chandler D, Burroughs N, Evans DJ, Ryabov EV (2011) Recombinants between Deformed wing virus and Varroa destructor virus-1 may prevail in Varroa destructor-infested honeybee colonies. J Gen Virol 92:156–161
24. Ongus JR, Peters D, Bonmatin JM, Bengsch E, Vlak JM, Van Oers MM (2004) Complete sequence of a picorna-like virus of the genus Iflavirus replicating in the mite Varroa destructor. J Gen Virol 85:3747–3755
25. Harboa JR, Harrisa JW (2005) Suppressed mite reproduction explained by the behaviour of adult bees. J Apic Res 44:21–23
26. Ibrahim A, Spivak M (2006) The relationship between hygienic behavior and suppression of mite reproduction as honey bee (*Apis mellifera*) mechanisms of resistance to Varroa destructor. Apidologie 37:31–40
27. Spivak M, Masterman R, Ross R, Mesce KA (2003) Hygienic behavior in the honey bee (*Apis mellifera* L.) and the modulatory role of octopamine. J Neurobiol 55:341–354
28. Spivak M, Reuter GS (2001) Resistance to American foulbrood disease by honey bee colonies *Apis mellifera* bred for hygienic behavior. Apidologie 32:555–565
29. Spivak M, Reuter GS (2001) Varroa destructor infestation in untreated honey bee (Hymenoptera: Apidae) colonies selected for hygienic behavior. J Econ Entomol 94:326–331
30. Parker R, Guarna MM, Melathopoulos AP, Moon KM, White R, Huxter E et al (2012) Correlation of proteome-wide changes with social immunity behaviors provides insight into resistance to the parasitic mite, Varroa destructor, in the honey bee (*Apis mellifera*). Genome Biol 13:R81
31. Guarna MM, Melathopoulos AP, Huxter E, Iovinella I, Parker R, Stoynov N et al (2015) A search for protein biomarkers links olfactory signal transduction to social immunity. BMC Genom 16:63
32. Barnett EA, Charlton AJ, Fletcher MR (2007) Incidents of bee poisoning with pesticides in the United Kingdom, 1994–2003. Pest Manag Sci 63:1051–1057
33. Rondeau G, Sanchez-Bayo F, Tennekes HA, Decourtye A, Ramirez-Romero R, Desneux N (2014) Delayed and time-cumulative toxicity of imidacloprid in bees, ants and termites. Sci Rep 4:5566
34. Rortais A, Arnold G, Halm MP, Touffet-Briens F (2005) Modes of honeybees exposure to systemic insecticides: estimated amounts of contaminated pollen and nectar consumed by different categories of bees. Apidologie 36:71–83
35. Schafer S, Bicker G (1986) Distribution of GABA-like immunoreactivity in the brain of the honeybee. J Comp Neurol 246:287–300

36. Roat TC, Carvalho SM, Nocelli RC, Silva-Zacarin EC, Palma MS, Malaspina O (2013) Effects of sublethal dose of fipronil on neuron metabolic activity of Africanized honeybees. Arch Environ Contam Toxicol 64:456–466
37. El Hassani AK, Dacher M, Gauthier M, Armengaud C (2005) Effects of sublethal doses of fipronil on the behavior of the honeybee (*Apis mellifera*). Pharmacol Biochem Behav 82:30–39

Chapter 14
Application of Proteomic Biomarkers in Livestock Disease Management

Ehsan Oskoueian, Peter David Eckersall, Elena Bencurova and Thomas Dandekar

Abstract The applications of proteomics in animal husbanding are broad and include monitoring proteome changes in the tissue and body fluids to interpret the physiological process during growth, development and production and in the detection and management of disease. The diversity of farm animal species from cattle, sheep, goats, chickens to fish and even invertebrate aquaculture species complicate the analysis and interpretation of proteome data. The recent technological advances in extraction and fractionation techniques along with platform sensitivity and data analysis have allowed discovery of next-generation biomarkers with high sensitivity, specificity and precision. These robust biomarkers are useful in monitoring health and well-being of animals, surveillance against animal pathogens, elucidating disease mechanisms, assessing pharmacologic response to therapeutic and directing genetic selection and breeding. A literature survey revealed that discovery of proteomic biomarkers in biological fluids (serum, plasma, urine, milk, exudates, tear, semen and genital secretion) provide readily

---

E. Oskoueian (✉)
Agricultural Biotechnology Research Institute of Iran (ABRII),
East and North-East Branch, P.O.B. 91735/844, Mashhad, Iran
e-mail: e.oskoueian@abrii.ac.ir

P.D. Eckersall
Institute of Biodiversity, Animal Health and Comparative Medicine,
College of Medical, Veterinary and Life Sciences, University of Glasgow,
Glasgow, UK
e-mail: David.Eckersall@glasgow.ac.uk

E. Bencurova · T. Dandekar
Department of Bioinformatics, Biocenter, University of Wuerzburg,
97074 Würzburg, Germany
e-mail: elena.bencurova@uni-wuerzburg.de

T. Dandekar
e-mail: dandekar@biozentrum.uni-wuerzburg.de

E. Bencurova
Laboratory of Biomedical Microbiology and Immunology,
Department of Microbiology and Immunology, University of Veterinary
Medicine and Pharmacy in Kosice, 041 81 Kosice, Slovakia

G.H. Salekdeh (ed.), *Agricultural Proteomics Volume 2*,
DOI 10.1007/978-3-319-43278-6_14

accessible sources of samples for non- or minimally-invasive and cost-effective diagnosis tools. This area of research is actively expanding and future research would profitably focus on applications of multiple biomarkers to increase the diagnosis precision in livestock disease management. Therefore this review is aimed to provide a brief overview on successful experiences in using proteomics biomarkers identified in biological fluids for livestock diseases management.

Keywords Proteomics · Mass spectrometry · Electrophoresis · Bio-monitoring · Farm animals · Diseases · Animal Husbandry · Next-generation biomarkers

14.1 Introduction

The proteins in the livestock are mainly involved in structural roles, growth, development, metabolism, signaling, production and disease response. For in-depth elucidation of protein function in animal health, a growing body of research has been conducted over the last decade to characterize the proteome of biological fluid, cells, tissues, and organs in specific conditions or in a comparative way [1–5]. The advances in proteomics methodologies, such as sample fractionation, mass spectrometry and protein arrays, together with the development of high-throughput technologies, have opened new avenues in the search for clinically useful biomarkers and provide deeper understanding in livestock disease management from diagnosis to treatment response [6]. For this purpose, numerous proteins have been explored in biological fluids as non-invasive or minimally invasive monitoring methods. Biological fluids, such as saliva, urine, milk, plasma, serum, tear, semen, genital secretions, respiratory exudates, egg yolk and egg white, contain a broad range of valuable biomarkers, which can be potentially used for monitoring health and well-being of animals, for surveillance against animal pathogens, to elucidate disease mechanisms, assessing pharmacologic response to therapeutic, vaccine development and implementing genetic selection in breeding (Fig. 14.1) [3, 6–12].

Biological fluids are considered as minimally invasive sources (serum, plasma and semen) or non-invasive (milk, urine, saliva and exudates) and readily accessible sources of real-time disease biomarkers. Comparative proteomic studies on biological fluids have resulted in the discovery of various biomarkers which next require sensitivity, specificity and validation analysis (Fig. 14.1) [13]. For instance, proteomic biomarkers have been identified for diagnosis of gastrointestinal, respiratory, cardiovascular, neurological, reproductive system diseases in livestock such as cattle, sheep, goats, chicken and swine as well as aquaculture species including trout, salmon, bream and shrimps [1, 8, 10, 13–17]. Overall the results showed that modern proteomics biomarkers could indicate disease trait (risk factor), disease state (preclinical or clinical), disease rate (progression), the response to therapy or even to monitor environmental stress [18–20]. The following sections will further highlight proteomics biomarkers identified in biological fluids for livestock diseases management.

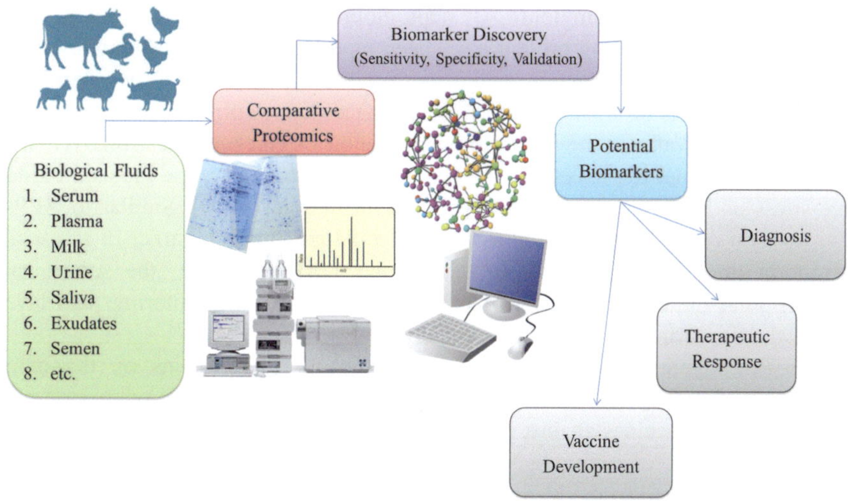

Fig. 14.1 A typical workflow for proteomics biomarker discovery and their applications in livestock diseases management

14.1.1 Blood Biomarkers

An animal's entire proteome can potentially be represented in the blood as has been shown for human blood proteomic studies [21]. The presence of disease induces pathological responses which cause quantitative or qualitative changes in tissue proteins that leak into blood altering the serum proteins profile which can be potentially useful as disease biomarkers [22]. Thereby the blood proteome appeared to be a favorable source of biomarkers for health status monitoring, prompt diagnosis, and evaluation of treatment response, and is generally considered as a primary source of biomarkers in farm animals.

Serum and plasma are generally used for minimally invasive clinical diagnosis of diseases. However, discovery and validation of high-specific protein biomarkers in the serum is hindered by the presence of certain high abundant proteins such as albumin, which interrupt detection of relatively less abundant but specific proteins. Selective affinity based elimination and prefractionation techniques help depleting high abundance proteins and detecting proteins with biomarker potential [23].

Among blood biomarkers, the acute phase proteins (APPs) are associated with innate immune response and are responsible for the systemic reaction to inflammation, including the opsonization of several antigens, the elimination of potentially toxic compounds and the overall regulation of different stages of inflammation process [2, 6, 7, 16]. It has been established that increases in the concentrations of APPs are valuable biomarkers of inflammation in the body. For instance, the expression of serum amyloid A, haptoglobin, α-1 acid glycoprotein and lipopolysaccharide binding protein have been reported in the case of endometritis, lung and pulmonary infection and digestive tract disorders in livestock [7]. Furthermore, it has been shown that heat stress can induce physiological responses and significantly affects the health and production of chicken [9]. Moreover, this group reported that the excessive temperature above comfort zone induced oxidative stress and resulted in alteration of APPs in the plasma [9]. Similar to APPs, transthyretin could be considered as a potent inflammation biomarker in assessing long-term moderate heat stress in dairy cows [24]. Moreover, the serum and cerebrospinal fluid concentration of transthyretin helped in monitoring pain level and evaluating drug response in cattle [25].

The investigation of immunogenic properties of the proteins on the outer membrane of pathogenic microorganisms helps in early detection of disease and development of vaccines against infection [26, 27]. As shown by Le Marechal et al. [28] the detection of immunogenic proteins of *Staphylococcus aureus* in sheep blood serum resulted in the identification of the organism's core seroproteome and accessory seroproteome which helped in early detection of mastitis in sheep. Furthermore, proteomic analysis of pathogenic bacteria such as *Bacillus anthracis, Listeria monocytogenes, Mycobacterium paratuberculosis* subsp. *avium*, and *Dichelobacter nodosus* has resulted in identification of coat-associated proteins with immunogenic properties. Application of these protein biomarkers in vaccine development improved the antibody production and efficacy of vaccine against anthrax, listeriosis, paratuberculosis and footrot diseases [29, 30].

In addition to pathogenic microorganisms, parasite burden is a significant threat to domestic animals which cause economic losses in farm animal industry. The development of resistance in parasites to antiparasitic drugs has challenged infection control. The proteomics approach can be used in searching for biomarkers that aid the development of new drugs and next-generation polyvalent vaccines [31]. In addition, biomarkers facilitated the discrimination of antigens, determination of developmental phases of pathogens, and monitoring the host response during infection. They also have been useful in identification of pathways and key regulators that could be therapeutically targeted for more effective treatments to a wide spectrum of animal hosts including cattle, sheep, goats and water buffalo [31, 32].

If physiological parameters deviate from normal, biomarkers are useful to determine the degree of physiological imbalance. Early detection reduces the risk of disease and improves production and reproduction performance in the livestock animals. For instance, pyruvate carboxylase and isocitrate dehydrogenase are serum based hepatic biomarkers to determine the physiological imbalance in the cows during early lactation while alcohol dehydrogenase-4 and methylmalonate-semialdehyde dehydrogenase act in a similar way for cows in mid lactation as apparent from MS investigations [33]. The changes in abundance of these proteins revealed physiological imbalance and thereby indicated the cows at risk for disease during lactation.

14.1.2 Milk Biomarkers

New biomarkers may be determined from milk in contrast to traditional biomarkers which mostly are measured in the blood. As milk collection is less invasive and readily available throughout a day, it provides an opportunity for the discovery of potential biomarkers at an affordable cost. Milk biomarkers are considered as new tool for animal health management [3]. Bovine mastitis is one of the major diseases in dairy industry that decrease the milk production and results in economic losses. The detection of bovine mastitis in an early stage provides quick recovery with no significant decrease in the milk production. However, detection of bovine mastitis in a late stage decreases the milk production accompanied with reproductive disorders in dairy cows [34]. The diagnosis of subclinical mastitis is difficult due to the absence of any visible clinical sign. Moreover, evaluation of electrical conductivity, somatic cell counts and lactate dehydrogenase activity are unable to detect a subclinical mastitis in a early stages [15]. Proteomics approaches introduced promising biomarkers to detect subclinical mastitis in early stages. Based on the available results changes in the abundance of α-lactalbumin and β-lactoglobulin proteins [35], complement C3 and C4 proteins [36], apolipoprotein A-I, cathelicidin-1, APA serum amyloid A [37], α-1-acid-glycoprotein [38] and prostaglandin D synthase [39] in the milk are considered as putative biomarkers to detect the mastitis in the early stages. Quantitative proteomic methods for monitoring a multiplex of milk proteins in mastitis have been proposed to increase diagnostic capabilities [40]. In addition investigation of low molecular weight peptides which increase in milk during mastitis due to protease activity, may also provide early signs of impending disease [41]. Biomarkers have also helped in early detection of mastitis in goats and sheep together with identification of virulent and host immune response studies [30, 42]. Overall, milk biomarkers are useful in development of new management strategies to improve the animal productivity and public health, through promoting early detection and prevention of disease.

14.1.3 Urinary Biomarkers

In recent years several have identified putative urinary biomarkers warranting their use in clinical trials. The discovered biomarkers were useful for early detection of disease, classification of disease, choice of therapeutic agents and monitoring efficiency of therapeutic regimen [10, 43–45].

The urine proteome has been studied by various proteomic techniques, such as using two-dimensional gel electrophoresis and characterization by liquid chromatography coupled tandem mass spectrometry (LC-MS/MS) [46] and SELDI-TOF-MS [47]. As a result, urinary biomarkers were investigated for the diagnosis, pathogenesis and monitoring therapy of various diseases such as kidney, liver, diabetes, neurodegenerative disorders and prion disease [10, 43, 46]. Notably, cathelicidin, clusterin, uroguanylin, Ig gamma-2 chain C region and protease resistant prion (infectious agent) isoform were reported as urinary biomarker proteins for diagnosis of the dangerous bovine spongiform encephalopathy [46, 48, 49]. The detection of pregnancy in the early stages has also being possible with the help of urinary biomarkers. The results of Zheng et al. [50] indicates that urinary uromodulin possesses the biomarker potential to be used for early detection of pregnancy in cattle. The evident progress in the field of urinary proteomics is demonstrate also by work of Shao et al. [51] which collected all published urinary protein biomarker with more than 400 entries (March, 2016) for various animals and that is available as the Urinary Protein Database (website: http://122.70.220.102/biomarker).

14.1.4 Salivary Biomarkers

The saliva is the fluid present in the oral cavity. It contains various proteins which vary among species reflecting diverse diets and types of digestion. Saliva as a non-invasive diagnostic fluid has facilitated the safe, efficient and low-cost collection of large numbers of samples [52]. The saliva is a valuable source to assess stress and systemic disease related biomarkers. Studies that characterize livestock salivary proteome have started very recently and potential biomarkers identified in goats, sheep [8, 30, 53, 54], pig [55] and cattle [56, 57]. Transportation stress, social stress, common management practices, and different environmental stressors increased the cortisol concentration in porcine saliva [5]. In addition to cortisol, IgA has been known as another biomarker present in the saliva which indicates the stress level in pigs. The IgA increase in the saliva appeared to be due to the activation of the sympathetic nerve system in the salivary submandibular and sublingual glands [52].

Salivary acute phase proteins can be applied to identify clinical and even sub-clinical disease as well as therapeutic response [58]. Proteins such as haptoglobin, C-reactive protein and serum amyloid A have been used in pigs to distinguish between healthy and diseased animals [59]. Furthermore, the salivary biomarkers such as antibodies enable the detection of viral diseases including swine fever, reproductive and respiratory syndrome virus or foot and mouth disease virus [5]. A comparative proteomic approach was used in the work of Fuentes-Rubio et al. [60] where albumin and odorant-binding proteins were downregulated in three groups of the pigs after stress exposure which suggested their potential to monitor stress in livestock.

14.1.5 Exudate Biomarkers

The proteomics studies of host infected tissues, respiratory exudate and respective pathogen aid in the discovery of biomarkers useful for accurate diagnosis, identification of virulence-associated antigens and vaccine development. In this regards,

Weldearegay et al. [61] reported pleuropneumonia in cattle caused by *Mycoplasma mycoides* as one of the most prevalent diseases in the cattle industry. *Mycoplasma mycoides* induced the accumulation of pleural fluid and the proteome characterization of fluid and respiratory exudates revealed the strong enrichment of proteins involved in antigen processing, platelet activation, degranulation and apoptosis together with increased abundance of acute phase proteins. The proteome analysis of mycoplasma indicated the presence of proteins involved in virulence-associated processes and capsule synthesis. The result suggested candidate biomarkers for accurate diagnosis of pleuropneumonia caused by *Mycoplasma mycoides*, monitoring the disease stage and introduced new valuable targets for vaccine development [61]. In another study conducted by Nanduri et al. [62] the proteome analysis of bronchoalveolar lavage of the lamb infected by *Mannheimia haemolytica* upon treatment by antibiotics revealed the down-regulation of biomarkers related to the *M. haemolytica* leucotoxin. This result provided valuable information on therapeutic progress and effectiveness.

14.1.6 Seminal Biomarkers

A large part of general costs in the dairy cattle industry is allocated to reproductive management and there have been proteomic investigation of seminal fluid and related biological fluids [63]. Low semen quality is responsible for a significant percentage of reproductive failure. The comparative proteome analysis of sperm of high- and low-fertility bulls revealed the differential expression in proteins involved in sperm-egg interaction and cell cycle regulation [64]. Hence, these proteins could be considered as potential biomarkers to differentiate high- and low-fertility sperms prior to artificial insemination [17, 65, 66]. In line with this study Boe-Hansen et al. [64] evaluated the seminal plasma proteins in *Bos indicus* Brahman bulls and identified candidate biomarkers associated with morphologically normal sperm. Elucidation of physiological functions and relationships between seminal plasma proteins and sperm attributes is an important step toward discovery of candidate biomarkers for early life prediction of male fertility and improving sperm preservation methods.

Furthermore, in vitro fertilization (IVF) has been used in modern breeding of farm animals, however, challenges with a low pregnancy rate due to the unknown viability of in vitro developed embryos, still remained unsolved. The morphological differentiation is inadequate and complementary methods are still required. However, new proteomics biomarkers enable the successful differentiation of viable embryos with high precision [67]. In line with this study Deutsch et al. [68] has also reported proteins involved in lipid metabolism such as apolipoprotein A1 as biomarkers to confirm the viability of bovine embryo in culture medium before transfer to the recipient cows.

14.1.7 *Tears Biomarkers*

Tears are a potentially interesting material for the discovery of novel biomarkers, however only few studies consider this body fluid. The natural function of tears is protection of the eye by lubricating the outer parts of eye, removing of foreign material and prevention against the infection. A reference map of the proteome of tears of dogs [69], cow, sheep and camel [70] has been reported and demonstrated differential protein expression among animal species. The protein analysis of rabbit tears revealed that after the mechanical abrasion of the cornea, two defensins, NP-1 and NP-2 were significantly increases during the curative process and thus may play a crucial role in the protection of the cornea against the microbial infection during the heling period [71].

14.2 Conclusion

The high sensitive MS-based detection of biomarkers is one of the most important aspects of the application of proteomics in livestock health management. The protein biomarkers from biological fluids are non- or minimally-invasive, readily accessible, real time and cost-effective sources for diagnosis purposes, monitoring therapeutic response and development of vaccines against various diseases. The diversity of farm animal species from cattle, sheep, goats, chickens to aquaculture species complicate the analysis and interpretation of their proteomes. However, the current technological improvements allow investigation of animal disease biology and next-generation of biomarkers with high sensitivity, specificity and precision. Building on these insights, a number of cheap and reliable tests using these proteomic biomarkers will soon be developed, for instance antibody-based colorimetric assays or similar robust sandwich assays to reduce risks of zoonoses, enhance farm animal welfare, improve quality and safety of animal products and decrease farmers financial losses.

References

1. Bendixen E, Danielsen M, Hollung K, Gianazza E, Miller I (2011) Farm animal proteomics—a review. J Proteomics 74:282–293
2. Eckersall PD, De Almeida AM, Miller I (2012) Proteomics, a new tool for farm animal science. J Proteomics 75:4187–4189
3. Roncada P, Piras C, Soggiu A, Turk R, Urbani A, Bonizzi L (2012) Farm animal milk proteomics. J Proteomics 75:4259–4274
4. Bassols A, Turk R, Roncada P (2014) A proteomics perspective: from animal welfare to food safety. Curr Protein Pept Sci 15:156–168
5. Gutierrez AM, Ceron JJ, Fuentes-Rubio M, Tecles F, Beeley JA (2014) A proteomic approach to porcine saliva. Curr Protein Pept Sci 15:56–63

6. Almeida AM, Bassols A, Bendixen E, Bhide M, Ceciliani F, Cristobal S et al (2015) Animal board invited review: advances in proteomics for animal and food sciences. Animal 9:1–17
7. Ceciliani F, Ceron JJ, Eckersall PD, Sauerwein H (2012) Acute phase proteins in ruminants. J Proteomics 75:4207–4231
8. Lamy E, Mau M (2012) Saliva proteomics as an emerging, non-invasive tool to study livestock physiology, nutrition and diseases. J Proteomics 75:4251–4258
9. Shakeri M, Zulkifli I, Soleimani AF, O'reilly EL, Eckersall PD, Anna AA et al (2014) Response to dietary supplementation of L-glutamine and L-glutamate in broiler chickens reared at different stocking densities under hot, humid tropical conditions. Poult Sci 93: 2700–2708
10. Bathla S, Rawat P, Baithalu R, La Yadav M, Naru J, Tiwari A et al (2015) Profiling of urinary proteins in Karan Fries cows reveals more than 1550 proteins. J Proteomics 127:193–201
11. Choi JW, Kim GJ, Lee S, Kim J, Demello AJ, Chang SI (2015) A droplet-based fluorescence polarization immunoassay (dFPIA) platform for rapid and quantitative analysis of biomarkers. Biosens Bioelectron 67:497–502
12. Ceciliani F, Eckersall D, Burchmore R, Lecchi C (2014) Proteomics in veterinary medicine: applications and trends in disease pathogenesis and diagnostics. Vet Pathol 51:351–362
13. Di Girolamo F, D'amato A, Lante I, Signore F, Muraca M, Putignani L (2014) Farm animal serum proteomics and impact on human health. Int J Mol Sci 15:15396–15411
14. Rodrigues PM, Silva TS, Dias J, Jessen F (2012) PROTEOMICS in aquaculture: applications and trends. J Proteomics 75:4325–4345
15. Abd El-Salam MH (2014) Application of proteomics to the areas of milk production, processing and quality control-a review. Int J Dairy Technol 67:153–166
16. O'reilly EL, Eckersall PD (2014) Acute phase proteins: a review of their function, behaviour and measurement in chickens. Worlds Poult Sci J 70:27–43
17. Codina M, Estanyol JM, Fidalgo MJ, Ballesca JL, Oliva R (2015) Advances in sperm proteomics: best-practise methodology and clinical potential. Expert Rev Proteomics 12: 255–277
18. Downing G (2001) Biomarkers definitions working group. Biomarkers and surrogate endpoints. Clin Pharmacol Ther 69:89–95
19. Boehmer JL, Olumee-Shabon Z (2011) Veterinary biomarker discovery: proteomic analysis of acute phase proteins. INTECH Open Access Publisher, Rijeka
20. Marco-Ramell A, De Almeida A, Cristobal S et al (2016) Proteomics and the search for welfare and stress biomarkers in animal production in the one-health context. Mol Biosyst 12:2024–2035
21. Anderson NL, Polanski M, Pieper R, Gatlin T, Tirumalai RS, Conrads TP et al (2004) The human plasma proteome: a nonredundant list developed by combination of four separate sources. Mol Cell Proteomics 3:311–326
22. Eckersall PD, Mclaughlin M (2011) Proteomics in Animal Health and Disease. In: Eckersall PD, Whitfield PD (eds) Methods in animal proteomics. Wiley, Chichester, pp 243–318
23. Marco-Ramell A, Bassols A (2010) Enrichment of low-abundance proteins from bovine and porcine serum samples for proteomic studies. Res Vet Sci 89:340–343
24. Min L, Zheng N, Zhao S, Cheng J, Yang Y, Zhang Y et al (2016) Long-term heat stress induces the inflammatory response in dairy cows revealed by plasma proteome analysis. Biochem Biophys Res Commun 471:296–302
25. Rialland P, Otis C, De Courval ML, Mulon PY, Harvey D, Bichot S et al (2014) Assessing experimental visceral pain in dairy cattle: a pilot, prospective, blinded, randomized, and controlled study focusing on spinal pain proteomics. J Dairy Sci 97:2118–2134
26. Hughes V, Garcia-Sanchez A, Smith S, Mclean K, Lainson A, Nath M et al (2012) Proteome-determined type-specific proteins of *Mycobacterium avium* subspecies paratuberculosis. Vet Microbiol 158:153–162

27. Hughes V, Denham S, Bannantine JP, Chianini F, Kerr K, May L et al (2013) Interferon gamma responses to proteome-determined specific recombinant proteins: potential as diagnostic markers for ovine Johne's disease. Vet Immunol Immunopathol 155:197–204
28. Le Marechal C, Jan G, Even S, Mcculloch JA, Azevedo V, Thiery R et al (2009) Development of serological proteome analysis of mastitis by Staphylococcus aureus in ewes. J Microbiol Methods 79:131–136
29. Hughes V, Bannantine JP, Denham S, Smith S, Garcia-Sanchez A, Sales J et al (2008) Immunogenicity of proteome-determined *Mycobacterium avium* subsp. paratuberculosis-specific proteins in sheep with paratuberculosis. Clin Vaccine Immunol 15:1824–1833
30. Katsafadou AI, Tsangaris GT, Billinis C, Fthenakis GC (2015) Use of proteomics in the study of microbial diseases of small ruminants. Vet Microbiol 181:27–33
31. De La Fuente J, Contreras M (2015) Tick vaccines: current status and future directions. Expert Rev Vaccines 14:1367–1376
32. Marcelino I, De Almeida AM, Ventosa M, Pruneau L, Meyer DF, Martinez D et al (2012) Tick-borne diseases in cattle: applications of proteomics to develop new generation vaccines. J Proteomics 75:4232–4250
33. Moyes KM, Bendixen E, Codrea MC, Ingvartsen KL (2013) Identification of hepatic biomarkers for physiological imbalance of dairy cows in early and mid lactation using proteomic technology. J Dairy Sci 96:3599–3610
34. Reinhardt TA, Lippolis JD, Nonnecke BJ, Sacco RE (2012) Bovine milk exosome proteome. J Proteomics 75:1486–1492
35. Hogarth CJ, Fitzpatrick JL, Nolan AM, Young FJ, Pitt A, Eckersall PD (2004) Differential protein composition of bovine whey: a comparison of whey from healthy animals and from those with clinical mastitis. Proteomics 4:2094–2100
36. Danielsen M, Codrea MC, Ingvartsen KL, Friggens NC, Bendixen E, Røntved CM (2010) Quantitative milk proteomics–host responses to lipopolysaccharide-mediated inflammation of bovine mammary gland. Proteomics 10:2240–2249
37. Smolenski G, Haines S, Kwan FYS, Bond J, Farr V, Davis SR et al (2007) Characterisation of host defence proteins in milk using a proteomic approach. J Proteome Res 6:207–215
38. Boehmer JL, Bannerman DD, Shefcheck K, Ward JL (2008) Proteomic analysis of differentially expressed proteins in bovine milk during experimentally induced *Escherichia coli* mastitis. J Dairy Sci 91:4206–4218
39. Baeker R, Haebel S, Schlatterer K, Schlatterer B (2002) Lipocalin-type prostaglandin D synthase in milk: a new biomarker for bovine mastitis. Prostaglandins Other Lipid Mediat 67:75–88
40. Bislev SL, Kusebauch U, Codrea MC, Beynon RJ, Harman VM, Rontved CM et al (2012) Quantotypic properties of QconCAT peptides targeting bovine host response to *Streptococcus uberis*. J Proteome Res 11:1832–1843
41. Mansor R, Mullen W, Albalat A, Zerefos P, Mischak H, Barrett DC et al (2013) A peptidomic approach to biomarker discovery for bovine mastitis. J Proteomics 85:89–98
42. Chiaradia E, Valiani A, Tartaglia M, Scoppetta F, Renzone G, Arena S et al (2013) Ovine subclinical mastitis: proteomic analysis of whey and milk fat globules unveils putative diagnostic biomarkers in milk. J Proteomics 83:144–159
43. Pisitkun T, Johnstone R, Knepper MA (2006) Discovery of urinary biomarkers. Mol Cell Proteomics 5:1760–1771
44. Il'yasova D, Scarbrough P, Spasojevic I (2012) Urinary biomarkers of oxidative status. Clin Chim Acta 413:1446–1453
45. Heilmann RM, Wright ZM, Lanerie DJ, Suchodolski JS, Steiner JM (2014) Measurement of urinary canine S100A8/A9 and S100A12 concentrations as candidate biomarkers of lower urinary tract neoplasia in dogs. J Vet Diagn Invest 26:104–112
46. Simon SL, Lamoureux L, Plews M, Stobart M, Lemaistre J, Ziegler U et al (2008) The identification of disease-induced biomarkers in the urine of BSE infected cattle. Proteome Sci 6:23

47. Liu Z, Yuan Z, Zhao Q (2014) SELDI-TOF-MS proteomic profiling of serum, urine, and amniotic fluid in neural tube defects. PLoS ONE 9:e103276
48. Shaked GM, Shaked Y, Kariv-Inbal Z, Halimi M, Avraham I, Gabizon R (2001) A protease-resistant prion protein isoform is present in urine of animals and humans affected with prion diseases. J Biol Chem 276:31479–31482
49. Ma D, Li L (2012) Searching for reliable premortem protein biomarkers for prion diseases: progress and challenges to date. Expert Rev Proteomics 9:267–280
50. Zheng J, Liu L, Wang J, Jin Q (2013) Urinary proteomic and non-prefractionation quantitative phosphoproteomic analysis during pregnancy and non-pregnancy. BMC Genom 14:777
51. Shao C, Li M, Li X, Wei L, Zhu L, Yang F et al (2011) A tool for biomarker discovery in the urinary proteome: a manually curated human and animal urine protein biomarker database. Mol Cell Proteomics 10(M111):010975
52. Muneta Y, Yoshikawa T, Minagawa Y, Shibahara T, Maeda R, Omata Y (2010) Salivary IgA as a useful non-invasive marker for restraint stress in pigs. J Vet Med Sci 72:1295–1300
53. Lamy E, Da Costa G, Santos R, Capela ESF, Potes J, Pereira A et al (2009) Sheep and goat saliva proteome analysis: A useful tool for ingestive behavior research? Physiol Behav 98:393–401
54. De Sousa-Pereira P, Cova M, Abrantes J, Ferreira R, Trindade F, Barros A et al (2015) Cross-species comparison of mammalian saliva using an LC-MALDI based proteomic approach. Proteomics 15:1598–1607
55. Gutierrez AM, Miller I, Hummel K, Nobauer K, Martinez-Subiela S, Razzazi-Fazeli E et al (2011) Proteomic analysis of porcine saliva. Vet J 187:356–362
56. Ang CS, Binos S, Knight MI, Moate PJ, Cocks BG, Mcdonagh MB (2011) Global survey of the bovine salivary proteome: integrating multidimensional prefractionation, targeted, and glycocapture strategies. J Proteome Res 10:5059–5069
57. Muthukumar S, Rajkumar R, Rajesh D, Saibaba G, Liao CC, Archunan G et al (2014) Exploration of salivary proteins in buffalo: an approach to find marker proteins for estrus. FASEB J 28:4700–4709
58. Rahman M, Müller U, Sauerwein H et al (2013) Investigation of salivary acute phase proteins in calves. In: Almeida A, Eckersall D, Bencurova E et al (eds) Farm animal proteomics 2013. Wageningen Academic Publishers,Wageningen, Netherlands, pp 213–216
59. Gomez-Laguna J, Gutierrez A, Pallares FJ, Salguero FJ, Ceron JJ, Carrasco L (2010) Haptoglobin and C-reactive protein as biomarkers in the serum, saliva and meat juice of pigs experimentally infected with porcine reproductive and respiratory syndrome virus. Vet J 185:83–87
60. Fuentes-Rubio M, Ceron JJ, De Torre C, Escribano D, Gutierrez AM, Tecles F (2014) Porcine salivary analysis by 2-dimensional gel electrophoresis in 3 models of acute stress: a pilot study. Can J Vet Res 78:127–132
61. Weldearegay YB, Pich A, Schieck E, Liljander A, Gicheru N, Wesonga H et al (2016) Proteomic characterization of pleural effusion, a specific host niche of *Mycoplasma mycoides* subsp. *mycoides* from cattle with contagious bovine pleuropneumonia (CBPP). J Proteomics 131:93–103
62. Nanduri B, Lawrence ML, Vanguri S, Burgess SC (2005) Proteomic analysis using an unfinished bacterial genome: the effects of subminimum inhibitory concentrations of antibiotics on *Mannheimia haemolytica* virulence factor expression. Proteomics 5:4852–4863
63. Souza CE, Rego JP, Lobo CH, Oliveira JT, Nogueira FC, Domont GB et al (2012) Proteomic analysis of the reproductive tract fluids from tropically-adapted Santa Ines rams. J Proteomics 75:4436–4456
64. Boe-Hansen GB, Rego JP, Crisp JM, Moura AA, Nouwens AS, Li Y et al (2015) Seminal plasma proteins and their relationship with percentage of morphologically normal sperm in 2-year-old Brahman (*Bos indicus*) bulls. Anim Reprod Sci 162:20–30
65. Gaviraghi A, Deriu F, Soggiu A, Galli A, Bonacina C, Bonizzi L et al (2010) Proteomics to investigate fertility in bulls. Vet Res Commun 34(Suppl 1):S33–36

66. Holland A, Ohlendieck K (2015) Comparative profiling of the sperm proteome. Proteomics 15:632–648
67. Rodgaard T, Heegaard PM, Callesen H (2015) Non-invasive assessment of in-vitro embryo quality to improve transfer success. Reprod Biomed Online 31:585–592
68. Deutsch DR, Frohlich T, Otte KA, Beck A, Habermann FA, Wolf E et al (2014) Stage-specific proteome signatures in early bovine embryo development. J Proteome Res 13:4363–4376
69. De Freitas Campos C, Cole N, Van Dyk D, Walsh BJ, Diakos P, Almeida D et al (2008) Proteomic analysis of dog tears for potential cancer markers. Res Vet Sci 85:349–352
70. Shamsi FA, Chen Z, Liang J, Li K, Al-Rajhi AA, Chaudhry IA et al (2011) Analysis and comparison of proteomic profiles of tear fluid from human, cow, sheep, and camel eyes. Invest Ophthalmol Vis Sci 52:9156–9165
71. Zhou L, Beuerman RW, Huang L, Barathi A, Foo YH, Li SF et al (2007) Proteomic analysis of rabbit tear fluid: defensin levels after an experimental corneal wound are correlated to wound closure. Proteomics 7:3194–3206

Index

G.H. Salekdeh (ed.), *Agricultural Proteomics Volume 2*,
DOI 10.1007/978-3-319-43278-6

MIX
Papier aus verantwortungsvollen Quellen
Paper from responsible sources
FSC® C105338
FSC
www.fsc.org

If you have any concerns about our products,
you can contact us on
ProductSafety@springernature.com

In case Publisher is established outside the EU,
the EU authorized representative is:
Springer Nature Customer Service Center GmbH
Europaplatz 3, 69115 Heidelberg, Germany

Printed by Libri Plureos GmbH
in Hamburg, Germany